International Conference on Emerging Trends in Engineering Research 2016

AA006233

IOP Conference Series: Materials Science and Engineering
Volume 183

Chennai, India
20 – 21 October 2016

Editors:

M. Chandrasekaran
S. Arun

ISBN: 978-1-5108-3771-3
ISSN: 1757-8981

Printed from e-media with permission by:

Curran Associates, Inc.
57 Morehouse Lane
Red Hook, NY 12571

Some format issues inherent in the e-media version may also appear in this print version.

Printed by Curran Associates, Inc. (2017)

For permission requests, please contact the Institute of Physics
at the address below.

Institute of Physics
Dirac House, Temple Back
Bristol BS1 6BE UK

Phone: 44 1 17 929 7481
Fax: 44 1 17 920 0979

techtracking@iop.org

Additional copies of this publication are available from:

Curran Associates, Inc.
57 Morehouse Lane
Red Hook, NY 12571 USA
Phone: 845-758-0400
Fax: 845-758-2633
Email: curran@proceedings.com
Web: www.proceedings.com

Table of contents

Volume 183

International Conference on Emerging Trends in Engineering Research
20–21 October 2016, Chennai, India

Accepted papers received: 28 February 2017
Published online: 10 March 2017

Advanced Manufacturing Technology

Renewable energy

Alternative fuels

IOP Conf. Series: Materials Science and Engineering **183** (2017) 011001 doi:10.1088/1757-899X/183/1/011001

FOREWORD

ICETER16 is the second International conference held by School of Engineering, VELS University, Chennai, India. The papers have been received from various emerging areas of research in Composites, Metals, Alloys, Manufacturing, Automotive Technology, Energy Engineering, Optimization techniques, Computer Algorithm, Construction and Industrial Engineering. This brings out Industrial experts to exhibit their proof of concept on various domains. This was an ideal platform for researchers and practioners to rediscover the new frontiers. One of the goals of the Conference is to share most up-to-date knowledge and to preserve and widen the working contacts between the Research scholars, Industrial People and Students. The program of conference included invited lectures given by keynote speakers from Malaysia and Kuwait.

The success of ICETER16 depends completely on the effort, talent, and energy of more than 200 researchers in the field of Engineering who have written and presented papers on a variety of topics. Praise is also deserved for the program committee members and external reviewers who have invested significant time in analyzing and assessing multiple papers, who hold and maintain a high standard of quality for this conference.

Guest Editors

Dr. M. Chandrasekaran, Dr. S. Arun

IOP Conf. Series: Materials Science and Engineering **183** (2017) 011001 doi:10.1088/1757-899X/183/1/011001

CONFERENCE PHOTO

Releasing e-proceedings of International Conference on Emerging Trends in Engineering Research (ICETER 2016) organized by School of Engineering, Vels University on 20th and 21st October, 2016.

ORGANIZING COMMITTEE

Chief Patron:

Dr.Ishari K.Ganesh, President

Patrons:

Dr.A.Jothi Murugan, Vice President / Planning & Development

Dr.V.Thamizh Arasan, Vice Chancellor

Dr.S.Sivasubramanian, Research Advisor / Chairman CARD

Dr.A.R.Veeramani, Registrar

Chairman:

Dr.P.Swaminathan, Dean / School of Engineering

Convenors:

Dr.M.Chandrasekaran, Director / Mechanical

Dr.C.Dhanasekaran, Coordinator / Engineering

Dr.S.Arun, Director / IQAC

Coordinators:

Dr.R.Rathan Raj, HOD / Civil

Ms.K.Kalaivani, HOD / CSE

Dr.V.Rajendran, HOD / ECE

Dr.R.Krishna Kumar, HOD / EEE

Dr.V.Muthuraman, HOD / Mechanical

Dr.P.Brindha Devi, HOD / Bio-Tech

Organising Secretaries:

Dr.T.Ilango,Assistant Professor / Civil

Dr.R.A.Karthika,Assistant Professor / CSE

Dr.S.Jerritta, Assosiate Professor / ECE

Dr.S.Murali, Assosiate Professor / Mechanical

IOP Conf. Series: Materials Science and Engineering **183** (2017) 011001 doi:10.1088/1757-899X/183/1/011001

INTERNATIONAL ADVISORY COMMITTEE

Dr.Jay N. Meegoda, Director, New Jersey Institute of Technology, USA

Dr.S.Neelamani, Kuwait Institute for Scientific Research, Kuwait

Dr.Bryan Barragan, CMRDI,Egypt

Dr.Muthu Krishnan, Kwazulu Nadal University, South Africa

Dr.Kumar Radhakrishnan, Coffey Geotechnics, Albury, Australia

Dr.Kenneth Sundaraj, Technical University of Malaysia, Malacca

Dr.R.Badlishah, Universiti Malaysia Perlis (UniMAP), Malaysia

Dr.Jeyakesavan Veerasamy, The University of Texas, Dallas, USA

Dr.John Cole, University of Texas, Dallas, USA

Dr.Nur Azah Binti Hamzaid, University of Malaya, Malaysia

Dr.Murugappan M, Kuwait College of Science and Technology,kuwait

Dr.Saidatul Ardeena Natie Binti Awang, University Malaysia Perlis (UniMAP), Malaysia

Dr.Md. Anamul Islam, Stony Brook University, New York

Dr.Nizam Uddin Ahamed, University Malaysia Pahang (UMP), Malaysia

Dr.Md. Asraf Ali, Daffodil International University, Dhaka

Dr.Tom Page, Loughborough University, UK

Dr.Eriki Ananda Kumar, Caledonian University, Oman

Dr.Subramanium Arunachalam, University of East London, UK

Dr. Lei LI, Hosei University, Japan

NATIONAL ADVISORY COMMITTEE

IOP Conf. Series: Materials Science and Engineering **183** (2017) 011001 doi:10.1088/1757-899X/183/1/011001

Dr.R.G.Robinson, Indian Institute of Technology Madras, Chennai

Dr.E.B.Perumal Pillai, Director, HRDC, Vel Tech University, Chennai

Dr.Satish Chandra, Director, CRRI, New Delhi

Dr.M.Harikrishna, National Institute of Technology, Calicut

IOP Conf. Series: Materials Science and Engineering **183** (2017) 011002 doi:10.1088/1757-899X/183/1/011002

Peer review statement

All papers published in this volume of *IOP Conference Series: Materials Science and Engineering* have been peer reviewed through processes administered by the proceedings Editors. Reviews were conducted by expert referees to the professional and scientific standards expected of a proceedings journal published by IOP Publishing.

IOP Conf. Series: Materials Science and Engineering **183** (2017) 012001 doi:10.1088/1757-899X/183/1/012001

Woven Hybrid Composites - Tensile and Flexural Properties of Jute Mat Fibres with Epoxy Composites

P Gopal[1], V K Bupesh Raja[2], M Chandrasekaran[3] and C Dhanasekaran[4]
[1]Research Scholar, Department of Mechanical Engineering, Vels University, Chennai, Tamil Nadu, India.
[2]Professor & Head, Department of Automobile Engineering, Sathyabama University, Chennai, Tamil Nadu, India.
[3] Professor, Department of Mechanical Engineering, Vels University, Chennai, Tamil Nadu, India.
[4]Associate Professor, Department of Mechanical Engineering, Vels University, Chennai, Tamil Nadu, India.

E-mail : bupeshvk@gmail.com

Abstract. The jute mat fibers are fabricated with several layers of fiber with opposite orientation in addition with coconut shell powder and resins. In current trends, metallic components are replaced by natural fibers because of the inherent properties such as light in weight, easy to fabricate, less cost and easy availability. This material has high strength and withstands the load. In this investigation the plates are made without stitching the fiber. The result of tensile strength and flexural strength are compared with nano material (coconut shell powder).

1. Introduction

The study or research on the influence of jute fiber without coconut shell powder is very less. Jute is one of the most cost effective and the toughest of all natural fibers [7, 8, 9, 10] .The epoxy and coconut shell powder added the aim of evaluating the efficiency of increasing the reinforcing mechanism able to improve the delamination resistance of laminates[5]. Mohan and Kishore in 1985 reported that jute performed well as a core material in jute-glass hybrid laminates. They investigated flexural properties and compressive properties of the jute-glass reinforced epoxy laminates concocted by a filament winding technique using flat mandrel [11, 12].

This is focused on two classes of cross ply stacking sequences [0/90]2s [90/0]2s arranged that the role of controlling damage progression of laminates and their capability to reduce the impact sensitivity. The flexural strength is reduced due to the addition of coconut shell powder. The impact energy is higher than threshold level and delaminations are sufficiently developed. [0/90]2slaminates on the other hand, stress concentration regions generated by the quality of good mixing.

In another study it was given good aesthetic nature and capable of preventing the initiation and spread of delamination, it induces a clear reduction of damage area when the fabrication is made by water abrasive jet machine. The plates are easily fabricated without any damage by water jet machine with fine accuracy. It is demonstrated that predicting the influence of adding coconut shell powder in different percentage pattern. The variable factors such as types of fiber, resin, lay up configuration and mixing ratio of powder, loading conditions, etc. without stitching the jute mat, the considerable tensile strength/flexural strength obtained.

2. Experimental Investigation

This paper discusses about the fabrication of jute mat composite with/without coconut shell powder and about the tensile strength/flexural strength obtained. The composite plate specimen with different orientation was fabricated by manned method. The tensile specimens are fabricated as per ASTM standard by abrasive jet machine. Fractographic analysis was conducted on the specimen after testing, which were investigated for failure mode by Scanning electron microscope (SEM).

3. Manufacturing methods

The aluminum plates were taken as mold size (500X500X10mm). The sizes of plate (300X300X3mm) rubber mold were placed over them. To avoid sticking on to the plate a small quantity of monsion polish was added. Then resin with hardener was applied in a ratio of 10:1 placing the first jute mat, then applying the brushing by rotter processing. Then the second jute mat was placed over this. Add resin gently, then remove the air bubble by pressing the surface by hand brushes and repeat the same process for 4 jute mat[6]. Placing the 100kg weight over the aluminium plate and allowed to cure for the period of 3 hours. Similarly the coconut shell plates are prepared (the mixing is an epoxy resin / hardener/coconut shell powder 15% of total weight taken for mixing).

4. Tensile testing

Tensile testing is a testing procedure done to calculate the material's property's ability to resist tensile forces, deformation and flexural failure. The test is conducted in UTM (Universal Testing Machine).

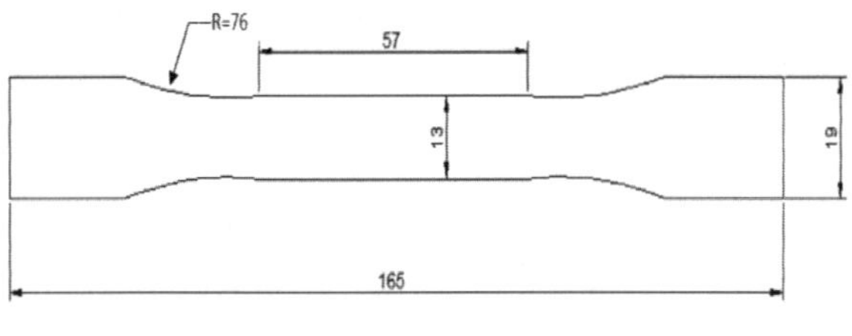

Figure.1 Tensile Specimen

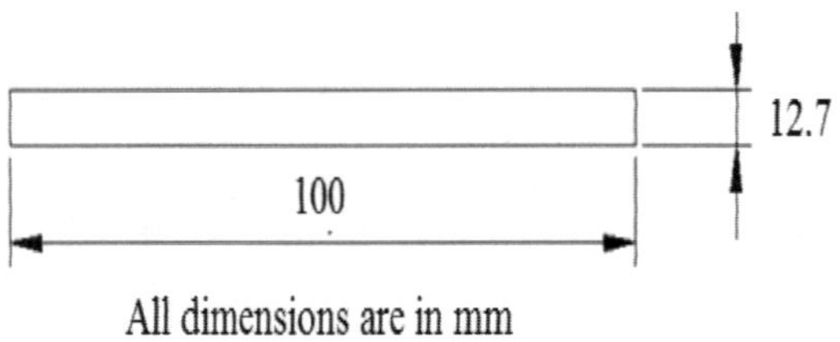

Figure.2 Flexural Specimen

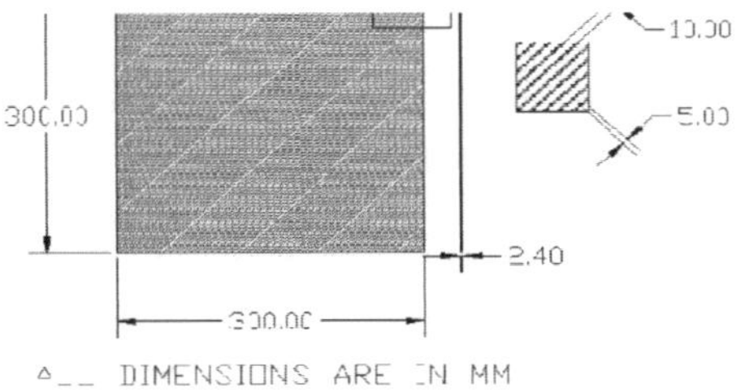

Figure.3 Flexural Specimen

Figure.4 Universal Testing Machine

The tensile test specimen/flexural specimen were fabricated in the abrasive water jet machine as per the standard as shown in figure 1. The tensile specimen subjected to uniaxial loading in a universal testing machine as shown in figure 4.

5. Results

The tensile test/flexural test results are tabulated as shown in the table. This indicates that the tensile and flexural test are increased in coconut shell powder added plates.

IOP Conf. Series: Materials Science and Engineering **183** (2017) 012001 doi:10.1088/1757-899X/183/1/012001

Table.1 Tensile & Flexural strength of specimen

Sl.no	Composite plate	Ultimate tensile strength	Flexural strength
1	Jute mat without coconut powder	1475	0.140
2	Jute mat with coconut powder	1848	0.170

Comparing the characterization of jute mat composite plate with coconut powder is higher tensile/flexural strength.

The fractography images obtained from the scanning electron microscope (SEM) shows lack of bonding and failure due to fiber pull and breakage. In this case the plate without nano particle is in lack of bonding properties. Whereas in the case of coconut shell powder and epoxy mixed with jute mat is a perfect bonding between resin and fiber and thereby no delamination occurs.

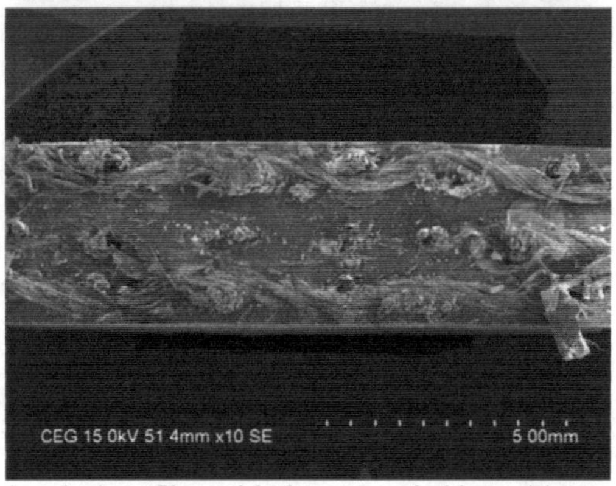

Figure.5 This figure shows that the jute fiber with the cross stitching with coconut shell powder particle as a function of SEM image.

6. Conclusion

The tensile test and flexural test were conducted and the results are compared. The composite plate without powder is less tensile and less flexural whereas proper mixing of coconut shell powder will give good tensile strength and flexural strength and further the laminate improve the tension and the delamination properties. The tensile specimen opted for fractographic studies will indicate the influence of nano particle will improve the tensile and flexural character of natural fibers. This is also a green composite of hybrids.

7. References

[1] V Lopresto 2006 Effect of stitches on the impact behaviour of graphite/epoxy composites. *Composites science and technology* **66** 206-214.

[2] Bupesh Raja V K 2012 Experimental Investigation on Tensile Strength of Cross Stitched and Un-stitched Fibre Glass-Epoxy Laminates in *International Journal on Design and Manufacturing Technologies (IJODM)* **6(1)**

[3] Bupesh Raja V K, et al., 2012 Experimental Investigation on Impact Strength of Cross Stitched and UN-stitched Fibre Glass-Epoxy Laminates In the *Processing of International Conference on Recent Advances in Mechanical Engineering(INCRAME'12).*

[4] Bupesh Raja V K., et al., 2012 Fracture Topography Investigation of Cross Stitched and Un-stitched Fiber Glass-Epoxy laminates *National Conference on Evolving Techniques in Mechanical Engineering (NCETME-2012).*

[5] Md. Faruk Hossen , Sinin Hamdan , Md. RezaurRahman, Md. Saiful Islam , FuiKiewLiew , Josephine Chang hui Lai , Md. MizanurRahman 2016 Effect of clay content on the morphological, thermo-mechanical and chemical resistance properties of propionic anhydride treated jute fiber/polyethylene/nanoclaynano composites; *Measurement* **90** 404–411

[6] M. Jawaid, H.P.S. Abdul Khalil, A. Abu Bakar 2011 woven hybrid composites: Tensile and flexural properties of oil palm-woven jute fibres based epoxy composites *Materials Science and Engineering A* **528** 5190–5195

[7] J A Khan and M A Khan 2015 The use of jute fibers as reinforcements in composites;*Biofiber Reinforcements in Composite Materials* 3–34

[8] B C Kundu , K C Basak and P B Sarkar 1959 Jute in India Indian Central Jute Committee Calcutta India.

[9] A Majumder , S Samajpati , P K Ganguly , D Sardar and P C Das Gupta 1980 Swelling of jute: Heterogeneity of crimp formation *Textile Research Journal* **50** 575 – 578

[10] A road map for jute *International Jute Study Group (IJSG) Dhaka, Bangladesh and International Trade Centre UNCTAD/WTO (ITC)* Geneva Switzerland.

[11] Kishore and Mohan R 1983 Compressive strength of jute-glass hybrid fibre composites. *J. Mater. Sci. Lett.* **2**, 99–102.

[12] Kishore and Rengarajan Mohan 1985 Jute-glass sandwich composites. *J. Reinf. Plast.Compos.* **4,** 186–194.

Characterization of Nylon 6 Nano Fiber/E-Glass Fiber Reinforced Epoxy Composites

T Vinod Kumar[1], M Chandrasekaran[2], V Santhanam[3] and N Udayakumar[4]
[1]Research Scholar & Assistant professor, Department of Mechanical Engineering, Vels University, Chennai, Tamil Nadu, India.
[2]Professor, Department of Mechanical Engineering, Vels University, Chennai, Tamil Nadu, India.
[3]Professor, Department of Mechanical Engineering, Rajalakshmi Engineering College, Chennai, Tamil Nadu, India.
[4]PG Student, Department of Mechanical Engineering, Rajalakshmi Engineering College, Chennai, Tamil Nadu, India.

E-mail: vinodkmrmech@gmail.com

Abstract. In the paper thermoplastic polymer Nylon-6 is generated in the form of Nanofibers by using an electro spinning method, and concentration of a solution is 4% as a constant then, by varying the process parameters such as flow rate (0.8ml/hr, 1ml/hr and 1.2 ml/hr) of the solution. The results indicated Nanofibers with 4% concentration and 1 ml/hr produced optimum fibers due to continuous fiber formation. Composites Plates are fabricated by using a Hand lay-up method with different volume fraction (0.5, 1, 2 % v/v) of Nanofibers ratio. Then, the optimum Nanofibers volume ratio (2 % v/v) is reinforced with E-glass fibers and epoxy resin as a matrix. In order to find Nanofibers effect, Mechanical properties like (Tensile, Flexural and Impact) is performed and evaluated.

1. Introduction

Fiber Reinforced Polymer (FRP) composites are used in almost every type of advanced engineering structure. E-Glass, Carbon and Aramid Fibers are used glass fibers are most commonly used as reinforcement [1,2]. The idea of using Nanofiber to reduce porous fiber diameter and high stiffness Nanofibers is used as a reinforcement in the form of fibers, tubes, whiskers, platelets, & particles. Nano-fibers are certain increases the mechanical properties, The beneficial of nano-fibers reinforcements is required for reinforcements to be very low [3,4]. Investigation of electro spun fibers for two different polymers Poly (methyl methacrylate) PMMA and Polyacrylonitrile (PAN) polymers is Generated by electro spinning. Then three difference combination Nylon-66 fibers are reinforced with PAN, PMMA and PMMA & PAN mixed together fabrication is made and conducted mechanical testing for all plates conducted. Among the three the PMMA and PAN it 5% higher than other two composite plates [5]. The Electro spun fibers as filler materials, Nylon 6 (NY) and polycaprolactone (PCL) used as reinforcements and matrix materials and different weight ratio of filler materials (1wt %, 3wt %, 5%). Compression molding is used in the fabrication process for the three different combination PCL/NY3 was prepared as sandwich of one layer of nylon mat between two PCL films. PCL/ NY5 was obtained from two layers of Nylon 6 mats alternated by PCL films. PCL/ NY8 was prepared interposing all the nylon fiber between two PCL layer (3wt %) is optimum among other the results is better than others [6]. Then, conducted Sixteen ply quasi-isotropic composite laminates interleaved AS4/3501-6 composite laminate clamped all around were the impact test is conducted and the improvement in impact resistance of composite laminates that have been interleaved by electro spun nylon 66 nano-fiber. The interleaved nano-fiber increase the impact force resistance and reduces damage from the impact forces [7]. In this Paper author conducted two different polymers Nylon 6 and Nylon 66 as a solute and Polycaprolactine as Solvent. Three different solution concentration and three

IOP Conf. Series: Materials Science and Engineering **183** (2017) 012002 doi:10.1088/1757-899X/183/1/012002

different viscosities for two polymers. Scanning Electron Microscope SEM images is an analysis of all samples and found low viscosity and concentrations goat fibers in Nanofibers [8]. Four different polymers are Polylactic acid PLA, Nylon 6 PA6, Polyethersulfone PES and TAC used, and kenaf fibers used for reinforcement, PA6 polymer tensile properties are increases than other [9]. The use of short electro spun fibers for the preparation of homogeneous polymer Nanofibers composites and the improvement in mechanical properties is amount of continuing long fibers required for achievement almost the same strength was much higher. I.e. 38 weight% in comparison to the 2 weight% required for the short fibers [10].Only Few researcher concentrated on Nylon-6 Nanofibers in different volume ration and reinforced with epoxy composites and hybrid E-glass composites .Mechanical properties are increased while using Nanofibers in composites.

2. Materials And Methodology

2.1. *Concept of Electro- spinning.*

An electro- spinning is one of the efficacious technique is used to generate polymers in Nanofibers structure. A simple working process behind the electro spinning methods. An Electro spinning apparatus is shown in figure 1, and an apparatus used in R&D (Rajalakshmi Engineering College) Chennai. A solution is filled in the syringe positive charge is connected and plate collector is placed at the bottom.

At high voltages is applied to solve. Then, the solution is ejected from the syringe, and collected in the top surface of the plate collector. The fiber diameter ranged from 300 to 550 mm.The schematic diagram of electro spinning is shown in figure 2. The electro spinning parameters used were: 4% Nylon-6 concentration, 20 KV applied voltage, 10 cm distance between the needle and collector, 1 ml/h flow rate. The Nanofibers weight percentage used in between 0.5% and 2 %.

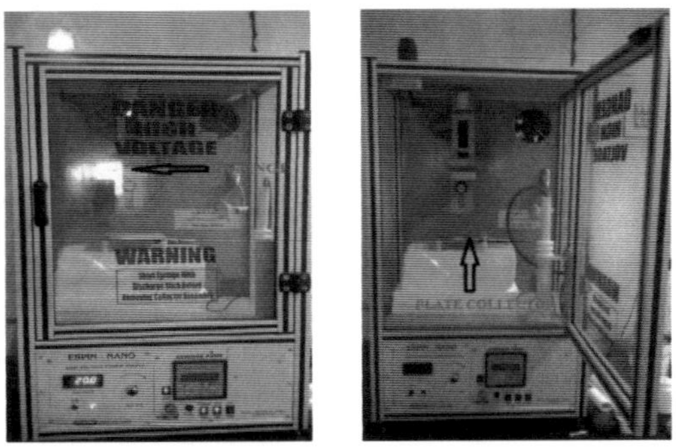

Figure 1. Represent Electro-spinning apparatus

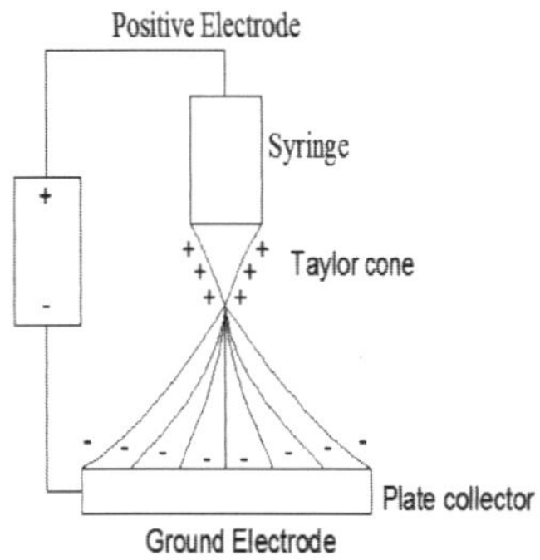

Figure 2. Schematic diagram of Electro-spinning

2.2. Materials.

The matrix used as Epoxy resin purchased from (Sakthi glass fiber) Chennai. A thermoplastic polymer Nylon 6 pellets used as sample and purchased from Sigma–Aldrich (USA). The solution 4% nylon 6 solutions in formic acid/acetic acid (80/20, wt/wt %). A fabrication of composites plates in different composition in order to find the effect of Nanofibers, in the Composites.

2.3. Fabrication Specimens.

Composites plates are fabricated and mechanical testing (Tensile, Flexural and Impact test) is conducted for all plates, in order to find the effects of Nanofibers in the composites. Fibers are chopped and stirred with Epoxy resin.Figure:3 Represents chopped Nylon-6 Nanofibers.

Figure 3. Chopping of Nylon-6 Nanofibers.

The dimension of the composites plates 200× 200 ×3 mm and according to the American Society for Testing and Materials (ASTM) Standard Tensile specimen is prepared length 165mm, width 19mm and thickness 3mm ASTM D638-03 Standard. Flexural specimen is prepared at length 100mm, width 12.7mm and thickness 3mm ASTM D790 and Impact specimen is prepared at length 65mm, width 12.7mm and thickness 3mm ASTM D256. Composites Plates are fabricated by using a Hand lay-up method with different volume fraction (0.5, 1, 2 % v/v) of Nanofibers ratio. Then, the optimum Nanofibers volume ratio (2 % v/v) is reinforced with E-glass fibers and epoxy resin as a matrix.

3. Results and Discussion

In this section effect of Nanofibers in epoxy resin and Mechanical test is conducted. NYLON 6 polymer solution was generated by dissolving the NYLON 6 in formic acid/acetic acid and the PA6 Nanofibers with different Flow rate (0.8ml/hr, 1ml/hr, 1.2 ml/hr) were tested. The temperature in the electro spinning apparatus 25-30 (^{0}c) respectively. Although Fibers were observed among the NYLON 6 Nanofibers in 4 % in 1 ml/hr is observed diameter range from 300 - 550nm approximately and 0.8ml/hr flow rate of NYLON 6 is observed air bubbles is formed in these fiber due to increasing the flow rate and despite the fact that fiber diameters range from 100 -200 NM but more air bubbles is formed. Then, 1.2 ml/hr is the NYLON 6 Nanofibers could not be formed from the High flow rate. The tensile, Flexural and Impact specimen are prepared based on ASTM standard, average of Five values Mechanical testing (tensile, Flexural and Impact) is evaluated. By increasing (0.5, 1, 2 % v/v) of Nanofibers Volume ratio in the composites. In the graph samples A is (100% v/v) Resin, samples B is (0.5 %v/v), samples is C (1% v/v), samples is D (2 % v/v), samples E is (30%v/v) E -Glass and samples F is 30% E- Glass and 2% Nanofibers. (c) Impact Energy.

3.1. Tensile properties

Tensile strength is comparatively improved in hybrid E-glass composites. A proper bonding between Matrix and E-glass fiber and Nanofibers in the composites that lead to increases in Tensile properties.

Tensile strength is increased in hybrid E-glass composite samples F is 152.9 (Mpa). The samples is D (2 % v/v) higher compare to A, B and C samples. By increasing the Nanofibers ratio and proper alignment in samples, that lead to tensile strength is increased for samples D 16.6 (Mpa). Tensile modulus is calculated from the stress - strain relation. Figure:4 Represents average values of Tensile Strength (MPa) and Tensile Modulus (GPA).

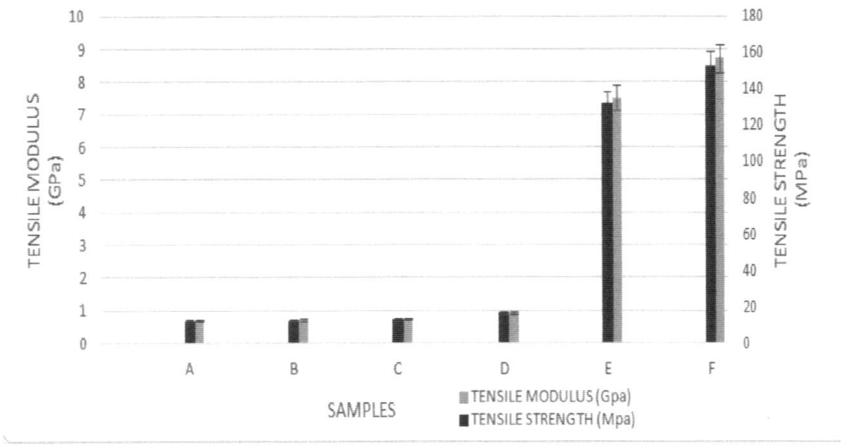

Figure 4. Represents average values of Tensile Strength (MPa) and Tensile Modulus (GPA).

3.2. Flexural properties

Flexural strength is comparatively improved in hybrid E-glass composites. A samples D is (2 % v/v) higher comparing to A, B and C Samples .Flexural strength of samples D is 83.33 (MPa), by increasing the Nanofibers ratio and proper alignment in samples, that lead to Flexural strength of hybrid composites is increase samples F is 143.58(Mpa). Flexural modulus is calculated. Figure 5. Represents average values of Flexural Strength (MPa) and Flexural Modulus (GPA).

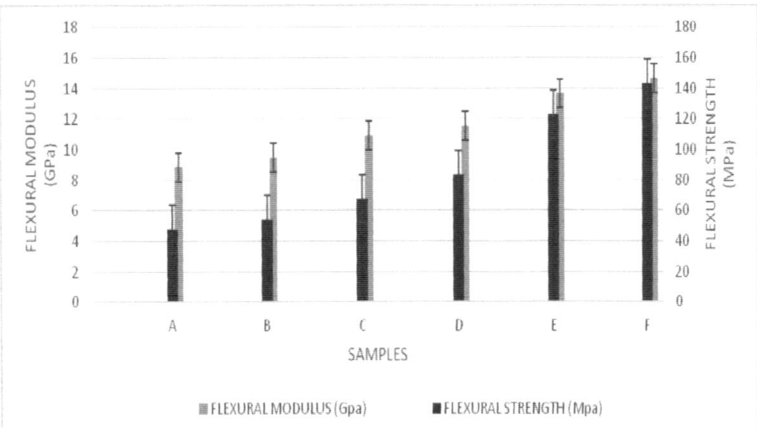

Figure 5. Represents average values of Flexural Strength (MPa) and Flexural Modulus (GPA).

3.3. Impact Energy (J)

Impact specimen is prepared based on ASTM standard .Then, samples F hybrid composites are higher because of proper alignment in samples F is 1.75(J) Impact Energy. Samples D is higher compare to A, B and C Samples, the result of sample D is 0.8 Impact Energy. Figure: 6 Represent the average values of Impact Energy(J).

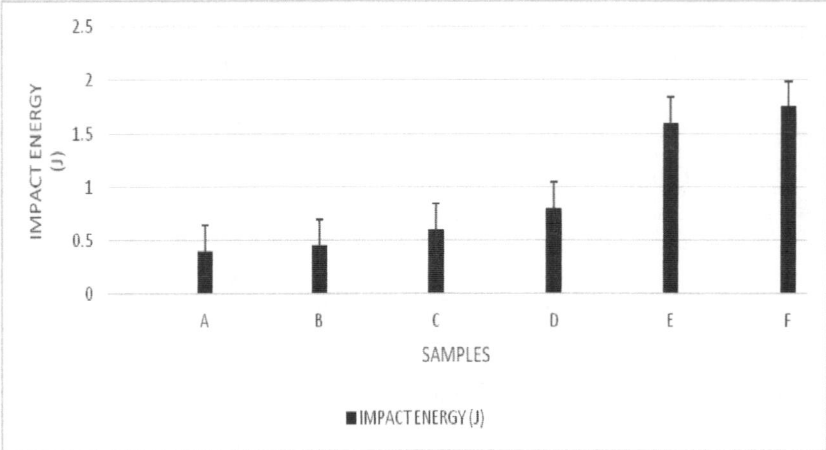

Figure 6. Average values of Impact Energy(J).

4. Conclusion

Thermoplastic polymer Nylon 6 can generate in the Electro spinning machine. Mechanical properties (Tensile, Flexural and Impact) are calculated for all samples. The Hybrid, E-Glass fiber reinforced with Nanofibers increases the Mechanical properties compared to other samples. Increasing the Nanofibers volume will result Proper Bonding between matrix and fiber.

5. References

[1] P K Malik Fiber *reinforced composites materials manufacturing and design* third edition 2007.

[2] Subbiah T, Bhat G S, Tock R W, Parameswaran S and Ramkumar S S 2005 Electrospinning of nanofibers *Journal of Applied Polymer Science* **96(2)** 557-569.

[3] Chronakis I S 2005 Novel nanocomposites and nanoceramics based on polymer nanofibers using electrospinning process—a review *Journal of Materials Processing Technology* **167(2)** 283-293.

[4] Lin S, Cai Q, Ji J, Sui G, Yu Y, Yang X and Deng X 2008 Electrospun nanofiber reinforced and toughened composites through in situ nano-interface formation *Composites Science and Technology* **68(15)** 3322-3329.

[5] Neppalli R, Marega C, Marigo A, Bajgai M P, Kim H Y and Causin V 2010 Poly (ε-caprolactone) filled with electrospun nylon fibres: A model for a facile composite fabrication *European Polymer Journal* **46(5)** 968-976.7.

[6] Goichi Ben, Yuri Kobayashi and Kazuhiro Sakata, Mechanical properties of kenaf composites coated with several nanofibers *Journal of Reinforced Plastics and Composites* 1–8.

[7] Akangah P, Lingaiah S and Shivakumar K 2010 Effect of Nylon-66 nano-fiber interleaving on impact damage resistance of epoxy/carbon fiber composite laminates, *Composite Structures*, **92(6)** 1432-1439.

[8] Gunavathi P, Ramachandran T and Chellamani K P 2012 Characterization of nanomembrane using nylon-6 and nylon-6/poly (e-caprolactine) blend *Indian Journal of Fibre and Textile Research* **37(3)** 211.

[9] Ben G, Kobayashi Y and Sakata K 2013, Mechanical properties of kenaf composites coated with several nanofibers Journal of Reinforced Plastics and Composites 0731684413494111.

[10] Jiang S, Duan G, Schöbel J, Agarwal S and Greiner A 2013 Short electrospun polymeric nanofibers reinforced polyimide nanocomposites *Composites Science and Technology* **88** 57-61.

Optimization of process parameters in drilling of fibre hybrid composite using Taguchi and grey relational analysis

B Vijaya Ramnath[1], S Sharavanan[2] and J Jeykrishnan[3]

[1]Professor, Department of Mechanical Engineering, Sri Sairam Engineering College, Chennai-600 044, India.
[2]Research Scholar, Department of Mechanical Engineering, Vels University, Pallavaram, Chennai-600117, India.
[3]Assistant Professor, Department of Mechanical Engineering, Sri Sairam Engineering College, Chennai-600 044, India.

E-mail: vijayaramnath.mech@sairam.edu.in

Abstract. Nowadays quality plays a vital role in all the products. Hence, the development in manufacturing process focuses on the fabrication of composite with high dimensional accuracy and also incurring low manufacturing cost. In this work, an investigation on machining parameters has been performed on jute-flax hybrid composite. Here, the two important responses characteristics like surface roughness and material removal rate are optimized by employing 3 machining input parameters. The input variables considered are drill bit diameter, spindle speed and feed rate. Machining is done on CNC vertical drilling machine at different levels of drilling parameters. Taguchi's L16 orthogonal array is used for optimizing individual tool parameters. Analysis Of Variance is used to find the significance of individual parameters. The simultaneous optimization of the process parameters is done by grey relational analysis. The results of this investigation shows that, spindle speed and drill bit diameter have most effect on material removal rate and surface roughness followed by feed rate.

1. Introduction

In present days, composites processed from natural fibres are increasingly useful and preferred in the field of automotive, aircraft and construction industries. Natural fibres provide an additional advantage over synthetic fibres due to their low cost, biodegradability and eco-friendly properties. Consumption of synthetic fibre has been reduced in aircraft and military application due to high cost even though they have good mechanical properties. El Tayeb [1] has investigated and acknowledged the fact that the composites reinforced with natural fibres and polymer matrices have better mechanical properties. Malkapuram et al. [2] studied the adhesion between the fibre and matrix surface and concluded that the mechanical properties can be improved by chemical modification methods and novel processing techniques. Various natural fibres like jute, flax, sisal, kenaf, abaca and acacia are used for commercial purposes [3]. An investigation shows that jute-flax hybrid composite has better tensile and shear properties than flax-fiber mono composite but flax composite is superior to hybrid composites in flexural and impact properties [4]. Srinivasan *et al* [5] have experimentally tested and proved that hybrid composite with glass fibre reinforced composite (GFRP) have better thermal properties and also shows higher resistance under impact and flexural loading when compared to single fibre composite. Vijaya Ramnath *et al* [6, 7] have fabricated abaca-jute hybrid composite and evaluated various mechanical properties. They concluded that hybrid natural fibre composites have better tensile, flexural, impact and shear properties when compared with mono fibre composite. Wong *et al* [8] has studied the fracture characteristics of short bamboo fibre reinforced polyester composite and concluded that fracture toughness of all types of composites are higher compared to mere polymer composite. Koenig *et al* [9] has studied the various factors affecting machining of FRP's and concluded fibre orientation also affects machining. Teti [10] and Abrao *et al* [11] have concluded that various characteristics of composite lamination such as non-homogenous, anisotropic, high abrasiveness make them difficult to be machined. Ali *et al* [12] has performed comparative study of drilling and milling process in hole making of GFRP composites. They concluded that milling process has been more suitable than drilling at higher cutting speed and lower feed rate. Sonbaty *et al* [13] have performed the drilling operation on GFRP with twist drills of various diameters with constant rotational speeds and found that with increase in drill diameter and feed rate, thrust force and torque has been increased. Khashaba [14] has studied the influence of drilling variables and material variables on thrust force and torque and concluded that the feeds and fiber volumes have direct effects on thrust forces and torques. Ariffin *et al* [15] has performed optimization of drilling process using design of experiments on aircraft composite. It has been concluded that minimum and maximum damage length of 0.05mm and 0.44mm was obtained when performed at 3000rpm and feed rate of 80.2 mm/rev for high speed steel. Tsao and Chiu [16] have investigated the influence of cutting velocity, feed rate and drill diameters on thrust force and found that cutting velocity, feed rate are the most influential among a group of five control factors. Yang *et al* [17] has conducted end milling on high purity graphite and optimized the various parameters such as cutting speed, feed rate and depth of cut. Strenkowski *et al* [18] has described a model for determining thrust force and torque in drilling. This model has been applicable for general drill geometries under various cutting condition. Hocheng and Tsao [19] have studied the effects of special drill bits on drilling induced delamination of composite material and reviewed the application of special drill bits and non-traditional machining methods and they have also concluded that the drilling thrust force varies with feed rate and has devised a feed rate strategy that avoids delamination due to high thrust force in drilling. Jin *et al* [20] has concluded that the thrust force have been influenced by thickness of drilling plates and also thrust force increases with feed rate and number of holes drilled. Marta *el al* [21] has studied drilling of carbon epoxy using one shot drill bit and the results showed the influence of tool wear and thickness of the work piece on the thrust force and torque in the drilling process. Duraoa *et al* [22] and Hocheng *et al* [23] have analyzed the effects of drilling parameters on composite plate's damage. The study inferred that lower feed rates and conservative cutting speed are most suitable for drilling of the composite plates. Jeykrishnan et al [24-28] have done experiments on non-traditional machining processes such as electro-discharge machining (EDM) and electro-chemical machining (ECM) and have optimized their input variables using traditional techniques like Taguchi method in obtaining the best feasible output characteristics such as material removal rate (MRR), tool wear rate (TWR) and surface roughness (R_a) in materials like steels, nickel alloys etc., and have also concluded that certain parameters like applied current, inter-electrode gap between tool and the workpiece impacts the responses, significantly.

However, the task of optimizing a particular parameter has been a tedious and time consuming process. Taguchi principles and grey relational analysis (GRA) are used nowadays to optimize the process parameters. Chen *et al* [29] has discussed the integration of Taguchi method with GRA for resolving multi-quality characteristic into grey relational grades and obtained the results by comparing these grades. Tzeng *et al* [30] has made use of Taguchi method and GRA to optimize the input parameters for CNC turning process and concluded that the depth of cut (DOC) and cutting speed has the maximum influence on roughness and roundness. Hasani *et al* [31] has performed GRA to find optimum process parameters in end spinning process of yarns and concluded that rotor diameter has most significant impact. Upinderkumar and Deepak Narang [32] have optimized the cutting parameters in high speed turning by GRA. They have employed Taguchi method and analysis of variance (ANOVA) for individual character optimization in the machining of medium carbon steel in dry conditions while varying the different input characteristics like cutting speed, feed and depth of cut for obtaining the optimal values of surface roughness and material removal rate (MRR). Sardinas *et al* [33] has performed a genetic algorithm (GA) based multi objective optimization technique on cutting parameters in turning process. The results were analyzed for various production conditions and also discussed the advantage of multi –objective optimization approach. Hassan *et al* [34] have used Taguchi's L27 orthogonal array (OA) for optimizing MRR in turning process and concluded that MRR was affected by cutting speed and feed rate, significantly.

2. Materials

Natural fibres are substances produced by plants and animals and they are environment friendly. The added key advantages are their strength, high stiffness to weight ratio, acoustic isolation, high damping and rapid production. The following natural fibers namely jute and flax have been used in this work.

2.1 Jute fibre

Jute which is called as "The golden fibre" mainly because of its colour and high cash value comes under the family name of Sparrmanniaceae. Jute fibres are 100% bio-degradable and recyclable. They are environmental friendly. They are the 2nd most available vegetable fibre. They also have good tensile strength and low extensibility.

2.2 Flax fibre

Flax is biologically named as Linum Usitatissimum belong to the family called Linaceae. Flax fibres are thermo regulating, non-allergenic, antistatic and anti-bacterial.

2.3 Glass Fibre Reinforced Polymer (GFRP)

They are made up of strands of silica glass fibre. Glass fibres have high strength, light weight, good resistance, low maintenance and high durability. They also have high strength to weight ratio. The main function of GFRP is to protect the fibre from atmosphere moisture.

2.4 Resin and Hardener

The main purpose of Resin and Hardener are to bind various layers of fibre. Epoxy resins have excellent mechanical strength, resistance to heat, adhesive strength and low curing contraction. These resin and hardener are used at room temperature to achieve better results. Epoxy LY558 Resin and Araldite HY951 hardener are used for fabrication of composite in this work.

3. Fabrication of composite

Hand layup process is used to fabricate Jute Flax based GFRP composite. Initially, the fibres are allowed to dry in sunlight for complete removal of moisture. Combing is done to separate small flakes present in jut fibre. A releasing agent (Poly Vinyl Alcohol) is spread over flat surface of mold for easy removal of composites. Resin hardener mixture is applied between layers of fibers for binding purpose. GFRP is used as laminate on outer side of composites so that it covers the composites on both sides. Here, the jute fibers are arranged in vertical orientation and flax fibers are arranged at 45° to jute fiber. The composite consists of 5 layers of fibers at different orientation as shown below.

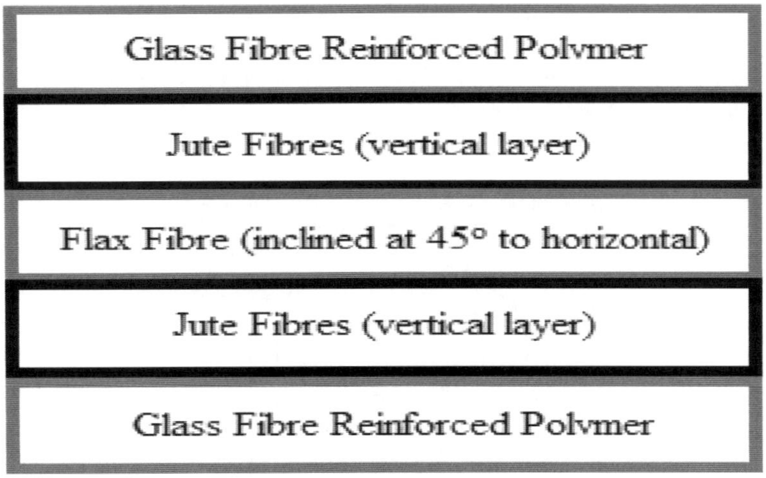

Figure 1. Schematic diagram of Jute+Flax+GFRP composite

4. Experimental setup

4.1 Machining setup

In this work, a hole was drilled on hybrid composite laminate. The following machines are used for drilling and also for measuring R_a.

4.1.1 Drilling machine

In this work, drilling was done on a vertical CNC machine, BMV TC24 manufactured by BFW as shown in figure 2. Initially, test drill was performed to ensure that the work piece is clamped properly. The drill was performed under varying combination of cutting parameters namely drill bit diameter, spindle speed and feed rate. The drill bit was made up of K20 Carbide.

Figure 2. CNC drilling machine

4.1.2 Surface Roughness testing machine

Surface roughness was measured using Surfcontour-1500 as shown in figure 3 has been manufactured by Zeiss THK (Tokyo Seimitsu, Japan). In order to achieve good results, the measurement was carried out at a speed of 0.3mm/s for a length of 4mm.

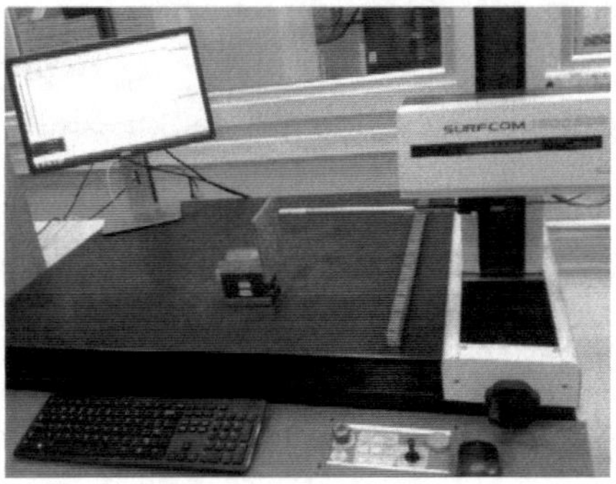

Figure 3. Surfcontour-1500

4.2 Measurement procedure

The following machined work piece parameters are considered in this work.

4.2.1 Material removal rate (MRR)

The rate at which the material is removed from the work piece is termed as Material Removal Rate. It depends on Diameter of drill (D), speed of the spindle (N), feed rate (f_r). Though high MRR rate reduces the cost of production and also the machining time, it increases the rate of tool wear as well as surface roughness. MRR is calculated from the formula as shown in eqn 1.

$$MRR = (\pi D^2/ 4)\ (N * f_r * 60) \quad (mm^3/sec) \tag{1}$$

4.2.2 Surface roughness

Surface roughness is defined as the degree of waviness present on the surface of the machined material. If the surface finish is good, the machined material is said to have a smooth surface. It is measured either in R_a or RMS. R_a is called as arithmetic mean roughness value. Since, the machined surface must have better finish, this work prefers the smaller values of surface roughness.

5. Optimization of machining parameters

5.1 Design of Experiment

Taguchi method is mainly used to study the effect of all the parameters with minimum possible number of experiments. Initially, a proper orthogonal array must be selected based on the number of parameters. Since, this work is aimed to analyze the effect of machining parameter more accurately, signal to noise ratio are calculated. The contribution of each parameter can be clearly obtained using ANOVA method.

5.2 Taguchi approach

Taguchi approach is a statistical method. The experiments were performed using Taguchi approach which is used to investigate the effect of different parameter affecting the mean and variance of a process performance characteristic that define the outcome of the product with less number of experiments.

There are 3 quality characteristic of Signal-to-Noise ratios for an optimization problem:

(I) Smaller the better

$$SN_s = -10 \log (\Sigma y^2/n) \quad (2)$$

(II) Larger the better

$$SN_L = -10 \log (\Sigma (1/y^2)/n) \quad (3)$$

(III) Target the better

$$SN_T = 10 \log (\bar{y}^2/s^2) \quad (4)$$

Where,

SN - Signal to Noise ratio

n -Number of observation in a trail

y - Observed value of the response

$\bar{y}$ - Average of observed value

Since, this work mainly focuses on finding the optimum parameters that are required to achieve maximum material removal rate and minimum surface roughness, the quality characteristic namely Larger the better and Smaller the better are used. In this work, multi-performance optimization of process parameters are also performed using GRA.

6. Experimental plan and details

Table 1. Levels and process parameters

Symbol	Drilling Parameter	Level 1	Level 2	Level 3
D	Drill bit Diameter (mm)	6	8	10
N	Spindle speed (rpm)	600	1200	1800
F_r	Feed rate (mm/rev)	0.1	0.2	0.3

Three major factors that are affecting the material removal rate and surface roughness are selected and each factor is designed to have three different levels to perform the process of optimization. The parameters and their levels are shown in table 1. After reducing the number of experiments with Taguchi method, statistical method ANOVA and GRA are applied to find optimal process parameter for drilling.

Table 2. Design of experiment and calculation

EX. No	D (mm)	N (rpm)	f_r (mm/rev)	MRR(mm^3/sec)	Ra (μm)
1	6	600	0.1	28.26	7.68
2	6	1200	0.2	113.04	10.052
3	6	1800	0.3	254.34	6.472
4	8	600	0.2	100.48	10.576
5	8	1200	0.3	301.44	4.749
6	8	1800	0.1	150.72	4.418
7	10	600	0.3	235.5	8.31
8	10	1200	0.1	157	7.012
9	10	1800	0.2	471	9.178

Table 2 shows the calculated value of MRR and surface roughness for all the experiments conducted based on L9 orthogonal array. These found values are analysed to find the optimal machining condition.

7. Results and discussion

In drilling process, the material removal rate and surface roughness are important features. In order to find the factors which significantly affect them, ANOVA method is used. Response table shows the optimum condition of machining.

7.1 Effect of process parameter on material removal rate

Table 3 and 4 shows the ANOVA and response table for MRR. It can be seen from the relation that MRR is directly depends on all the three parameters. Table 3 shows the amount of contribution of each parameter on MRR. From table 3, it is seen that the speed contributes about 31.75%, while the diameter of drill and the feed rate has approximately equal amount of contribution. Response table 4 ranked the parameters based on their contribution.

Table 3. ANOVA table for mean of MRR

Source	Degree of Freedom (DF)	Sum of Squares (SS)	Mean of Squares (MS)	F-test	P-test	Contribution (%)
D	2	37797	18899	1.95	0.339	27.159
N	2	44186	22093	2.28	0.305	31.750
F_r	2	37797	18899	1.95	0.339	27.159
Error	2	19386	9693			13.930
Total	8	139167				

Table 4. Response table for MRR

Level	Drill bit Diameter	Spindle speed	Feed rate
1	131.9	121.4	112
2	184.2	190.5	228.2
3	287.8	292	263.8
Delta	156	170.6	151.8
Rank	2	1	3

From table 4 and table 1, it is concluded that the optimal machine settings for maximum MRR are drill bit diameter 10 mm, spindle speed 1800 rpm and feed rate 0.3 mm/rev.

7.2 Effect of process parameter on surface roughness

Table 5 and 6 shows the ANOVA table and response table for surface roughness. It can be seen from the table 5 that the feed rate has the major contribution among the three parameters of 64.81%. It is also clear that the drill diameter does not play a significant role in the outcome of surface roughness.

Table 5. ANOVA table for surface roughness

Source	DF	SS	MS	F	P	C
D	2	4.735	2.368	4.72	0.175	12.547
N	2	7.54	3.77	7.51	0.117	19.978
fr	2	24.462	12.231	24.38	0.039	64.816
Error	2	1.004	0.502			2.659
Total	8	37.741				

Table 6. Response table for surface roughness

Level	Drill bit Diameter	Spindle speed	Feed rate
1	8.068	6.855	6.37
2	6.581	7.271	9.935
3	8.167	6.689	6.51
Delta	1.586	2.166	3.565
Rank	3	2	1

From table 6 and table 1, it is concluded that the optimum machine settings for getting good surface finish are drill bit diameter of 8mm, spindle speed of 1800 rpm and feed rate of 0.1 mm/rev.

8. Grey relational analysis

Since, the experimental results are not unique and optimum, grey relational analysis (GRA) is employed to find the unique optimal machine setting. GRA uses a specific concept of information where situation with zero information is called as black and concept with required information called as white. Situation between these extremes are termed as Grey. In GRA, the quality characteristic are first normalized ranging from zero to one in which zero being the worst condition and one being the best condition. The procedure of GRA is shown in figure 4 as flow chart.

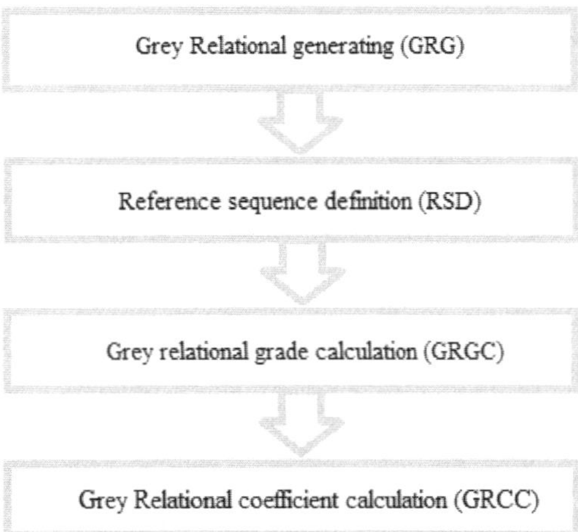

Figure 4. GRA flow chart

The S/N ratio, GRG and GRGC are calculated for each experiment and furnished in table 7. In this paper the distinguishing coefficient was set at 0.5, from which the required grey relation coefficient can be calculated.

Table 7. Grey relational analysis calculations

Ex. no.	Mean Values		S/N ratio		GRG		GRGC	
Xo	MRR	Ra	MRR	Ra	MRR	Ra	MRR	Ra
1	28.26	7.68	29.02	-17.707	0	0.47	0.33	0.49
2	113.04	10.052	41.06	-20.045	0.19	0.085	0.38	0.35
3	254.34	6.472	48.11	-16.221	0.51	0.666	0.51	0.6
4	100.48	10.576	40.04	-20.486	0.16	0	0.37	0.33
5	301.44	4.749	49.58	-13.532	0.62	0.946	0.57	0.90
6	150.72	4.418	43.56	-12.905	0.28	1	0.41	1
7	235.5	8.31	47.44	-18.392	0.47	0.368	0.48	0.44
8	157	7.012	43.91	-16.917	0.29	0.579	0.41	0.54
9	471	9.178	53.46	-19.255	1	0.227	1	0.39

Table 8 shows the grey relational coefficient calculated after giving equal weight age to each parameter. S/N ratio was calculated for the mean of GRG and they are ranked based on larger being the best.

Table 8. Grey relational grade calculation

Ex. No.	GRGC		Grade		Rank
	MRR	Ra	Mean	S/N ratio	
1	0.33	0.49	0.41	-7.76	7
2	0.38	0.35	0.37	-8.69	8
3	0.51	0.60	0.55	-5.15	4
4	0.37	0.33	0.35	-9.03	9
5	0.57	0.90	0.73	-2.68	1
6	0.41	1.00	0.70	-3.04	2
7	0.48	0.44	0.46	-6.69	6
8	0.41	0.54	0.48	-6.41	5
9	1	0.39	0.70	-3.14	3

8.1 Combined influence of the process parameter

Table 9 and 10 are made after the process of converting multi objective into single objective result. Table 9 represents the ANOVA for the mean of grey relational grade which shows that the maximum contribution among the three parameters is the speed at which the spindle is working which accounts about 49.02%. The next most influencing factor is the diameter of the drill bit which is 20.61%. The response table 10 shows the overall ranking of all the three parameter and their amount of influence.

Table 9. ANOVA for mean of grey relational grade

Source	DF	SS	MS	F	P	C
D	2	0.037	0.018	1.03	0.49	20.61
N	2	0.088	0.044	2.45	0.29	49.02
F_r	2	0.018	0.009	0.51	0.66	10.25
Error	2	0.035	0.018			20
Total	8	0.179				

Table 10. Response table for means of grey relational grade

Level	Drill bit Diameter	Spindle speed	Feed rate
1	0.443	0.409	0.531
2	0.598	0.527	0.473
3	0.546	0.651	0.584
Delta	0.154	0.242	.111
Rank	2	1	3

From table 10 the optimum process parameter to have maximum MRR and minimum surface roughness are drill bit diameters 8mm, spindle speed 1800rpm and feed rate 0.3mm/rev.

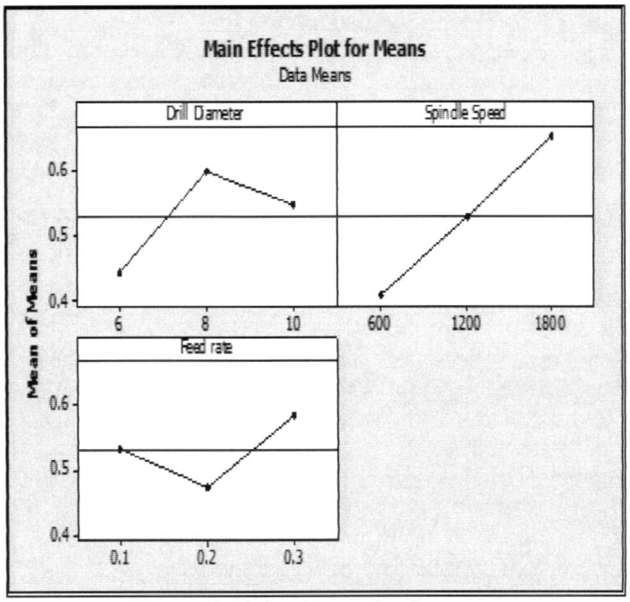

Figure 5. Main effects plot for means

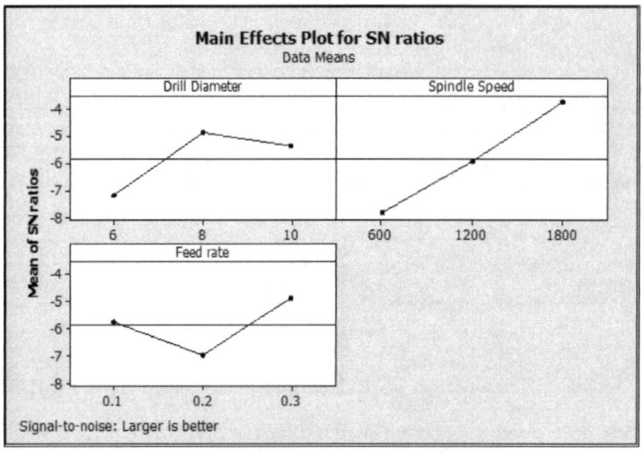

Figure 6. Main effects plot for SN ratio

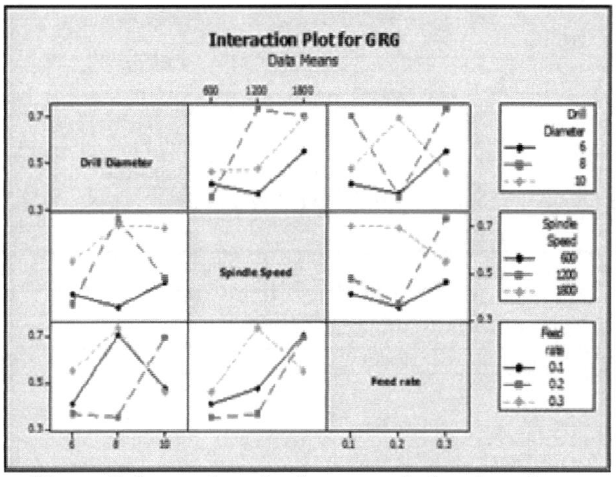

Figure 7. Interaction plot for grey relational grade

8.2 Main effects plots and interaction plot

A "main effect" is the effect of one of the selected independent variable on the dependent variable, ignoring the effects of all other independent variable. Figure 5 shows the main effects plot for mean and figure 6 shows the main effects plot for S/N ratio. S/N ratio is always high at optimum condition. So, from the figure 7 it can be seen that the optimum combination is diameter 8mm, spindle speed 1800 rpm and feed rate 0.3 mm/rev.

An interaction is present when the combined effect of two or more factors is bigger or smaller than the sum of their individual effect. It means that the factors are affecting each other as well as the product of process performance. This Interaction plot of grey relational grade and the three process parameters which are considered is clearly shown in the figure 6.

9. Conclusion

In this work, the three machining parameter like drilling, spindle speed, feed rate and their interactions are evaluated using ANOVA and also with GRA. The following conclusions are arrived in this study;

Spindle speed contributes more for MRR and surface roughness than the other two parameters.
MRR increased with drill diameter, feed rate and spindle speed.

The influence of drill bit diameter is very less when compared to the feed rate for surface roughness. GRA is used to find combined output and optimum machine setting.
By GRA, it is found that drill diameter of 8 mm, spindle speed of 1800 rpm and Feed rate of 0.3 mm/rev should be preferred in drilling of Jute –Flax fibre reinforced composite for getting optimum MRR and surface roughness.

10. References

[1] El-Tayeb NSM 2008 Development of low-cost polymeric composite materials *Mater. Des.* **30** 1151-1160.
[2] Malkapuram R, Kumar V and Negi Y S 2009 A review on mechanical behavior of natural fibre based hybrid composites *Reinf. Plast.Compos.* **28** 1169.
[3] Murali Mohan Rao K and Mohana Rao K 2007 Extraction and tensile properties of natural fibres: Vakka, date and bamboo *Composite structures* **77** 288-295.
[4] VijayaRamnath B, Elenchezian C, Nirmal P V, PremKumar G,Santosh Kumar V, Karthick S, Rajesh S and Suresh K 2014 Experimental investigation of mechanical behavior of jute-flax based glass fiber reinforced composite *Fibres and polymers* **15** 1251-1262.
[5] Srinivasan V S, RajendraBoopathy S, Sangeetha D and Vijaya Ramnath B 2014 Evaluation of mechanical and thermal properties of banana-flax based natural fibre composite *Materials and Design* **60** 620-627.
[6] Vijaya Ramnath B, Manickavasagam V M, Elanchezhian C, Vinodh Krishna C, Karthik S and Saravanan K 2014 Determination of mechanical properties of intra-layer abaca–jute–glass fibre reinforced composite *Materials and Design* **60** 643–652.
[7] Vijaya Ramnath B, JunaidKokan S, Niranjan Raja R, Sathyanarayanan R, Elanchezhian C, Rajendra Prasad A and Manickavasagam V M 2013 Evaluation of mechanical properties of abaca–jute–glass fibre reinforced epoxy composite *Materials and Design* **51** 357–366.
[8] Wong K J, Zahi S, Low K O and Lim C C 2010 Fracture characterization of short bambooFibre renforce polyester composites *Mater. Des.* **31** 4147–4154.
[9] Koenig W, Wulf C, Grass P and Willerscheid H 1985 Machining of fibre reinforced plastics *CIRP Annals-Manufacturing Technology* **34** 537-548.
[10] Teti R 2002 Machining of composite materials *CIRP Ann–Manuf. Technol.* **51** 611–634.
[11] Abrao A M, Faria P E, Campos Rubio J C, Reis P and Davim J P 2007 Drilling of fibre reinforced plastics: A review *J. Mater. Process. Technol* **186** 1–7.
[12] Hussein M Ali, Asif Iquball and Li Liang 2013 A comparative study on the use of drilling and milling processes in hole making of GFRP composite *Sadhan* **38 4** 743–760.
[13] El-Sonbaty I, Khashaba U A and Machaly T 2004 Factors affecting the machinability of GFR/epoxy composites *Compos. Struc.* **63** 329–338.
[14] Khashaba U A, Seif M A and Elhamid MA 2007 Drilling analysis of chopped composites *Composites Part A: Applied Science and Manufacturing* **38** 61–70.
[15] Mohd Ariffin M K A, Mohd Ali M I, Sapuan S M S M and Ismail N 2009 An optimized drilling process for an aircraft composite structure using design of experiments *Sci. Res. Essay.* **4** 1109–1116.
[16] Tsao C C and Chiu Y C 2011 Evaluation of drilling parameters on thrust force in drilling carbonfibre reinforced plastic (CFRP) composite laminates using compound core-special drills *International Journal of Machine Tools and Manufacture* **51** 740-744.
[17] Yang Y Y, Shie J R and Huang C H 2006 Optimization of dry machining parameters for high purity graphite in end-milling process *Mater. Manuf. Process.* **21** 832–837.
[18] Strenkowski J S, Hsieh C C, and Shih A J 2004 An analytical finite element technique for predicting thrust force and torque in drilling *International Journal of Machine Tools and Manufacture* **44** 1413–1421.
[19] Hocheng H and Tsao C C 2005 The path towards delamination-free drilling of composite

materials *Journal of Materials Processing Technology* **167** 251–264.

[20] Jin P J, Geun W K and Kang Y L 2005 Critical thrust force at delamination propagation during drilling of angle-ply laminates *Compos. Struct.* **68** 391–397.

[21] Marta F and Chris C 2006 Drilling of carbon epoxy using a one shot drill bit Part I: five stage representation of drilling and factors affecting maximum force and torque *Int. J. Mach. Tools Manufac.* **46** 70–75.

[22] Durãoa L M P, Magalhãesa A G, Marquesb A T, João Manuel R S and Tavaresb 2007 Effect of drilling parameters in composite plates damage *Proc. of Int. conference on high speed Industrial Manufacturing Processes* 1-8.

[23] Hocheng H, Puw H and Yao K C 1992 *Machining Composite Materials Symposium ASM Materials Week* 127-138.

[24] Jeykrishnan J, Maharaja K, Vijayarajan P and Priyadarshi Dutt 2017 Optimization of turning process variables in inconel 625 alloy using Taguchi techniques *International journal of engineering technology research and management* **1 (2)** 14-17.

[25] Jeykrishnan J, Vijaya Ramnath B, Jude Felix A, Rupan Pernesh C and Kalaiyarasan S 2016 Parameter optimization of electro-discharge machining (EDM) in AISI D2 die steel using Taguchi technique *Indian journal of science and technology* **9 (43)** 1-4.

[26] Jeykrishnan J, Vijaya Ramnath B, Sureshrajan G, Siva Bharath M, Hervin Savariraj X and Akilesh S 2016 Effects of Die sinking electro-discharge machining parameters on surface roughness in inconel 825 alloy *Indian journal of science and technology* **9 (41)** 1-5.

[27] Jeykrishnan J, Vijaya Ramnath B, Akilesh S, Pradeep Kumar R P 2016 Optimization of process parameters on EN24 tool steel using Taguchi technique in electro-discharge machining (EDM) *IOP conf. series: Materials science and engineering* **149** 012 – 022.

[28] Shebin Thomas, Ramesh S, Jeykrishnan J 2016 Parametric optimization of powder mixed electro-discharge machining using Taguchi technique *International journal of latest trends in engineering and technology* **7 (3)** 61-66.

[29] Chen C B, Lin C T, Chang C W, Ho C P 2000 Grey relation for solving multi quality characteristics problems of Taguchi method *Journal of Technology* **15** 25-33.

[30] Chorng-JyhTzeng, Yu-Hsin Lin, Yung-Kuang Yang and Ming-Chang Jeng 2009 Optimization of turning operation with multiple performance characteristics using Taguchi method and Grey Relational Analysis *Journal of Materials Processing Technology* **209** 2753-2759.

[31] Hossein Hasani, Somayeh Akhavan Tabatabaei and Ghafour Amiri 2012 Grey Relational Analysis to Determine the Optimum Process Parameters for Open-End Spinning Yarns *Journal of Engineered Fibres and Fabrics* **7** 81-86.

[32] Upinderkumar and Deepak Narang 2013 Optimization of Cutting Parameters in High Speed Turning by Grey Relational Analysis *International Journal of Engineering Research and Applications* **3** 832-839.

[33] Ramón QuizaSardiñas, Marcelino Rivas Santana and Eleno Alfonso Brindis 2006 Genetic algorithm-based multi-objective optimization of cutting parameters in turning processes *Engineering Applications of Artificial Intelligence* **19** 127 – 133.

[34] Hassan K 2012 Experimental investigation of Material removal rate in CNC turning using Taguchi method *International Journal of Engineering Research and Application* **2** 1581-1590.

IOP Conf. Series: Materials Science and Engineering **183** (2017) 012004 doi:10.1088/1757-899X/183/1/012004

Design and Fabrication of E-Glass /carbon/graphite epoxy hybrid composite leaf spring

T Gopalakrishnan[1], M Raja[2], V M Jothi Prakash[3] and C Gnanavel[4]
[1, 3, 4] Assistant Professor, Department of Mechanical Engineering,
VELS University, Pallavaram, Chennai, Tamil nadu, India
[2]Assistant Professor, Department of Mechanical Engineering
Government College of Engineering, Salem, Tamil nadu, India

E-mail: [1]gopalakrish185@gmail.com

Abstract. The Automobile Industry has shown increase interest for replacement of steel leaf spring with that of composite leaf spring. Substituting composite materials for conventional metallic materials has many advantages because of higher specific stiffness, strength and fatigue resistance etc. This work deals with the replacement of conventional steel leaf spring with a hybrid Composite leaf spring using E -Glass/Carbon/Graphite/Epoxy. The hybrid composite is obtained by introducing more than one fiber in the reinforcement phase. The hybrid composite is fabricated by the vacuum bag technique. The result shows that introduction of carbon and graphite fiber in the reinforcement phase increases the stiffness of the composite.

1. Introduction

Leaf spring is a simple type of suspension element commonly used in wheeled vehicles. Steel leaf spring is generally high weight so the whole suspension system becomes bulky. Due to the high weight over steer and directional instability may occur during cornering of vehicle so the automobile industry was always looking for better material instead of conventional material. Composite is a material composed of two phases, matrix and reinforced phase here the two phases are distinctly identifiable constituents. The Composite material is not only used in automobile but all other industries like aerospace, marine, civil etc. The composite material gives more advantage than the conventional materials the main advantages of composite material is the weight reduction most of the industry looking to reduce the weight but not compromise the strength so composite is the suitable alternate for the current requirement. The suspension leaf spring is one of the potential items for weight reduction. The main function of the leaf spring is to absorb and store energy and then release it slowly. So, increasing the energy storage capability of a leaf spring makes it more compliant. The amount of elastic energy that can be stored by a leaf spring is given by [1] $S = \frac{1}{2}\frac{\sigma^2}{E}$ where σ is the max allowable stress induced into the spring and E is the modulus of elasticity, both in the longitudinal direction. The equation shows that a material with maximum strength and minimum modulus of elasticity in the longitudinal direction is the most suitable material. Fortunately composite have these characteristics. There are many researcher analyses the composite material for the leaf spring application Rajendran I and Vijayarangan [2] presents an artificial genetic approach to optimize the

IOP Conf. Series: Materials Science and Engineering **183** (2017) 012004 doi:10.1088/1757-899X/183/1/012004

composite leaf spring. The design variable thickness and width are optimized by the use of GA the result makes the impact on weight of the leaf spring 8% to steel leaf spring and 23.4% to composite spring. Quershi H.A Al [1] presents a paper which describes mechanical and geometrical properties of a single leaf, variable thickness spring of glass fiber reinforced plastic (GFRP).E.Mahdi and A.M.S Hamouda[3] described the effect of ellipticity ratio on the leaf spring performance and carbon-glass/epoxy hybrid composite leaf spring is fabricated and the result shows that adding four layers of carbon fiber control the load carrying capacity. Shishay Amare and Gebremeskel[4] explains the the E-glass epoxy composite leaf spring with the fatigue life of $221.16*10^3$ cycles. Satish K G et al.[5] presents a paper on Characterization of In-Plane Mechanical Properties of Laminated Hybrid Composites. Composite material behavior always depend on the various parameters like types of reinforcement, types of resin, fiber orientation , fiber thickness, fiber stacking sequence. Therefore the main objective of the present work is to analyze the effect of stacking sequence in laminated hybrid composite for a leaf spring application.

2. Leaf spring design

The firm is a single source to the world leading automobile manufacturer. The existing capacity of the supplier is 4800 units per month. But the manufacturer increases his demand to 6000 units per month as increases of their customer demand. Hence there is 41% gap to meet the demand.Here weight and initial measurement of four wheeler Mahindra Bolero Pik-up vehicle are taken

Weight of the vehicle	= 1000 kg
Maximum load carrying capacity	= 1150 kg
Total weight	= 2150
Taking factor of safety	= 2.4
Acceleration due to gravity	= 9.8
	= 2150 *2.4*9.8
	= 50568 kg

Since the vehicle is four wheelers, a single leaf spring corresponding to one of the wheels takes up one fourth of the total weight

$$F = \frac{50568}{4} = 12642\ N$$

Design load	= 12642 N
Straight length of the leaf spring (L)	=1680 mm
Design stress	= 490 MPa

Strength of material concept [4]

$$\sigma_{\max} = \frac{6FL}{bh^2} \quad (1)$$

$$\delta_{\max} = \frac{4FL^2}{Ebh^2} \quad (2)$$

Solving equation ot the above for 'h' (thickness of the leaf spring)

$$h = \frac{\sigma_{\max}}{E\delta_{\max}} \quad (3)$$

Rearranging the equation to get width 'b'

$$b = \frac{6F(\frac{L}{2})}{\sigma_{max}h^2} \quad (4)$$

Therefore breath is taken as b = 104mm

Thickness = 20 mm

Camber Calculation [6]

$$\frac{C}{L} = 0.089 \quad (5)$$

Camber = 149 mm

The present work concentrates only a constant cross section because the accommodation of reinforcement fibers must be continuous and uniform distribution of resin is essential for analysis. Design model with the above dimensions are done with the help of CATIA.

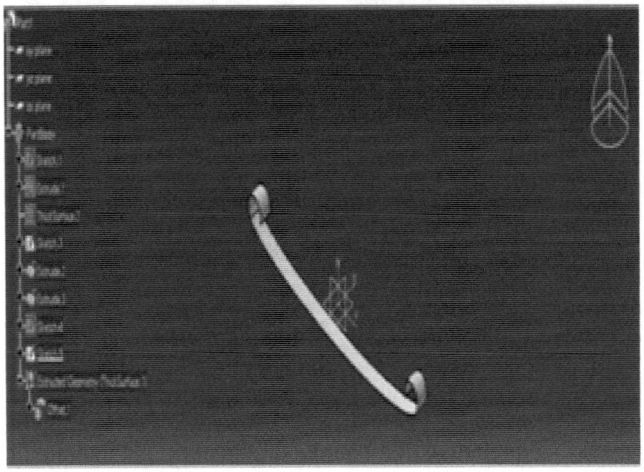

Figure 1. Leaf spring CATIA model

3. Hybrid composite

Hybridization in composite is obtained by introducing more than one fiber in the reinforcement phase in which two or more high-performance reinforcements are combined. The main aim to combine the properties of the different fibers is to get an average property, or even to create new or improved properties. The different reinforcement lamina is stacked with bottom to top order.

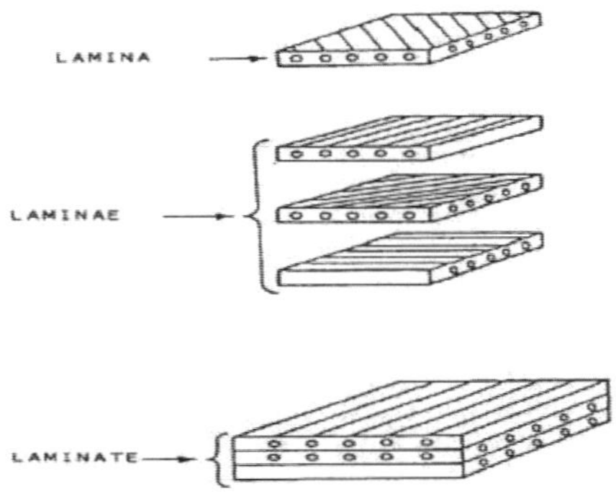

Figure 2. Lamina arrangements

4. Experimental Procedure

4.1 Polymers

In a polymer matrix composite structure the main function of matrix is binding the reinforcement in place and even distribution of stresses among the fibers under the applied force Epoxy resins are reasonably stable to chemical attacks and are excellent adherents having slow shrinkage during curing and no emission of volatile gases. Here L12 (3202) grade epoxy resin used as a matrix material.

Figure 3. Epoxy resin

4.2 Hardener

Hardener is mainly used to cure the epoxy resin which causes a chemical reaction without changing its own composition. The catalyst initiates the chemical reaction of the unsaturated polyester and monomer ingredient from liquid to a solid state. When used as a curing agent, catalysts are referred to as catalytic hardeners. Hardeners are mainly used to increase the polymerization process. The curing time mainly depends on the hardener and epoxy mixing ratio. Here the mixing ratio is 100:10. This makes the fabrication process as per our requirements.

4.3 Mould preparation

The wooden mould is prepared for the fabrication process. The main difficulty in the wooden mould preparation is making the eye design here the loose pattern type wooden roll is used to make the eye design.

4.4 Vacuum bag moulding

The composite leaf spring is fabricated by vaccum bag moulding. First the woden mould is prepared and the E glass, carbon, Graphite fibers are placed on mould layer by layer in different sequence. The main advantage of vaccum bag moulding is the void formation within the composite can be avoided.

Figure 4. wooden mould

5. Result and discussion

The deflection for the load application in various stacking sequence is tabulated. Initial analysis for the first sequence is arranged as first 10 layers of glass fiber and other 10 layers of carbon from bottom to top layer.

Table 1 Deflection for Various Lamina Stacking Sequence

Sl. no	Layer sequence	Deflection mm
1	1-10 glass ,10-20 carbon	133.7
2	1-10 glass ,10-20 graphite	125.9
3	1-10 carbon ,10-20 graphite	100
4	1-5 Glass,6-10 carbon,11-15 glass, 16-20 carbon	141.9
5	1-5 Glass,6-10 graphite, 11-15 glass, 16-20 graphite	135.7
6	1-5 Carbon, 6-10 glass, 11-15 carbon, 16-20 glass	132.5
7	1-5 Graphite, 6-10 glass, 11-15 graphite, 16-20 glass	124.488
8	1-5 graphite, 6-10 carbon, 11-15 graphite, 16-20 carbon	104.653
9	1,4,7,10,13,16,19 –glass, 2,5,8,11,14,17,20 – carbon, 3,6,9,12,15,18 – graphite	125.335
10	1,4,7,10,13,16,19 – carbon, 2,5,8,11,14,17,20 – graphite, 3,6,9,12,15,18 – glass	106.297
11	1-3 ,10-12,19,20 glass, 4-6, 13-15 – carbon, 7-9, 16-18 –graphite	145.49
12	1-3, 10-12,19-20 – carbon , 4-6, 13-15 – graphite, 7-9, 16-18 glass	107.97
13	1-3, 10-12, 19,20 – graphite, 4-6, 13-15 – glass, 7-9,16-18 – carbon	104.02
14	1,2,7,8,13,14,19,20 – glass, 3,4,9,10,15,16 – carbon, 5,6,11,12,17,18 -Graphite	126.435
15	1,2,7,8,13,14 – carbon, 3,4,9,10,15,16 – graphite, 5,6,11,12,17,18 – glass	105.74

The deflection of the leaf spring is shown in fig. The glass fiber having the youngs modulus value of 43000 MPa[7] in the x-direction but when its compared to carbon 177000 MPa so it makes the laminate as stiff as compared to laminate containing only glass fiber. Here the woven fibers are used so it makes the composite with minimum deflection range. The vacuum bag technique also used to remove the excessive resin that will avoid the matrix cracking and voids in the composite is reduced.

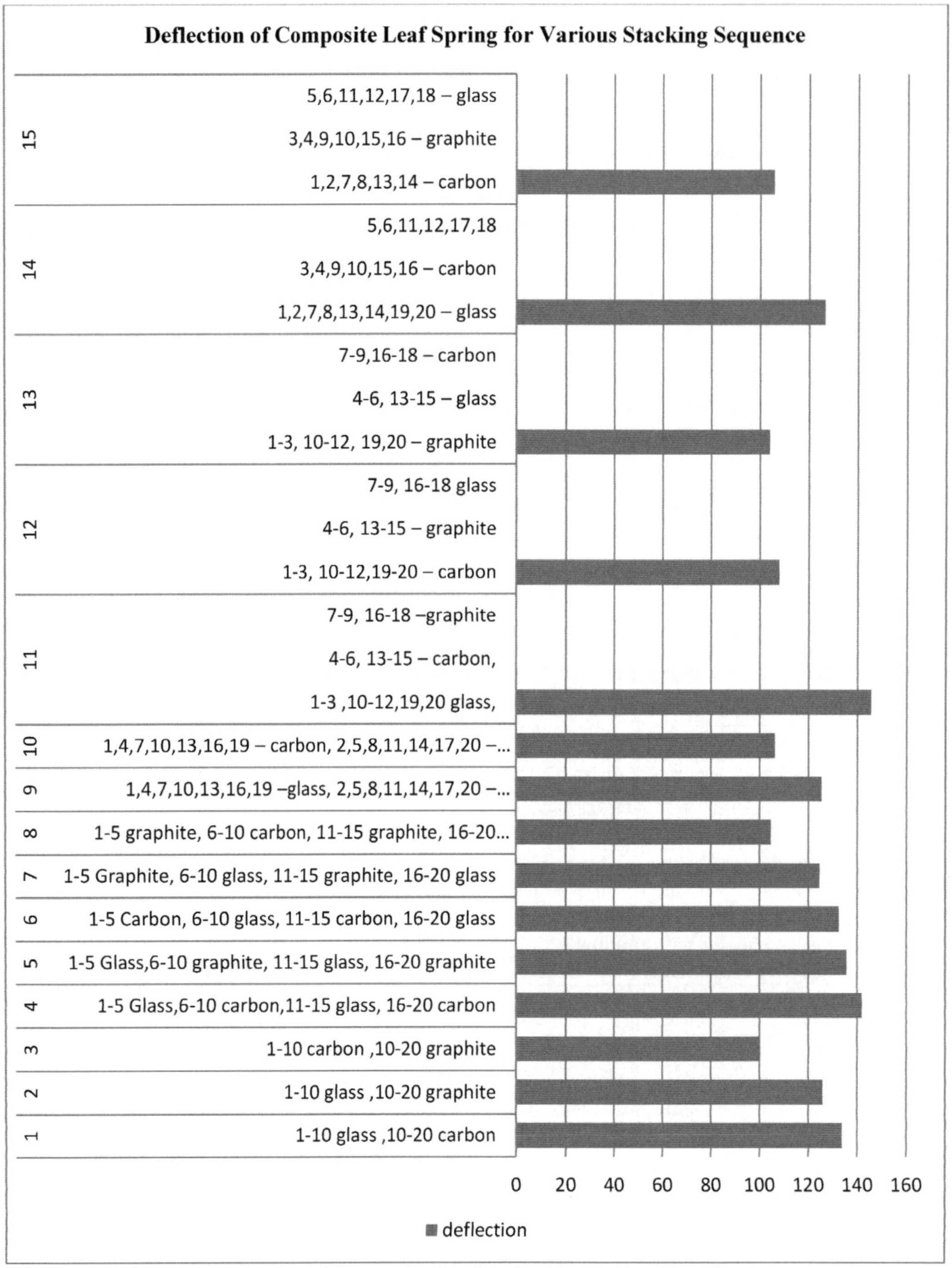

Figure 5. Deflection of Composite Leaf Spring for Various Stacking Sequence

7. Conclusion

The result shows that the introduction of carbon and graphite fibers in the reinforcement phase increases the stiffness of the composite leaf spring. The composite of $[\{0/90/0\}^{gr}\{90/0/90\}^{e\text{-}gl}\{0/90/0\}^{ca}\{90/0/90\}^{gr}\{0/90/0\}^{e\text{-}gl}\{0/90/0\}^{ca}\{90/0/90\}^{gr}]$ stacking sequence gives the more appropriate result over other and it shows that carbon and graphite fiber in the principle direction makes more effective instead of shear direction.

8. References

[1] Al-Qureshi H A 2001 Automobile leaf springs from composite materials *Journal of Material Processing Technology* **118(1)** 58-61.

[2] Rajendran I and Vijayarangan V 2001 Optimal design of composite leaf spring using genetic algorithm *computers and structures* **79(11)** 1121-1129

[3] Mahdi E, Hamouda A M S 2013 An experimental investigation into mechanical behaviour of hybrid and nonhybrid composite semi elliptical springs *Materials and design* **52** 504–513.

[4] Shishay Amare Gebremeskel 2012 Design,Simulation and prototyping of single composite leaf spring for light vehicle *Global Journal of Researches in Engineering Mechanical and Mechanics Engineering* **12(7)** 0975-5861.

[5] Satish K G, Siddeswarappa B, and Mohamed Kaleemulla K 2010 Characterization of In-Plane Mechanical Properties of Laminated Hybrid Composites *Journal of Minerals & Materials Characterization & Engineering* **9(2)** 105-114.

[6] ManasPatnaik, Koushik L P and Manoj Mathew 2012 Determination of Camber and Leaf Span of a Parabolic Leaf Spring for Optimized Stress and Displacement Using Artificial Neural Networks *International Journal of Modern Engineering Research* **2(11)**

[7] PankajSaini, AshishGoel and Dushyantkumar 2013 Design and analysis of composite leaf spring for light vehicles *International journal of innovative research in science, Engineering and technology* **2(5)** 2319-8753

Experimental Investigation of Thermal Properties in Glass Fiber Reinforced with Aluminium

S Joseph Irudaya raja[1] , T Vinod Kumar[2], R Sridhar[3] and P Vivek[4]
[1]Assistant Professor, Department of Mechanical Engineering, Vels University, Chennai, T.N, India.
[2]Assistant Professor, Department of Mechanical Engineering, Vels University, Chennai, T.N, India.
[3]Assistant Professor, Department of Mechanical Engineering, Vels University, Chennai, T.N, India.
[4]Assistant Professor, Department of Mechanical Engineering, Vels University, Chennai, T.N, India.

E-mail: joerajj78@gmail.com

Abstract. A test method of a Guarded heat flow meter are used to measure the thermal conductivity of glass fiber and filled with a aluminum powder epoxy composites using an instrument in accordance with ASTM. This experimental study reveals that the incorporation of aluminum and glass fiber reinforced results in enhancement of thermal conductivity of epoxy resin and thereby improves its heat transfer capability. Fiber metal laminates are good candidates for advanced automobile structural applications due to their high categorical mechanical and thermal properties. The most consequential factor in manufacturing of these laminates is the adhesive bonding between aluminum and FRP layers. Here several glass-fiber reinforced aluminum were laminates with different proportion of bonding adhesion were been manufactured. It was observed that the damage size is more preponderant in laminates with poor interfacial adhesion compared to that of laminates with vigorous adhesion between aluminum and glass layers numerically calculated ones and it is found that the values obtained for various composite models using experimental testing method.

1. Introduction

The most significant among them is their response to impact loading. A structure is subjected to an impact force when a foreign object hits it [1]. For instance, the loads imparted by dropped tool on the bonnet cover of car body, bird hit and runway debris on an aircraft engine are typical example of impact loads [6]. In its most rudimentary form a composite material is one, which is composed of at least two elements collaborating to engender material properties that are different to the properties of those elements on their own [2]. In practice, most composites consist of a bulk material (the 'matrix'), and a reinforcement of some kind, integrated primarily to increment the vigor and stiffness of the matrix [7]. This reinforcement is conventionally in fiber form. It withal avails to achieve directional properties. Reinforcements may be in the form of fibers, particles or flakes [3]. The fiber factors which contribute to the mechanical performance of a composite are length, orientation, shape and material. Matrix gives shape and forfends the reinforcement from the environment[4]. It also makes the individual fibers of the reinforcement act together and provides transverse shear strength and stiffness to the laminated composites[5].

2. Materials and Methods

In this laminate, Reinforcement such as Glass Fiber Reinforcement Plastic and E- Glass with Epoxy are utilized. The epoxy is having the correct ratio of resin and hardener is 10:1 where Resin: LY556 and Hardener HY951.The glass fibers come in variety of forms predicated on silica (SiO2) which is amalgamated with other elements to engender specialty glass [8]. Fibers of glass are engendered by extruding molten glass, at a temperature around 1200°C through apertures in a spinneret with diameter of 1 or 2 mm. Aluminum has a unique and unvanquishable cumulation of properties that make it into a multifarious, highly utilizable and alluring construction material

Figure.1 Glass Fiber

2.1 Laminate Preparation

Glass fiber with aluminum bidirectional type as shown in figure 2. is expected to withstand more stress. Albeit the method has been superseded with automated techniques, the lay-up of preimpregnated material by hand is the oldest and most mundane fabrication method for advanced composite structures such as laminated glass fiber as shown in figure.

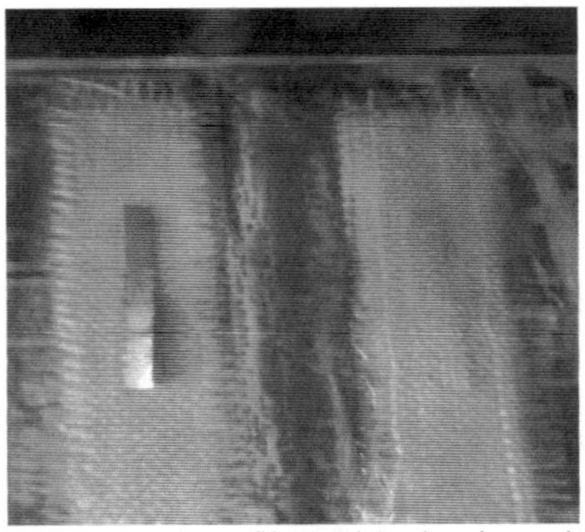

Figure.2 Glass fiber with aluminum bidirectional type

Figure.3 Glass Fiber Laminated

3. Experimental Procedure

In the present work is to investigate the mechanical properties such as Flexural, Tensile and Impact Vigor of glass fiber epoxy laminate with and without Aluminum alloy. Tensile load applied to a composite as shown in figure 4. The replication of a composite to tensile loads is very ward on the tensile rigidity and vigor properties of the reinforcement fibers, since these are far higher than the resin system on its own. Test was carried out with the avail of UTM (Macrocosmic Testing Machine).

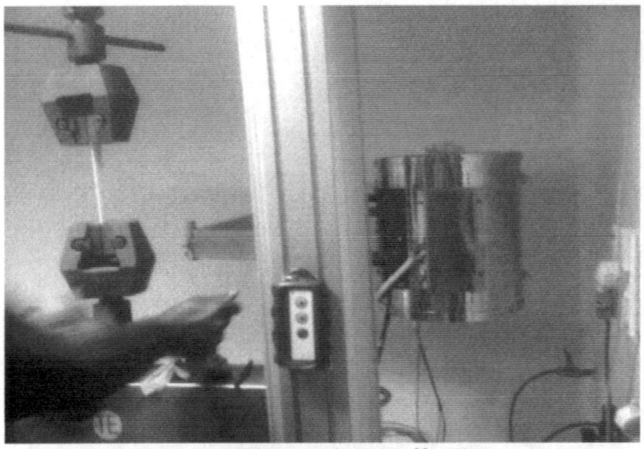

Figure 4. Tensile Test

A beam has an unpredictable cross section area where subjected to a crosswise loads the beam will bend as shown in figure 5. In addition to bending the other effects such as crumple and buckling may occur, and to scrutinize a problem that includes all the compound effects of crumple, bending and buckling could become a convoluted one. Thus we are paying attention to scrutinize the bending property alone we have to put certain restrictions on the geometry of a beam and the manner of loading.

Figure 5. Compression Test

Static tension tests of the un-notched specimen's do not always divulge the sensitivity of metal to brittle fracture. This paramount factor is purposeful in impact test. In impact tests we utilize the notched specimen's as shown in figure 6. This specimen is settle on its fortifies on anvil so that gust of the striker is antithesis to the notch an impact vigor is defined as the energy A, required to break the specimen, Impact Vigor = A / f Where f = A cross sectional area of the specimen in cm^2 at crack & conspicuously at notch. The main purport of jagged – bar tests is to study the collateral effect of stress concentration and high velocity load application Impact test are of the most astringent type and facilitate brittle friction.

Figure 6. Impact Test

4. Results and Graphs

Ultimate tensile load (KN) : 6.81
Ultimate tensile strength (MPa (or) N/mm^2) : 90.87

4.1. Tensile Test

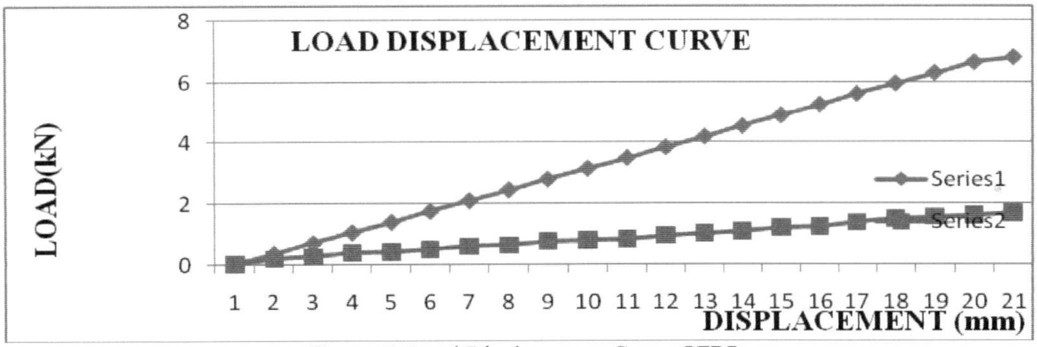

Figure 7. Load Displacement Curve GFRP

The Tensile test curve shows when the load applied on the specimen increases, the displacement also made of Glass Fiber Reinforced Polymer increases. The figure 7. indicates the load and displacement on tensile test for two different specimens.

4.2. Stress Vs Strain Graph

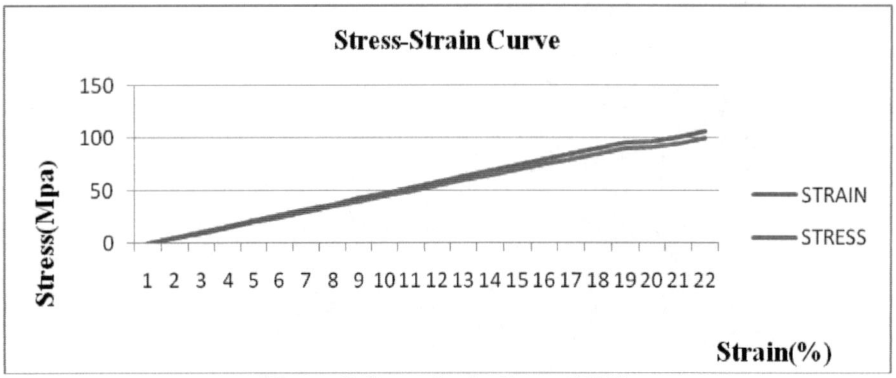

Figure 8. Stress Strain Curve GFRP

Figure 8. shows that when the stress of the specimen increases the strain also increases on glass fiber reinforced polymer.At the final stage stress is reduced compared to the strain curve.

4.3 Compression Test

Ultimate compressive load (KN) : 0.43
Ultimate compressive strength (MPa (or) N/mm2) : 5.733

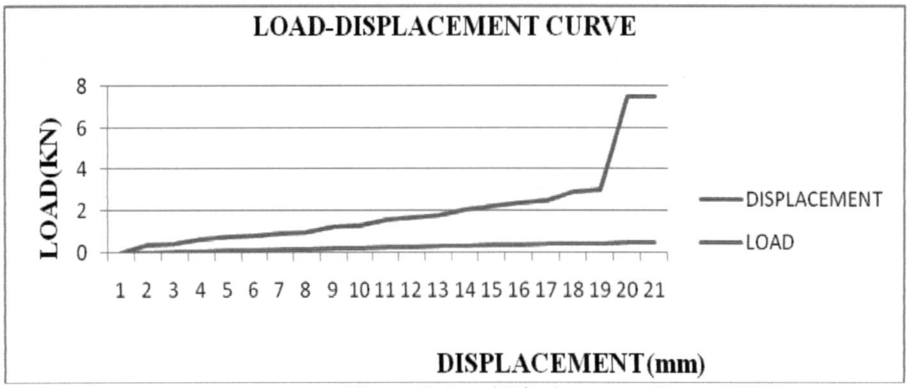

Figure 9. Load Displacement Curve GFRP

Figure 9. Shows the plot between load and displacement, initially when the load increases displacement also increases gradually and further when load reaches the maximum limit then the displacement will also reaches maximum. The load curve implies a linear distribution.

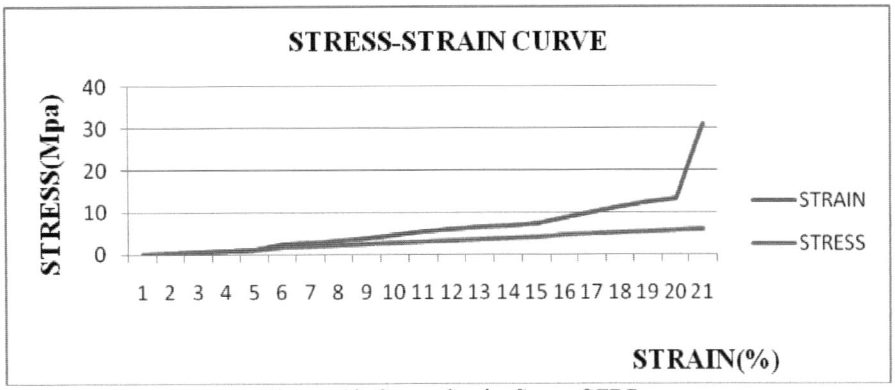

Figure 10. Stress Strain Curve GFRP

The plot between stress and strain for Glass Fiber Reinforced Polymer as shown in figure 10. represents when the stress increases strain gradually increases. Up to certain limit stress increases proportional to the strain, at the final stage stress limited to a minimum value.

Table 1. Impact Test GFRP

Test parameters	Sample 1	Sample 2	Sample 3	Average
Absorbed Energy-Joules	6	6	6	6

5. GFRP- Al Result

5.1 Tensile Test

Ultimate tensile load (KN) : 7.90
Ultimate tensile strength (MPa (or) N/mm2) : 105.33

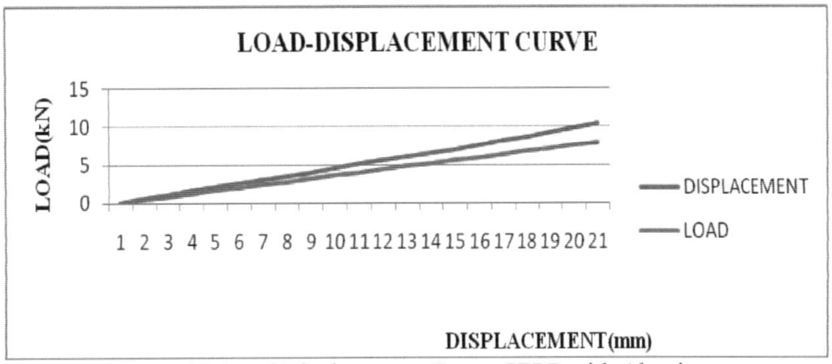

Figure 11. Load Displacement Curve GFRP with Aluminum

The figure 11. Shows the plot between load and displacement for GFRP with Aluminum where load absorbing capacity is very high compared to glass fiber reinforced polymer.

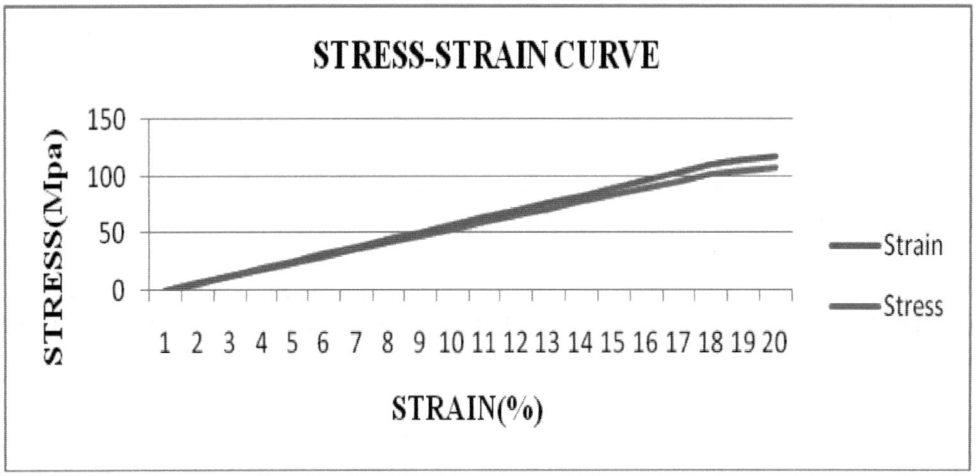

Figure 12. Stress Strain Curve GFRP with Aluminum

The figure 12. shows the plot between stress and strain for GFRP with Aluminum where the stresses is reduced due to reinforcement with aluminum.

5.2 Compression Test

Compressive Load KN : 0.66
Compressive strength in Mpa : 8.80

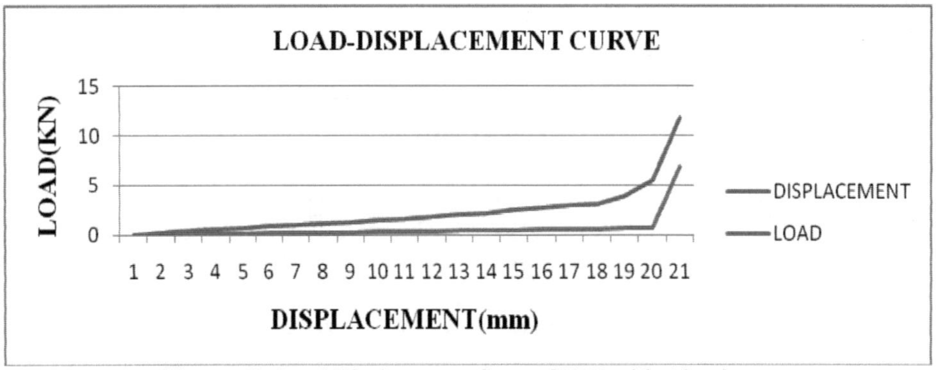

Figure 13. Load Displacement Curve GFRP with Aluminum

The figure 13. shows the plot between load and displacement for GFRP with Aluminum where displacement capacity is very high compared to glass fiber reinforced polymer.

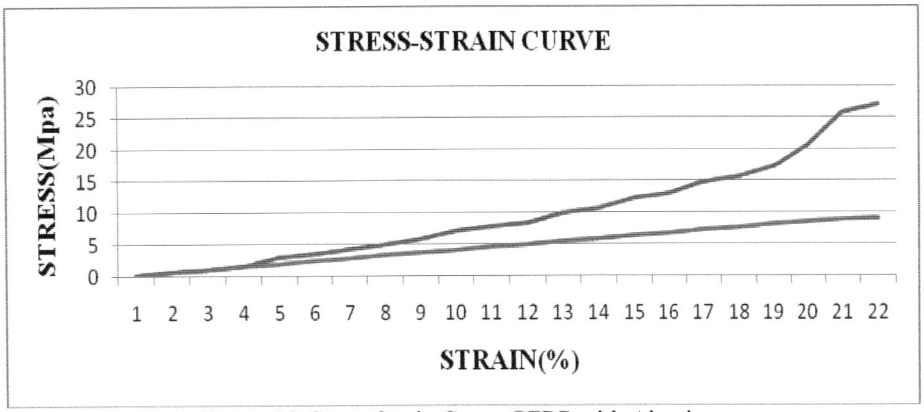

Figure 14. Stress Strain Curve GFRP with Aluminum

Table 2. Impact Test GFRP with Aluminum

Test parameters	Sample 1	Sample 2	Sample 3	Average
Absorbed Energy-Joules	8	6	8	7.33

6. Conclusion

The following are concluded from the above experimental analysis

- Impact Test performed for GFRP with Aluminum shows that higher energy absorbing capacity than GFRP.
- From the obtained results, it is found that the GFRP reinforced with aluminum is providing better thermal and mechanical properties.
- Further it will improve the performance of the pressure vessel and domestic applied gas cylinder also.
- It is also found that the stresses are reduced due to reinforcement with aluminum.

7. References

[1] Botelho E C, Silva R A, Pardini L C and Rezende M C 2006 A review on the development and properties of continuous fiber/epoxy/aluminum hybrid composites for aircraft structures. *Mater Res* **9(3)** 247–56.

[2] Bernhardt S, Ramulu M and Kobayashi A S 2007 Low-velocity impact response characterization of a hybrid titanium composite laminate. J Eng Mater Technoly **129** 220–6.

[3] Villanueva G R and Cantwell W J 2004 The high velocity impact response of composite and FML-reinforced sandwich structures *Compos Sci Technoly* **64** 35–54.

[4] Beumler T, Pellenkoft F, Tillich A, Wohlers W and Smart C 2006 Airbus costumer benefit from fiber metal laminates *Airbus Deutschland GmbH* Ref. no: L53pr0605135 **1** 1–18.

[5] Alderliesten R C and Benedictus R 2007 Fiber/metal composite technology for future primary aircraft structures In *48th Aiaa/Asme/Asce/Ahs/Asc structures, structural dynamics, and materials conference* 1–12.

[6] Cortes P and Cantwell WJ 2006 The prediction of tensile failure in titanium-based thermoplastic fiber–metal laminates *Compos Sci Technoly* **66** 2306–16.

[7] Asundi A and Choi Alta Y N 1997 Fiber metal laminates: *an advanced material for future aircraft* J Mater Process Technoly **63** 384–94.

[8] Vogelesang LB and Vlot A 2000 Development of fibre metal laminates for advanced *J Mater Process Technoly* **103** 1–19

Effect of Coupling Agent on Mechanical Properties of Palm Petiole Nanofiber Reinforced Composite

T Vinod Kumar[1], M Chandrasekaran[2], V Santhanam[3] and V D Sudharsan[4]
[1]Research Scholar& Assistant professor, Department of Mechanical Engineering, Vels University, Chennai, Tamil Nadu, India.
[2]Professor, Department of Mechanical Engineering, Vels University, Chennai, Tamil Nadu, India.
[3]Professor, Department of Mechanical Engineering, Rajalakshmi Engineering College, Chennai, Tamil Nadu, India.
[4]PG Student, Department of Mechanical Engineering, Rajalakshmi Engineering College, Chennai, Tamil Nadu, India.

E-mail:vinodkmrmech@gmail.com

Abstract. Composites have replaced conventional materials due to their advantages such as low cost, low density, high strength, etc. recently the research has been focused on natural fibers as a suitable replacement for synthetic fibers for reinforcement in composites. Hence, in this work, natural fiber is extracted from palm petiole and this fiber is investigated for reinforcement in epoxy polymer. The fiber is subjected to different chemical treatments for enhancing the surface wetting and Nano cellulose will also be extracted from the fiber. The composite will be fabricated by using handlay-up method by using 1%, 2%, 3% volume fraction of cellulose nanofiber in the epoxy resin. It is observed that the increase of the nanofiber in 3% volume fraction gives the good tensile, flexural and impact strength. So they obtained chemically treated cellulose nanofiber was treated with the amino propyl triethoxysilane and 3% of the silanesilane treated nano-fiber fabricated. Then the mechanical testings of this is compared with the 3% of chemically treated nano-fiber composite. The silane treated nanofiber gives the best results.

1. Introduction

The natural composite have a superior composites compare to the conventional composites [1]. The natural fibers which include banana, coconut, jute, kenaf, palm, sisaletcare widely used as a filler in the composite materials [2]. The applications of the composites are automotive, electronics, packaging, etc... The Cellulose fibers will be low in density, low in the cost, renewable. The cellulose fibers are best in strength and stiffness. The cellulose fibers have a many advantages while compare to the synthetic fiber,they can reduce the environmental problem caused by non-biodegradable materials. The disadvantage of the cellulose fiber is moisture absorption, low thermal stability and poor compatibility with the hydrophobic polymer matrix [3]. In India, Palm trees have a more productive so they are easily available. They also good in the biodegradable.The nano sized cellulose nano-fibers have a good strength and stiffness. They also have good interfacial adhesion with polymer matrices while comparing to the microfibrills [4]. The many processes are used to extract the cellulose nano-fiber and chemical treatment is the one among them. The chemical treatment has a three steps. They are Alkali treatment, Bleaching process and acid hydrolysis process [8]. The commonly used acids are NaOH, NaOCl and H2SO4.The chemical treatment will remove the hemicelluloses, lignin. They can break the hydrogen bonds between the hydroxyl groups and they will improve the surface roughness. The silane coupling agent which bonds with organic and the inorganic materials. They will give a good adhesion, surface modification and a mechanical strength to the composite while comparing to the non silane treated composite materials [6,7]. Here amino propyl triethoxysilane is used for coupling agent.

IOP Conf. Series: Materials Science and Engineering **183** (2017) 012006 doi:10.1088/1757-899X/183/1/012006

2. Experimental

2.1*Materials, Extractions and Fabrications of Cellulose Nanofibers*

Palm petiole fiber was extracted naturally from own source(Pulicat, Tamil Nadu, India), Sodium hydroxide used for alkali treatment (donated by chemistry department) , Sodium hypochlorite (umascientifics) and acetic acid (donated by chemistry Dept) was used for the bleaching process. Sulphuric acid (donated by chemistry Dept) was used for the acid hydrolysis process. Amino propyl triethoxysilane used for a coupling agent (VinayakAuxiChem, Bangalore, India).The natural palm petiole fiber was cut from the tree and it is kept in the room temperature to dry its moisture content for 2 – 4 days. Then the fiber was extracted easily. In the extraction process of the cellulose nanofibers, the alkaline, bleaching and acid hydrolysis process is used. The palm petiole fiber was chopped and sieved. In the alkaline treatment, Thethefiber is dispersed in the NaOHto remove the hemicelluloses content present in the fiber and it is stirred for 1hr at the 350C. Then it is centrifuged for several times by adding the distilled water until the fiber reaches the neutral stage. This treatment can remove the hemicelluloses in the fiber and also used to remove the intermolecular hydrogen bonding. This treatment will repeat once again after the bleaching process. The bleaching treatment, this process is carried out after the alkali treatment. The lignin content was removed during this process and also it uses to remove the color of the fiber. The alkali treated fiber was treated with the sodium hypochlorite and aqueous acetic acid.It is stirred for 1 hr at 350C, then it is centrifuged many times by adding distilled water until the fiber reaches pH.

The alkali treatment is repeated once again after the bleaching process and it washed still it turns neutral. The acid hydrolysis treatment followed. This fiber is dispersed in the sodium hydroxide and it is stirred for 3 hrs at 350C. Then it is centrifuged for 15mins at 7000rpm. Then it is sonicated in the ultrasonic cleaner to obtain the nano sized fiber [5]. The obtained cellulose nano-fiber was treated with aminopropyltriethoxysilane for 8hrs which acts as a coupling agent [9]. They will improve the thermal properties of the nano-fibers.

In the fabrication, the 1% of chemically treated nano-fiber was incorporated with the 99% of the epoxy resin. The2% of chemically treated nano-fiber was incorporated with the 98% of the epoxy resin. The3% of chemically treated nano-fiber was incorporated with the 97% of the epoxy resin. After obtaining the better mechanical properties [10] in the above plates. The silane treated nano-fiber is fabricated in that ratio.

3. Result and Discussion

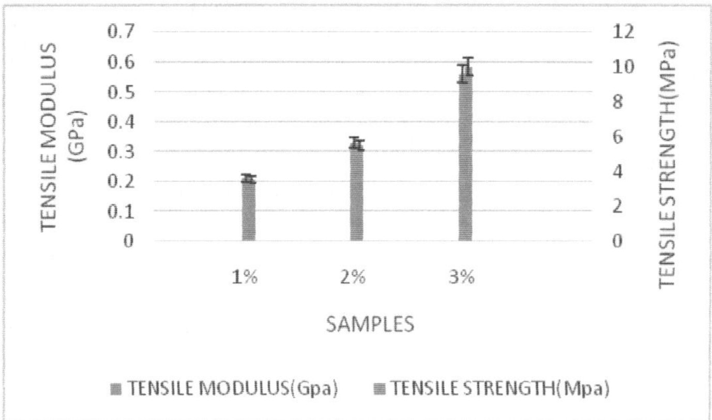

Figure 1. Graph of tensile strength and tensile modulus.

In the case of tensile strength and tensile modulus. Figure 1, shows that the 3% chemically treated nano-fiber reinforced composite have a better tensile strength and modulus while compared to the 1% and 2% chemically treated nano-fiber reinforced composites. It has 9.615 MPa tensile strength and 0.584GPa of tensile modulus. While 1% and 2% of chemically treated nano-fiber reinforced composites have 3.615MPa, 5.641MPa of tensile strength and 0.206GPa, 0.321GPaof tensile modulus respectively.

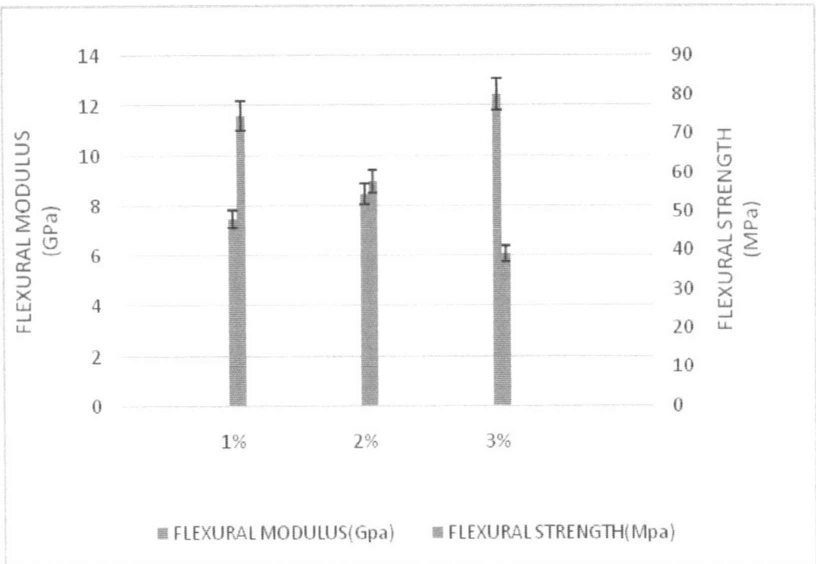

Figure 2. Graph of flexural strength and flexural modulus.

From the figure 2, The composite with the 1% and 2% of the chemically treated nano-fiber have a flexural strength 48.076 Mpa and 54.48 Mpa. Here we can see the rise in the flexural strength due to the addition of nano-fibers. In the 3% of the nano-fiber composite, they give 80.12 Mpa of the flexural strength. Which is greater than the 1% and 2% of the nano-fibers.In the flexural modulus, the 1% of the fiber composite have 11.61 GPa and 2% of the nano-fiber composite. The 3% of the nano-fiber composite have very low flexural modulus of 6.07% compared to the 1% and 2% of the nano-fiber composites. Here we can see the fall in the flexural modulus due to the addition of the nano-fibers.

IOP Conf. Series: Materials Science and Engineering **183** (2017) 012006 doi:10.1088/1757-899X/183/1/012006

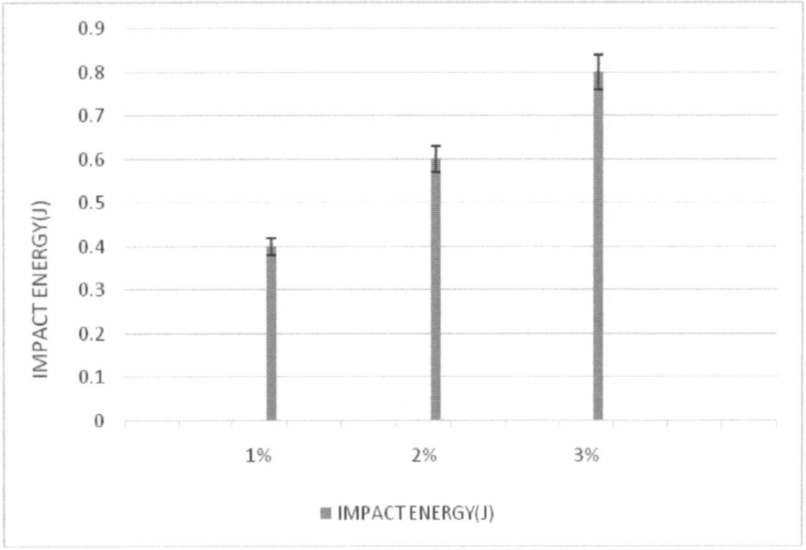

Figure 3. Graph of the impact energy.

In the case of impact energy from the figure 3, it shows that 1% and 2% of the chemically treatednano-fiber composites have a low impact energy of 0.4 J and 0.6 J while compare the 3% of the nano-fiber composites which have 0.8 J.

From this above study, we obtain that 3% of the chemically treated palm petiole nano-fiber have a good mechanical properties while compared to the 1% and 2% of the nanofiber composites. So it is compared with the silane [11]coupling agent treated nano-fiber in the same 3% of the coupling agent treated nano-fiber.

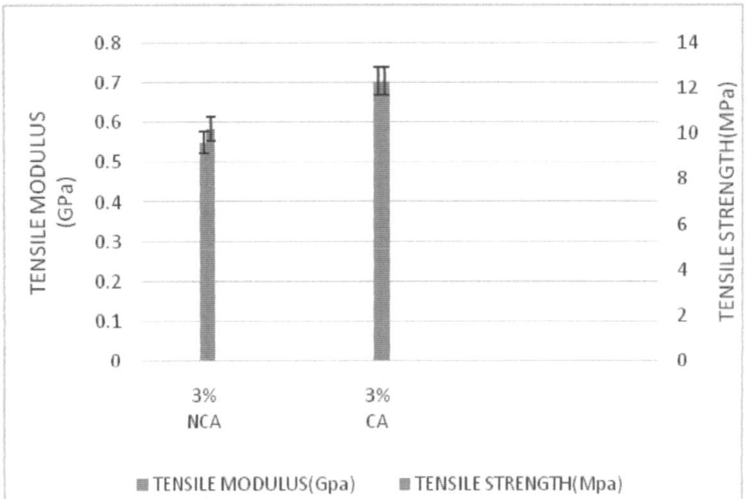

Figure 4. Comparison of NCA and CA for tensile modulus & tensile strength.

From the figure 4, we can see that the chemically treated nano-fiber after treated with the coupling agent (CA) have an improvement in the tensile properties while compare the non coupling agent(NCA) composite.

The tensile strength and tensile modulus of the 3% of the coupling agent nano-fiber composite have 12.308 MPa and 0.703GPa. But the 3% of non coupling agent composite have 9.615Mpa and 0.584Gpa of tensile strength and tensile modulus.

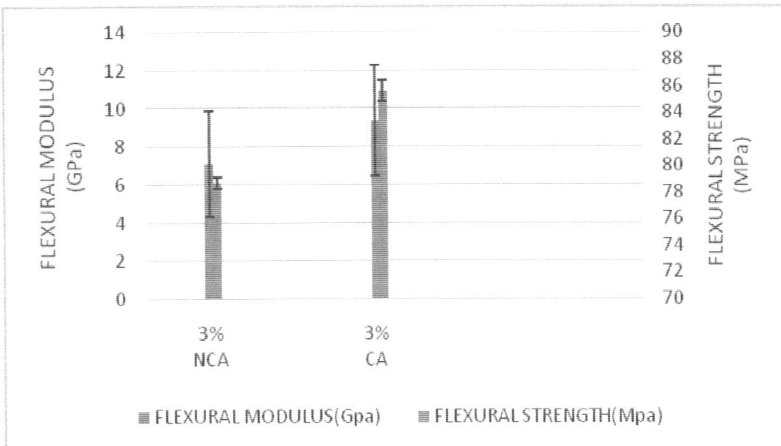

Figure 5. Comparison of NCA and CA for flexural strength and flexural modulus.

In the figure 5, it shows that 3% of the coupling agent composite have 83.33maps of flexural strength and 10.93GPa of flexural modulus. In the 3% of the non coupling agent composite have a 80.12MPa and 6.07GPa of flexural strength and flexural modulus. The coupling agent gives better adhesion with fiber and resin which helps in improving its flexural properties.

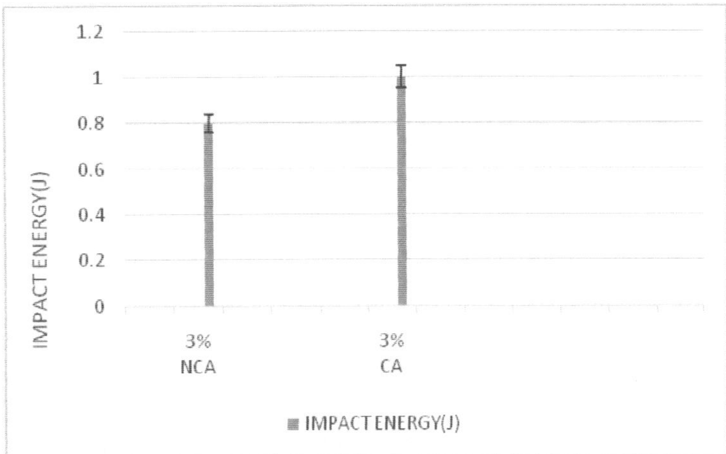

Figure 6. Comparison of NCA and CA for impact energy.

In the figure 6, impact testing also we can see that coupling agent treated composite have an impact energy of 1J while the impact energy of the non coupling agent composite has 0.8J. Which is lesser than the coupling agent treated composite.From this above study, we can observe that the silane coupling agent treated composite have a good mechanical properties while compare to the non silane treated coupling agent.

4. Conclusion

The cellulose nano-fiber was extracted through the chemical process. It is fabricated with the epoxy resin with various amounts of the nano-fiber in 1%, 2% and 3% .For the 3% of the chemically treated nano-fiber composite gives tensile strength and tensile modulus of 9.615MPa and 0.584GPa respectively. In the flexural testing they have 80.12maps of flexural strength and 6.07GPA of flexural modulus. They have a 0.8J of the impact energy. The 3% of nano-fiber composite gives better mechanical while compared to 1% and 2% of the nano-fiber composites. But 3% of the silane coupling agent treated nano-fiber gives a 12.308MPa tensile strength and 0.703GPa tensile modulus. In the flexural testings, they give 83.33maps of flexural strength and 10.93GPa of flexural modulus. In the impact testing they give 1J of impact energy. From this, we can obtain that 3% of silane coupling agent treated composite gives a better improvement in tensile, flexural and impact testings.

5. Reference

[1] Sharma N K and Kumar V 2013 Studies on properties of banana fiber reinforced green composite *Journal of Reinforced Plastics and Composites* 0731684412473005.

[2] De Oliveira P F and Maria de Fátima V M 2013 Comparison between coconut and curauafibers chemically treated for compatibility with PP matrixes *Journal of Reinforced Plastics and Composites* 0731684413516392.

[3] Wicaksono R, Syamsu K, Yuliasih I, Nasir M and Street K 2013 Cellulose nanofibers from cassava bagasse: Characterization and application on tapioca-film *Cellulose* **3(13)**79-87.

[4] Chen W, Yu H, Liu Y, Hai Y, Zhang M and Chen P2011 Isolation and characterization of cellulose nanofibers from four plant cellulose fibers using a chemical-ultrasonic process *Cellulose* **18(2)** 433-442.

[5] Kim HH, Kim S Y, Kim D H, Oh C Y and Jo N J 2014 Effect of Silane Coupling Agent on the Flexural Property of Glass Fiber Reinforced Composite Film *Journal of Materials Science and Chemical Engineering* **2(10)** 38.

[6] Threepopnatkul P, Kaerkitcha N and Athipongarporn N 2009 Effect of surface treatment on performance of pineapple leaf fiber–polycarbonate composites *Composites Part B: Engineering* **40(7)** 628-632.

[7] Takagi H, Nakagaito A N and Bistamam M S A 2013 Extraction of cellulose nanofiber from waste papers and application to reinforcement in biodegradable composites *Journal of Reinforced Plastics and Composites* **32(20)** 1542-1546.

[8] Singh A A and Palsule S 2013 Coconut fiber reinforced chemically functionalized high-density polyethylene (CNF/CF-HDPE) composites by Palsule process *Journal of Composite Materials* 0021998313513045.

[9] Sathishkumar T P, Nananeethakrishnan P, Shankar S and Rajasekar R 2013 Investigation of chemically treated longitudinally oriented snake grass fiber-reinforced isophthallic polyester composites *Journal of Reinforced Plastics and Composites* 0731684413495321.

[10] Deák T, Czigány T, Tamás P and Németh C 2010 Enhancement of interfacial properties of basalt fiber reinforced nylon 6 matrix composites with silane coupling agents *Express PolymLett* **4(10)** 590-598.

Estimation of fracture parameters for Al-SiC and Al-Fe_2O_3 metal matrix composites

S Jacob[1], R Sridhar[2] and S Joseph Irudaya Raja[3]
[1]Assistant professor, Department of Automobile Engineering,Vels University Chennai, Tamilnadu, India
[2]Assistant professor,Department of Mechanical Engineering, Vels University, Chennai, Tamilnadu, India
[3]Assistant professor,Department of Mechanical Engineering, Vels University, Chennai, Tamilnadu, India

E-mail: jacobthermal@gmail.com

Abstract. In this paper silicon carbide and iron oxide is reinforced with aluminum matrix. The prepared composite provide high strength than the commercial composite. It provides high elastic properties and high service temperature. Since this material have great scope in space application it has been chosen for estimation of fracture toughness. Hence grain size of the samples increased this is due to the effect of reinforcement bonding with aluminum matrix. The CMOD test has been carried out for this Purpose. Microscopic examination has shown that with increase in of reinforcement test confirmed that with increase in reinforcement ratio the fracture toughness also increased. In this work specimens have been preferred with and without addition of SiC and Fe_2O_3 to compare the influence. The results are compared and it has been studied that Fracture Toughness increased significantly by addition of inclusions.

1.Introduction

Metal matrix composites offer the ameliorated concrete properties as well as incremented elevated temperature performance relative to their unreinforced alloys. These are critical impute for many aero engine applications, making composites alluring contempt their higher cost. For brittle matrix systems such as inter-metallic, reinforcement can withal be utilized as an expedient of ameliorating toughness. In this paper, a wide range of long fiber and discontinuously reinforced composite systems are considered from an aero engine perspective predicated on Aluminium, titanium, and inter-metallic matrices peregrinating from moderate-temperature systems towards sultry ones. Processing methods felicitous to each type are described [2]. Properties such as concrete stiffness, strength, creep, and fatigue performance are considered. Another paramount aspect is the predisposition to engender residual stresses as a function of the discrepancy in thermal expansion between the phases [3]. In the final section a range of potential aero engine applications is examined. In this paper, stir casting process was adopted for preparing Aluminum matrix composite which is reinforced with various weight fractions (1% and 3%) and investigations are carried out by using SEM, EDAX, and CMOD [4]

2.Methodology

In this work stir casting is been followed for preparing the composites. In stir casting process uniform distribution of reinforcing particles is ensured by adding 5gms of reinforcement with 1 kg of Aluminum. The Aluminum is melted in a crucible by heating it in a muffle furnace at 800°C for one hour. The grain size of the Aluminum oxide particles used is 150 μm. The furnace temperature was first raised above the liquidus temperature of Aluminum near about 950 °C to melt the Al alloy thoroughly [6].A pink colored powder (Scum powder) Hexochloroethane is added to the melted Aluminum to remove the Slag from the melt. A degassing flux is added to remove the bubbles from the melt. Then the melted Aluminum is cooled down just below the liquidus to keep the slurry in Semi solid state.Automatic stirring was carried out with the avail of radial stirrer operated for about 10 minutes at stirring rate of 125 rpm. In order to minimize gas stored in the molten matrix during stirring, the stirring speed and the stirrer location was placed at a concrete location [7]. At this stage, the preheated SiC and Fe_2O_3 particles were integrated manually to the vortex. In the final commixing processes the furnace temperature was controlled within 700 ± 10°C. After stirring process the melt was poured in the sand mould to get desired shape of specimen in Figure 1.

Figure 1. Fabrication of composites

3. Results and Discussion

3.1. Optical Microscopic Examinations

Sample: 1 Al

Imported model in ANSYS workbench

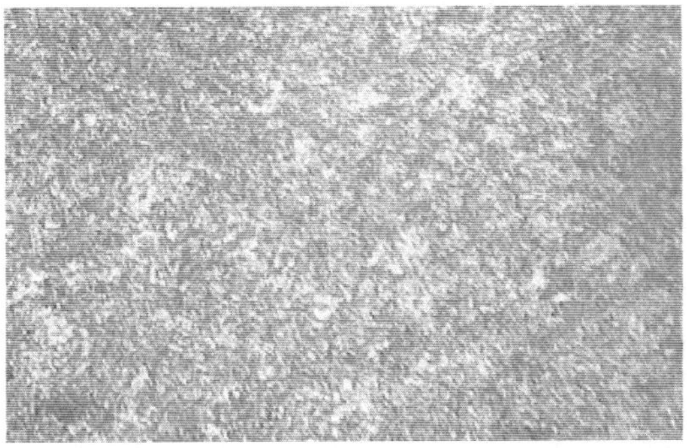

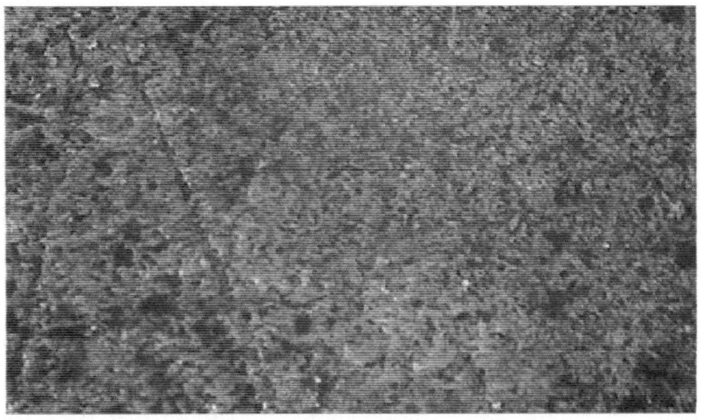

Magnification: 100X Etchant: H.F. soln

The metal matrix shows fine grains of Al-Si eutectic in aluminum solid solution. The constituents of phases present in the matrix are $Mg_2Si$and some $Cu\text{-}Al_2$ phases that are present in random distribution. As it is chill cast, the dendritic pattern of the grains is cored towards the direction of the chilled plate. No directional orientation could be seen in castings. The scanning of the plate in cross-section shows freedom from casting defects like shrinkage and micro porosities. The grains are finer due to the additives used and being a small area volume casting.

Sample:2 Al-SiC (1%)

Magnification: 100X Etchant: H.F. soln

The metal matrix shows fine grains of Al-Si eutectic in aluminium solid solution. The constituents of phases present in No directional orientation could be seen in castings. The scanning of the plate in cross-section shows freedom casting defects like shrinkage and micro porosities. The grains are finer due to the additives used and being a small area /volume casting. The presence of SiC in the matrix is uniform and present as dark particlesthe matrix are $Mg_2Si$and some $Cu\text{-}Al_2$ phases that are present in random distribution. As it is chill cast, the dendritic pattern of the grains is cored towards the direction of the chilled plate

Sample: 3 Al-Fe_2O_3 (1%)

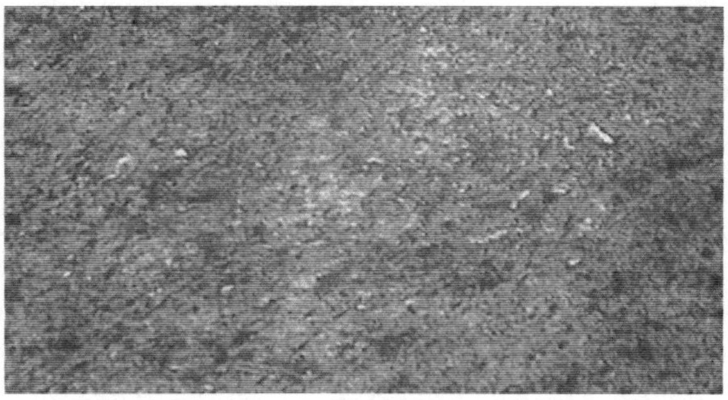

Magnification: 100X Etchant: H.F. soln

The metal matrix shows fine grains of Al-Si eutectic in aluminium solid solution. The constituents of phases present in the matrix are $Mg_2Si$and some $Cu\text{-}Al_2$ phases that are present in random distribution. As it is chill cast, the dendrite pattern of the grains is cored towards the direction of the chilled plate. No directional orientation could be seen in castings. The scanning of the plate in cross-section shows freedom casting defects like shrinkage and micro porosities. The grains are finer due to the additives used and being a small area /volume casting. The presence of Fe_2O_3 could be seen but cannot be confirmed. The microstructure by using SEM followed by EDAX can give the composition of the particles.

3.2. Crack mouth opening displacement test
In the CMOD test the specimens are proportional. If the thickness is represented by 'A', then the depth will either be 'A', for a square cross section, or, '2A' for a rectangular cross section with the standard length being '4.6A'.To prepare a specimen for a CMOD test, a incision on the surface is machined in the centre of the specimen and then an authentic fatigue crack is conscientiously induced at the base of the notch. The crack must be long enough to pass through any area showing plastic deformity caused by the machining process

Figure 2.CMOD Test Specimen

Figure.3.CMOD Testing Sample

3.3. EDAX Test

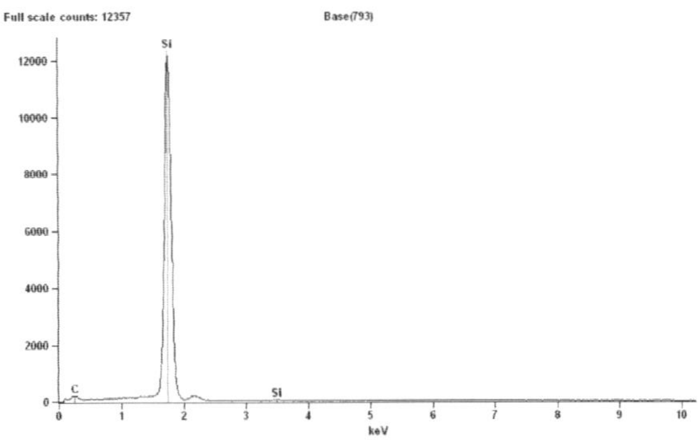

Figure 4. Quantitative Results for: Base(793)

IOP Conf. Series: Materials Science and Engineering **183** (2017) 012007 doi:10.1088/1757-899X/183/1/012007

EDAX Test Results For Fe_2O_3

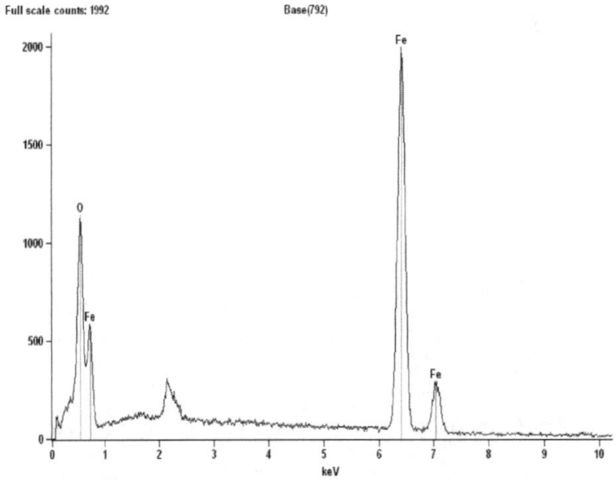

Figure.5. Quantitative Results for: Base(792)

Table 1. Quantitative Results for: Base(792)

Element	Net Counts	Weight %	Atom %
O	10354	14.66	37.49
Fe	35993	85.34	62.51
Fe	6387	---	---
Total		100.00	100.00

Table 2. J Integral test values for Al

Rec No	Max Load	Max Disp	Compliance	Crack Length	Kmax	Jmax	Jel	Jpl	Area Total	Area Elastic	Area Plastic
	kN	mm	mm/kN	mm	MPa/m	kJ/sq-m	kJ/sq-m	kJ/sq-m	kN-mm	kN-mm	kN-mm
1	0.278	0.089	0.220259	14.526	6.244	0.632	0.504	0.008	0.014	0.012	0.017
2	0.283	0.164	0.227537	14.684	9.256	1.935	1.256	0.666	0.034	0.034	0
3	0.256	0.242	0.254812	14.793	10.254	3.985	1.778	2.256	0.076	0.036	0.052
4	0.267	0.344	0.288645	15.251	11.235	5.325	1.987	4.985	0.112	0.054	0.089
5	0.281	0.432	0.342564	15.625	11.48	7.325	1.856	5.652	0.165	0.014	0.145
6	0.279	0.534	0.234561	16.234	10.825	9.365	1.652	7.562	0.152	0.016	0.15
7	0.269	0.634	0.477352	16.256	8.852	9.256	0.787	7.652	0.952	0.017	0.195
8	0.280	0.614	0.692652	17.679	7.765	10.462	0.658	9.762	0.298	0.013	0.215
9	0.258	0.802	0.820123	18.065	5.984	10.658	0.985	10.225	0.478	0.008	0.225

J integral vs Crack increment

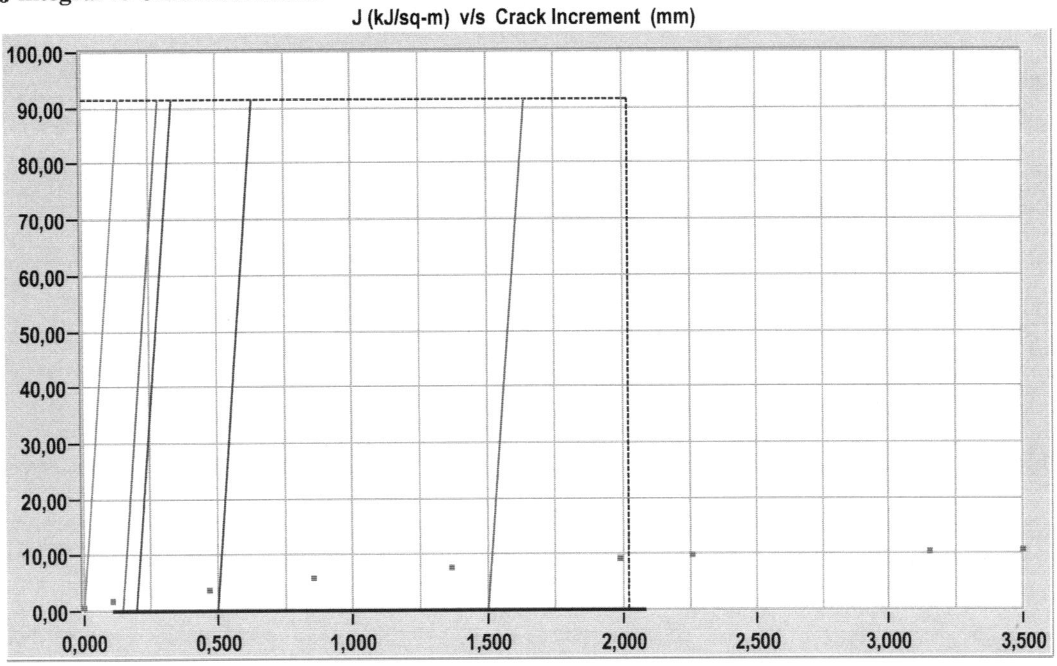

Figure 6. J integral vs Crack increment

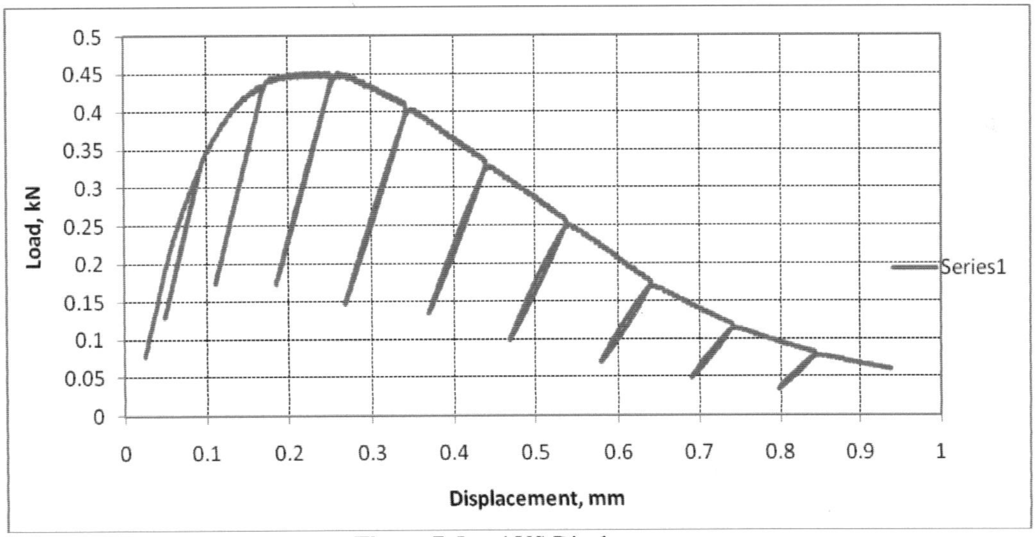

Figure 7. Load VS Displacement

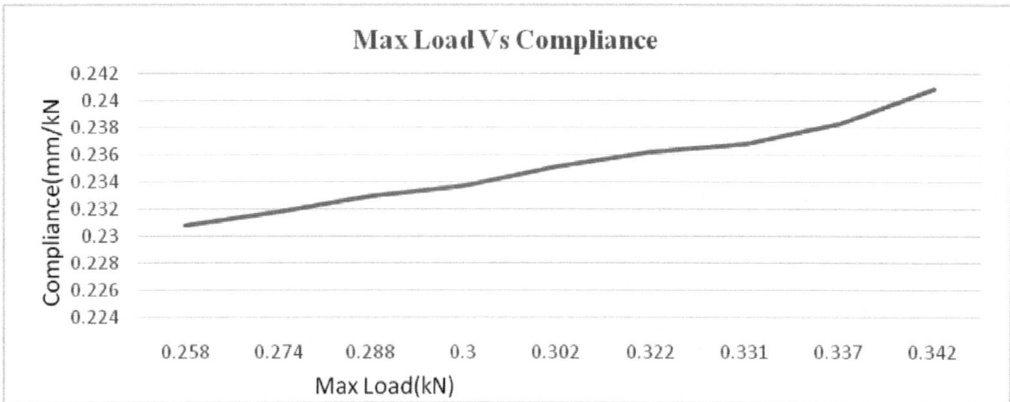

Figure 8. J- Integral test report for Al

Validation condition

1.Thickness, B>25*JQ/Yield stresses

2.Initial ligament,bo>25*JQ/Yield stresses

3.Slope of power law regression line,dj/da, evaluated at delta aQ is less than the yield stress.

Stress Intensity Factor

The stress intensity factor, K, is calculated to estimate the "stress intensity" near the tip of a crack generate by a distant load or leftover stresses. It is a theoretical construct conventionally applied to a homogenous, linear elastic material and is peripheral for providing a failure criterion for brittle materials. The concept can withal be applied to materials that exhibit minute scale yielding at a crack tip. The magnitude depends on sample geometry, the size and location of the crack and magnitude and modal distribution of loads on the material. Linear theory presages that the stress distribution (σij) near the crack tip, in polar co-ordinates(r, ϴ) with origin at the crack tip, has the form.It is denoted K_{Ic} and has the units of

K_{IC}=Y (3PS√a) / 2TW^2

Y=1.9-3.07 (a/w) +14.53 (a/w) 2-25.11 (a/w) 3+25.80 (a/w) 4

Where

P=Critical load for crack propagation S=Length of the span

a=Crack length T=Thickness , W=Width

Y=Non dimensional shear factor

J- Integral

The J integral is a line integral (path-independent) around the crack tip. It has huge consequentiality in elastic-plastic fracture mechanics.In materials science, fracture toughness is a property which describes the facility of a material containing a crack to resist fracture, and is one of the most paramount properties of any material for many design applications. The linear-elastic fracture toughness of a material is resolute from the stress intensity factor ( ) at which a thin crack in the material commences to grow. Plastic-elastic fracture toughness is denoted by JIC, with the unit of

J/cm2 or lbf-in/in2, and is a expression of the energy required to grow a thin crack.

JIC ≈ K_{IC}^2 / E

Where

K_{IC} = Stress intensity factor

E = Young's modulus of the material

Table: 3 Stress intensity factor and J- Integral value of specimen

S. No	Samples	K_{IC} (MPa m)	J_{IC} (Pa m)
1	Al (pure)	4.18	249.61
2	Al+ 3%SiC	6.931	686.23
3	Al+3%Fe_2O_3	9.932	1409.23

4. Conclusion

The following are concluded from the above analysis

- It has been studied that Fracture Toughness increased significantly by addition of reinforcements
- The addition of Fe_2O_3 has more significant fracture toughness when compared to SiC.

So by all the tests carried out, it has been concluded that the specimen with 3% reinforcement is having superior properties.

5. References

[1] Krupiński M, Labisz K, Dobrzański. L. A, and Rdzawski A, 2010 Image analysis used for aluminium alloy microstructure investigation *Journal of achievements in materials and manufacturing engineering* **42** 58-65.

[2] Loizaga A, Niklas, Fernandez-Calvo A I and Lacaze J 2009 Thermal analysis applied to estimation of solidification kinetics of Al-Si aluminium alloys *International Journal of Cast Metals Research* **22** 345-352.

[3] MadevaNagaral, Auradi V and Ravishankar M K 2013 Mechanical Behaviour Of Aluminium 6061 Alloy Reinforced With Al_2O_3 and Graphite Particulate Hybrid Metal Matrix Composites *International Journal Of Research In Engineering & Technology* **1** 193-198.

[4] Muhammad Hayat Jokhio, Muhammad Ibrahim Panhwar and Mukhtiar Ali Unar, 2010 Manufacturing of Aluminum Composite Material Using Stir Casting Process *Mehran University Research Journal Of Engineering & Technology* **30** 53-64.

[5] Naher S, Brabazon D and Looney L 2003 Simulation of the stir casting process *Journal of Materials Processing Technology* **24** 567-571.

[6] Narasimalu Srikanth, KhineKhine TUN and Manoj Gupta , 2003 Finite Element Based Energy Dissipation Studies of Al-SiC Composites *Journals of composite materials* **35** 1385-1410.

[7] Neelima Devi, Mahesh V, Selvaraj N 2011 Mechanical Characterization Of Aluminium Silicon Carbide Composite *International Journal Of Applied Engineering Research* **1** 793-799.

[8] Pardeepsharma, Gulshanchauhan, Neerajsharma 2003 Production of AMC by stir casting *International journal of contemporary practices* **2** 23-46.

[9] Phool Kumar, Azada A M, Vinod Kumar 2013 Frictional and Wear Characteristics of Stir-Cast Hybrid Composite Aluminium Al6061Reinforced with SiC Particulates *International Journal of Engineering Research & Technology* **2** 403-410.

[10] Bayraktar. E, Katundi. D, 2010 Development of a new aluminium matrix composite reinforced with iron oxide (Fe3O4) *Journal of achievements in materials and manufacturing engineering* **38** 7-14.

Reciprocating Sliding Behaviour of Solid Lubricant Coating over Modified Titanium Alloy Surfaces

V M Jothi Prakash[1], S Sathish[2], T Gopalakrishnan[3] and S Venugopal[4]
[1,3,4]Assistant Professor, Department of Automobile, Vels University, Chennai, India
[2]Head of the Department, Department of Automobile, Vels University Chennai, India

E-mail: jothiprakashsvce@gmail.com

Abstract. Tribological behaviour of contacting surfaces rigid sphere is using flat plate the with influence of normal and tangential loading (shear traction) is analysed using FEA model and surfaces being coated on flat plate by Titanium Alloy, Aluminium Alloy Molybdenum Di-sulphide. The finite element model facilitates to Evaluating the surface variables like contact stress distribution with the surface level and surface, contact pressure, shear stress and displacement. The finite element solution is validated through the hertz solution and on the successful verification.

1. Introduction

Contact is vital in mechanical power transmission. It is the science of microscopic surface geometry under friction, lubrication. The mechanics of contact is two bodies is very intricate one [1]. The determination of the stress fixing among two bodies when they are loaded together is important in the design of mechanical components [2]. Adhesive wear is the most common wear phenomenon in reciprocating sliding and it exposes the underlying material.Sliding-materials and sliding-distance and sliding-velocity, also operating environment, surface roughness, and hardness of material are the various parameters that affect the wear behaviour of the substrate [3]. These factors make the problem non-linear and thus the designer is faced with the difficulty of selecting a method to prolong the life of the components and thus surface coatings plays a vital role here. Under sliding, the surface coating tends to fail by delaminating, due to needful material similarity leading to poor bonding strength, thus affecting the adhesion strength. At the current scenario, scratch test is proving to be a better tool in determining the adhesion strength of the surface coating and this helps in determining the suitable material for that operating environment [4]. A coating is a cover that the applied to the surface of an object, usually referred to as the substrate. Narrow surface coatings are increasingly which are used for developing the tribological fulfilment and life-time of the metal and wood cutting tools, as these coatings provide high hardness and oxidation resistance for the tools [5].

The tribological impinging among two stressed surfaces in relative motion is a very complex system that is not easy to neither realize nor predict. The system becomes even more complex when coatings are introduced on the surfaces [6]. The basic idea in the finite element method is to find the solution of complicated problem by dividing it into simpler elements. Then the each individual element is added together under the given boundary condition to get the approximate solution.

Since the actual problem is restored by an easiest one in determining the solution, we will be able to determine only an estimate solution instead than the actual solution [7]. Developed a FEM to find the wear depth and volume of material displaced laterally among two contacting asperities during sliding.

Three-D hemispherical asperities with varying overlaps are modeled, the normal force acted on the asperity, shear force and the overall friction coefficient are determined for different E/Y ratios [8].

2. Experiments and Methods

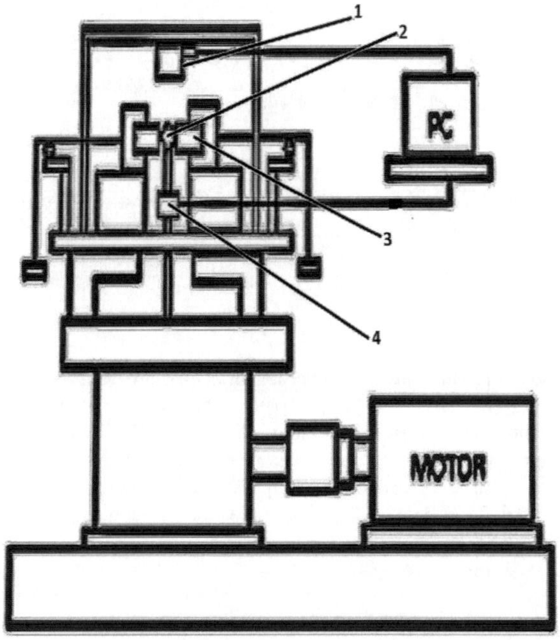

Figure.1. Schematic View of the Experimental Setup

The Figure 1 shows the experimental setup and it is placed at the central axis of the setup below the ball and it monitors the friction force of the contact pair and data is gathered in real-time using dewesoft. The ball is of 25.4mm in diameter and it is made up of AISI 52100 Bearing steel andthe selection of a particular type of bearing steel for an application involves the consideration of factors like corrosion resistance of the alloy and mechanical properties. Table 1 show the Mechanical Properties of AISI 52100 bearing steel.

Table.1 Mechanical Properties of AISI 52100 Bearing steel

Properties	Value
(a) Yield Strength	2.03 Gpa
Modulus of Elasticity	208 Gpa
Poisson's ratio	0.3
Density	8E-6kg/mm^3

The plate dimension is25 x 25 x 10 mm and it is made up of Ti-6Al-4V (Titanium grade-5) alloy and the material composition and mechanical property is shown in Table 2.

IOP Conf. Series: Materials Science and Engineering **183** (2017) 012008 doi:10.1088/1757-899X/183/1/012008

Table.2 Mechanical Properties of Titanium Alloy Material

Properties	Flat Plate (Ti Alloy)
(b) Yield Strength	880 MPa
Modulus of Elasticity	113.8 GPa
Poisson's ratio	0.3
Density	4.43E-6 kg/mm^3

The high power electric arc discharged at the target material blasts away some into highly ionized vapour to be deposited onto the work piece. The industrially synthesized used in most TIAlN coatings alloy targets with specific percentages of aluminium and titanium it is possible to generate TIAlN coatings with pure Al and Ti targets using a cathodic arc discharge technique. Titanium aluminum nitride is one of the widely used surface coatings in drill bits, dry metal-cutting tools, and high speed metal-machining to improve the wear resistance. Coating developers is put as much aluminum as possible into the mixing for this various reason. But there is a limit, adding high aluminum to the titanium-based nitride composition, and at long last the crystal structure of the coating will alter from cubic to hexagonal lattice

The rigid body is tended as the "master" surface and the deformable body as "slave", hence the sphere is tended as the "master" and the flat surface as "slave". The type of contact defined is a surface-to-surface contact.

The co-efficient of rolling friction value is defined based upon the material being employed. Any gap that exists between sphere and flat plate should be closed in order to avoid impact loading.

3. Result and Discussion

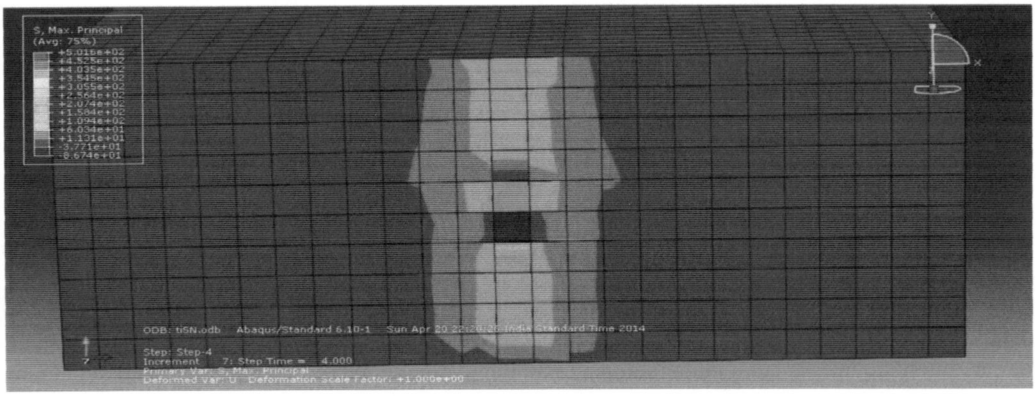

Figure 2. Maximum Principal Stress at 5N Load for Titanium Aluminium Nitride coating

The number of component to be chosen for idealization is related to the veracity desired. Although an enhance in the number of components generally means more exact results, there is a limit to number of elements that can be handled by ABAQUS beyond which the accuracy cannot be improved and distortion of the model will be resulted while meshing.

IOP Conf. Series: Materials Science and Engineering **183** (2017) 012008 doi:10.1088/1757-899X/183/1/012008

Figure (2) shows the Maximum Principal Stress at 5N Load for Titanium Aluminium Nitride coating region of contact is extremely fine meshed for better handling the high stress gradients and for accurate solution. Other zones are coarser meshed to reduce the computational time.

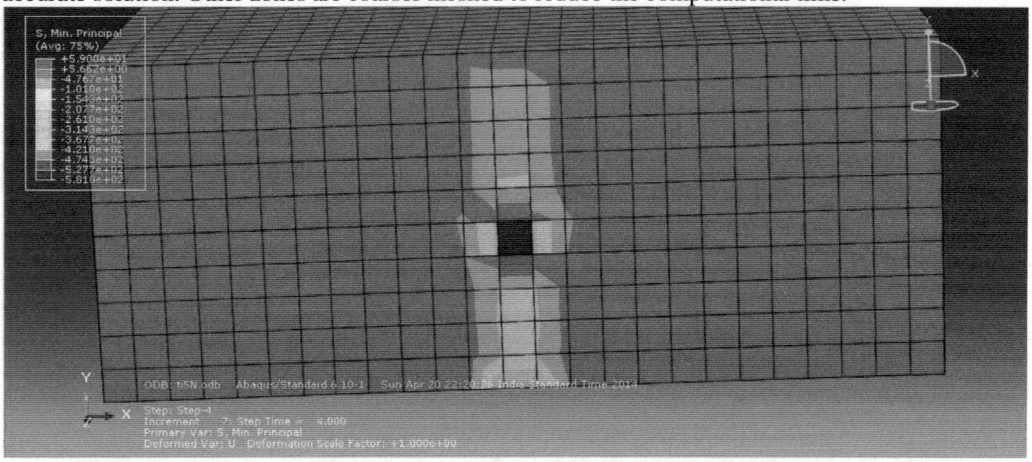

Figure 3. Minimum Principal Stress at 5N Load for Titanium Aluminum Nitride coating

Tribology of titanium and its alloy have found great interest among efficient and due to possible applications of titanium alloys with improved tribological properties in many potential areas. The Figure (3) shows the Minimum Principal Stress at 5N Load for Titanium Aluminum Nitride coating.

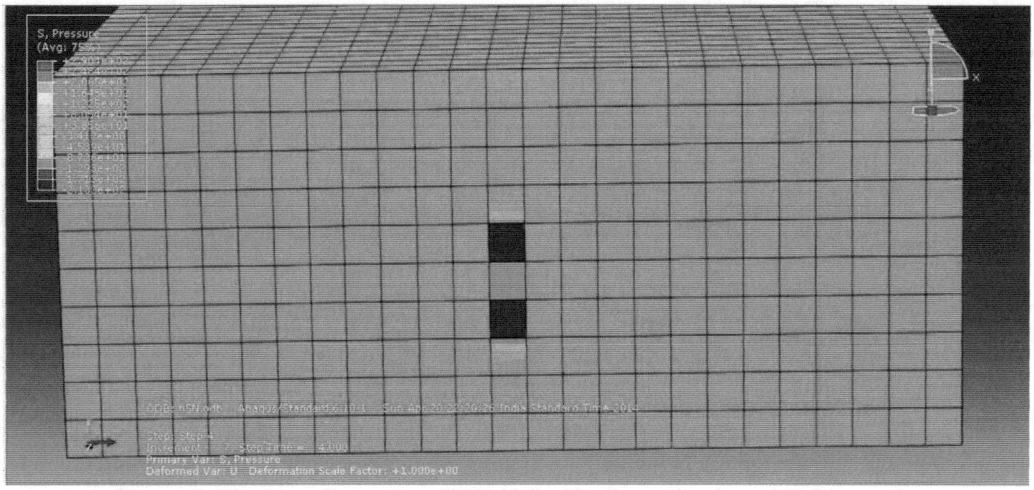

Figure 4. Maximum Pressure at 5N load for Titanium Aluminum Nitride coating

The Figure (4) shows the Maximum Pressure at 5N load for Titanium Aluminum Nitride coating here the polymer has outstanding physical and mechanical properties. For its excellent wear resistance, this coating has potential to provide a solution for tribological limitations of titanium alloys in biomedical and industrial applications.

4. Conclusion

The experimental results indicate that tribological performance of TiAlN coating is better compared to AlCrN and MOS2 based on friction force evolved over the time period of the experiment. The ploughing action due to normal load in the ball on flat reciprocating tribometer enhance the ploughing component of friction, thus has a material pile up result in front of the sliding tip that results in high tensile stresses in front of the tip, but also compresses material right in front of the tip below the coating. The compensating effect of this material compression can be seen as minimized first principal stresses right in front of the tip.

5. References

[1] Helena Ronkainen, AnssiLaukkanen, Kenneth Holmberg 2007 Friction in a coated Surface deformed by a sliding sphere *Wear* **263** 1315–1323.
[2] Michael Lovell 1997 Analysis of contact between transversely isotropic coated Surface and development of stress and displacement relationships using FEM *Wear* **214** 165-174.
[3] Kenneth Holmberg, Allan Matthews and Helena Ronkainen 1998 Coatings tribology-contact Mechanisms and surface design *Tribology International* **31 1–3** 107–120.
[4] Shankar .S and Mayoral M.M 2008 Effect of strain hardening in elastic-plastic Transition behaviour in a hemisphere in contact with a rigid flat *International Journal of Solids and Structures* **45** 3009–3020.
[5] Johnson K L 1985 Contact Mechanics *Cambridge University* **p**- 202-233
[6] Robert L. Jackson, Ravi S, Duvvuru, HasnainMeghani and Manoj Mahajan 2006 An Analysis of elasto-plastic sliding spherical asperity interaction *Wear* **262** 210–219.
[7] ASM International 1992 **18** Friction Lubrication and Wear Technology.
[8] Wei Peng, Bharat Bhushan *2001* Three Dimensional Contact Analysis of layered elastic/plastic With rough surfaces *Wear* **249** 741–760
[9] Dinia .D, Sackfieldb A Hills D.A 2008 an Axi-symmetric Hertzian Contact Subjected to Cyclic Shear And *Severe* Wear *Wear* **265** 1918–1922.

IOP Conf. Series: Materials Science and Engineering **183** (2017) 012009 doi:10.1088/1757-899X/183/1/012009

Design, Manufacture and Analysis of Al/SiC MMCs for Connecting Rod

S. Ramasubramanian[1], M.Chandrasekaran[2], R.Sridhar[3] and K.Karunakaran[4]

[1,3,4]Research Scholar & Assistant Professor, [2]Professor & Director
[1]Department of Automobile Engineering, Vels University, Chennai, TN, India.
[2,3,4]Department of Mechanical Engineering, Vels University, Chennai, TN, India.
E-mail: raams87@gmail.com

Abstract. The Metal matrix composite materials is particularly are widely used in automotive engineering applications. In this work discuss with production of Al/SiC composite and investigate it's suitability for the application on Connecting Rod. The results were compared with conventional material C70 Steel. The connecting rod is manufactured using the stir casting method. The basic properties were obtained by mechanical testing. The test results were used for FEA. The connecting rod is designed and meshed using solid works and hyper mesh, and then analysis is done used by ANSYS work bench 14.0 software. The results show that the composite connecting rod is more efficient than the conventional ones.

1. Introduction

In modern automotive IC engine, the connecting rod designed in steel. It's made up of titanium or aluminum for enhance the performance of engines. It is not a rigidly fixed at either end, so that the angle of contact at each end varies. The bigger end of the connecting rod is attached to the shaft and experiences tremendous stress from the reciprocating load takes place by the piston, and the load increases with every cycle and there is an enhanced speed of the engine. Generally the connecting rod is generally manufactured by forging from powder metal. Sudershan Kumar [1] suggested the aluminum boron carbide for carbon steel for connecting rod application. Since its factor of safety nearly theoretical value and improved the stiffness by about 48.55% and reduced stress by about 10.35%. Vivek [2] investigated about the induced stresses, and maximum stress at small end than the higher end. Since, the possibility of failure may occur the connecting rod is the fillet section of both ends. This work analyzed by pro-e wild fire 4.0 software and ANSYS using work bench 11.0 software. Pushpendra Kumar Sharma [3] reported that Weight is reduced using changing the material of the current forged steel connecting rod to crack able forged steel (C70). The dynamic loading based analysis on Al alloy using Finite Element Analysis by Ram Bansalin[4]. Pranav et al[5] conducted the static analysis and fatigue analysis and explored weight reduction possibility for a making of connecting rod and achieved the 9.24% maximum weight is reduction by suggesting the forged steel. In this study analysis used in ANSYS work bench 9 suggested by Pravardhan et al[6]. Rasekh et al [7] suggested to reduce the weight (10% lighter) and cost (25% less expensive) of the forged steel connecting rod. Conducted stress analysis on Mf-285 Connecting Rod using ANSYSv9, software. Hassan et al [8] is suggested the analysis using F-285connecting rod and reported that the higher pressure Stress is between the pin end and rod linkages, bearing cup and connecting rod Linkage and also the maximum tensile stress is incured in lower half of pin end and between Pin end and connecting rod linkages. The highest tensile stress was obtained in down part half of pin end and between Pin end and rod linkages. The wear resistance was achieved by Ghosh et al[9] by reinforcing SiC in Al2O3. In this research the Aluminum is reinforced with SiC by appropriate combination and to test its compatibility for the application.

2. Al/Sic MMCs for Connecting Rod

It is one of the powerful parts of a modern racing engine. The fault can come from any number of places, but the failure often occurs at the connecting rod because of the stresses involved and change over the linear motion of the piston to the rotary motion of the crankshaft. The each stroke, the connecting rod is stretched and pressed. In this pressure, plus another elements, can make the connecting rod to terminate. The broken rod is move through the engine block completely and run the engine, a condition known as "throwing a rod." The connecting rod often fails due to either of Over Reviving, Pin Failure, Stress Failure, Fatigue Failure or Hydro-lock. Manufacture of Al/SiC by Stir Casting methods shown in Figure 1.The liquid Infiltration Process involved in infiltration of a fibrous reinforcement, it's preformed using a liquid metal. Liquid-phase infiltration of MMCs is not straight forward; it's mainly trouble with wetting, ceramic reinforcement used as the molten metal.

The percolation of a fiber material to perform come about readily, reactions among the fiber and the molten metal may fall out which implication degrade the main properties of the fiber materials. The fiber coating is used to apply prior the implication, it is developing the leak and allows the control of interface reactions, which is improved and also generating some promotive results. This study even if, the weakness is that the fiber coating essential mot be unprotected the air prior to infiltration because surface oxidation of the coaing fall out. One liquid infiltration process is involving particulate reinforcement, known as Duralcan process it has been quite powerful and efficient. Ceramic particles and ingot-grade Al is blended and melted. The melt is agitated slightly above the liquids temperature around (600-700°C). The solidified is also under the secondary processing by rolling. The Duralcan process of producing the particulate composites by a liquid metal casting route involved and were 8-12 μm particles is used. For particles that are very small size (2-3 μm), result shows the very high interface region and a verying the viscous melt. In foundry-grade Metal Matrix Composite, high Si aluminum alloys (e.g., A356) is used to prevent the formation of the brittle compound Al_4C_3, which is formed from the interfacial reaction between Al and SiC, Al_4C_3 is particularly damaging to mechanical properties, toughness and corrosion resistance.

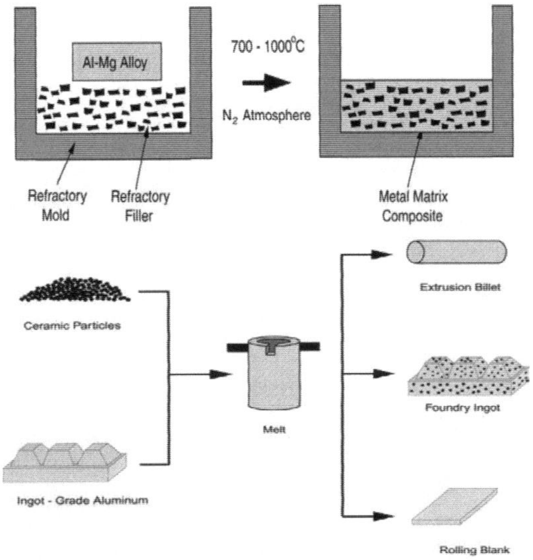

Figure 1. Stir casting technique for Al/SiC MMC

In this work the Aluminum reinforced using SiC and the ratios of Al 60% / SiC 40%, Al 75% / SiC 25% and Al 70% / SiC 30%. The properties is evaluated and furnished in Common physical properties in Table 2, Comparative Mechanical properties in table 3. The Dimensional parameter is shown in Table 4. The Lab Test results of Mechanical Properties were furnished in Table 5.

Table 1. Physical Properties of Al/SiC MMCs

Description of Properties	Quantity
Density (kg/m^3)	2784
Young's Modulus (MPa)	99974
Poisson Ratio	0.292
Co- efficient of thermal expansion (C^{-1})	16.002
Tensile Strength (MPa)	615.63
Compressive Strength (MPa)	400.3
Thermal Conductivity (Wm^{-1}K^{-1})	130
Specific Heat (Jkg^{-1}k^{-1})	919

Table 2. Mechanical Properties of Al/SiC MMCs

Composition (Vol.%)	Al75/ SiC25	Al70 / SiC3 0	Al60/ SiC40
Density (g/cm3)	2.8	2.8	2.9
Young's Modulus (GPa)	115	125	150
Young's Modulus/Density (GPa-cm3/g)	41	45	52
Flexural Strength (MPa)	-	-	-
Co- efficient of thermal expansion	15	14	13
Thermal Conductivity (W/mK)	145	150	155

Table 3. Paramaters of Connecting Rod

Description of Parameters	*Quantity*
Length of connecting rod	124mm
Outer Diameter of Big end	40mm
Inner Diameter of Big end	30mm
Outer Diameter of Small end	18mm
Inner Diameter of small end	12mm

Table 4. Laboratory test report on C70 Steel Vs Al/SiC

Properties	C70 Steel	AlSiC:60/40
Tensile Strength(MPa)	966	107.8
Hardness	156 HV	125 HV
Temperature (Melting point)	632oC	858oC

3. Compatibility Analysis of Al/SiC MMCs

3.1 Modeling of Connecting Rod:

The modeling was done by using ANSYS workbench 14.0 software (refer Figure 2). The stress, Strain, deformation and Thermal analysis is carried out. They are furnished for Al/Sic and for C70 respectively from the Figure 3 to Figure 10

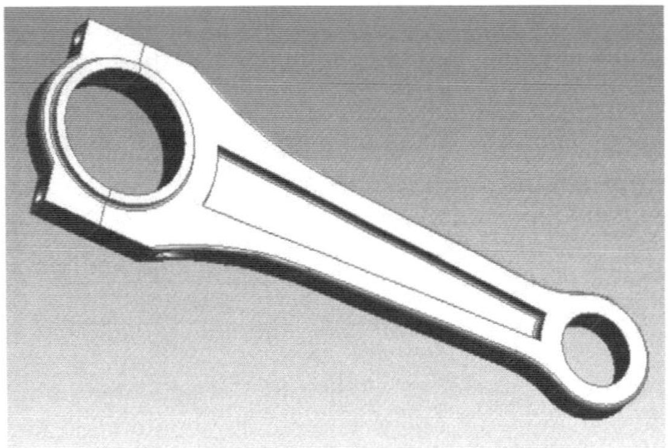

Figure 2. Model of Connecting Rod

3.2 Stress analysis

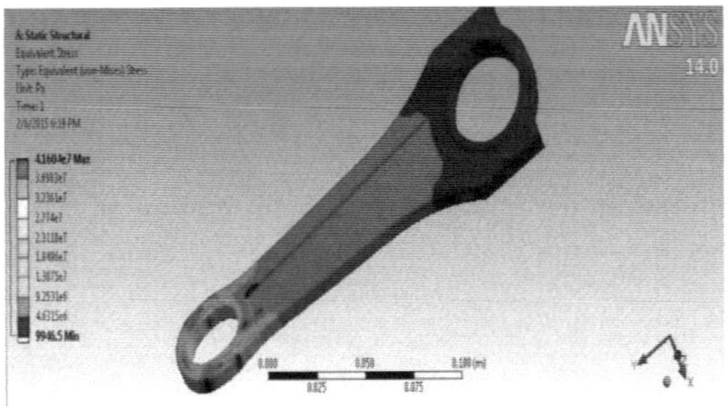

Figure 3. Stress analysis of Al/SiC

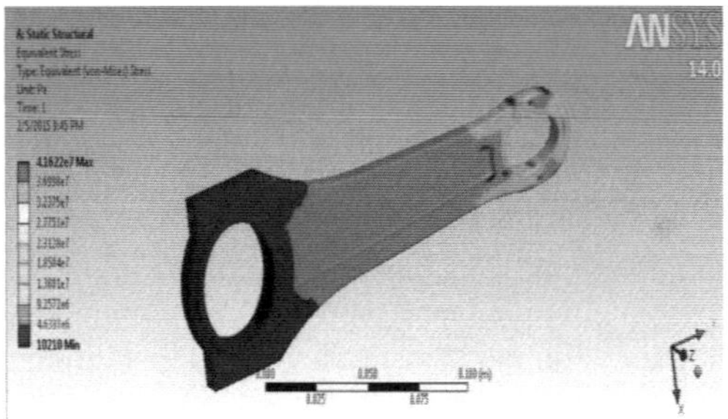

Figure 4 Stress analysis of C70 Steel

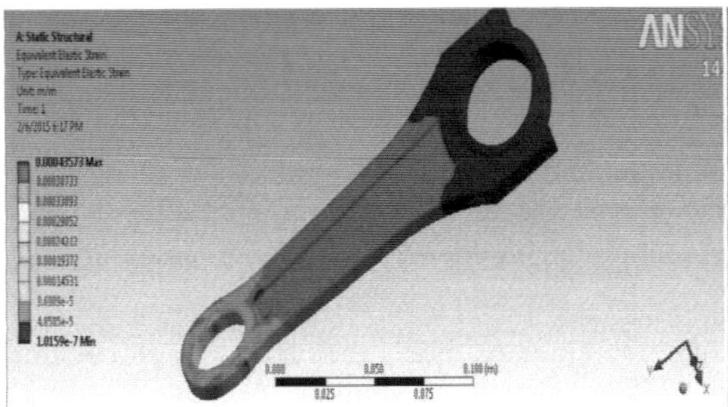

Figure 5 Strain analysis of Al/SiC

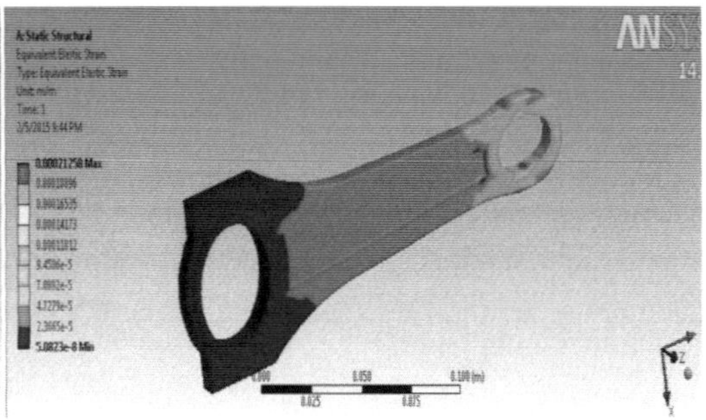

Figure 6 Strain analysis of C70 Steel

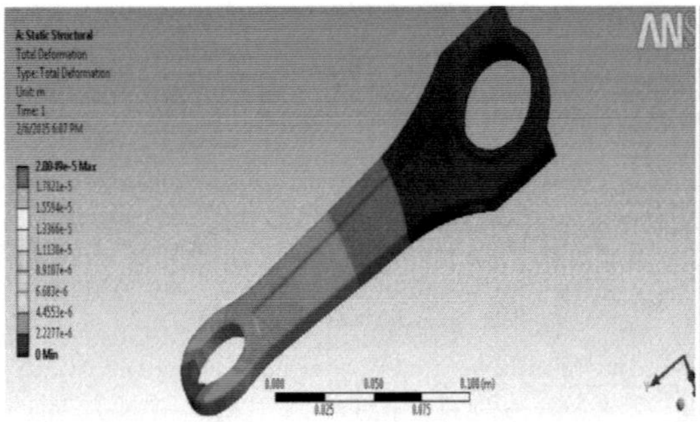

Figure 7 Deformation analysis of Al/SiC

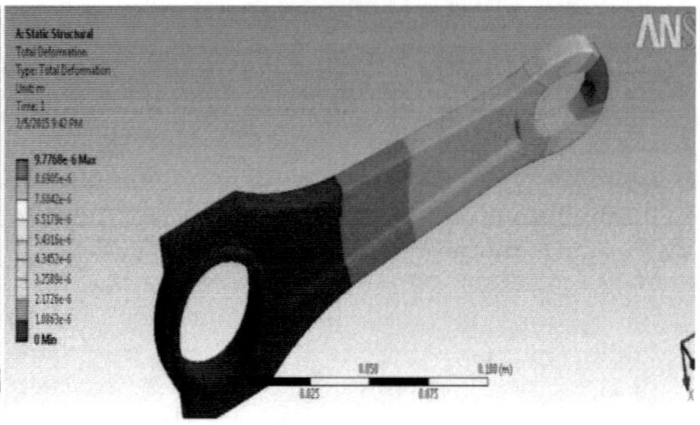

Figure 8 Deformation analysis of C70 Steel

3.3 Temperature Analysis

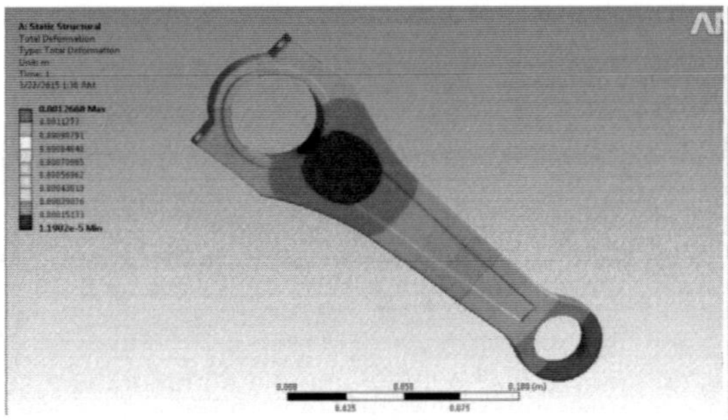

Figure 9 Temperature analysis of Al/SiC

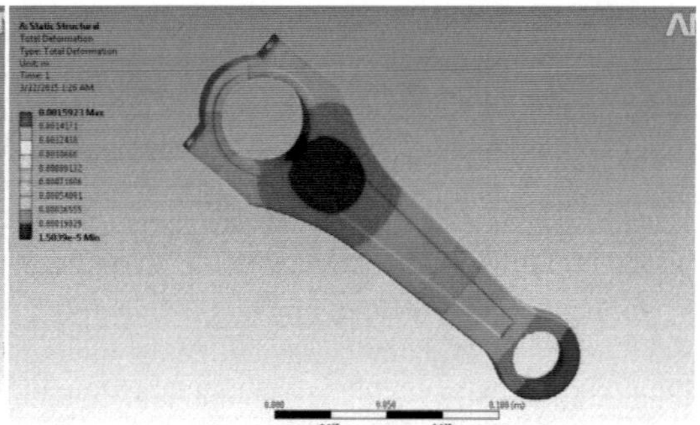

Figure 10 Temperature analysis of C70 Steel

3.4 Stress Analysis

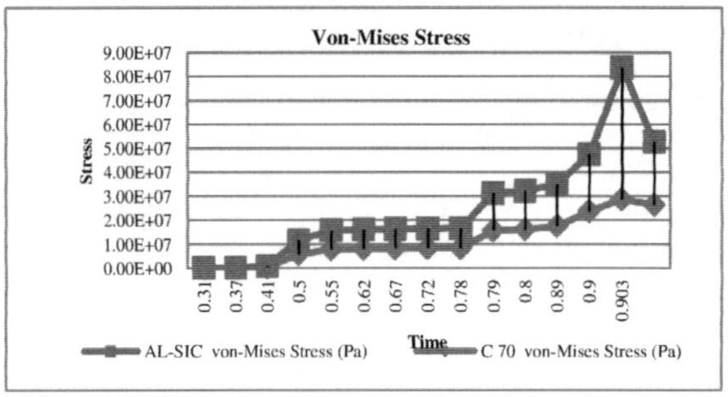

Figure 11. Strain Vs Time on Al/SiC and C70 Steel

3.5 Strain Analysis

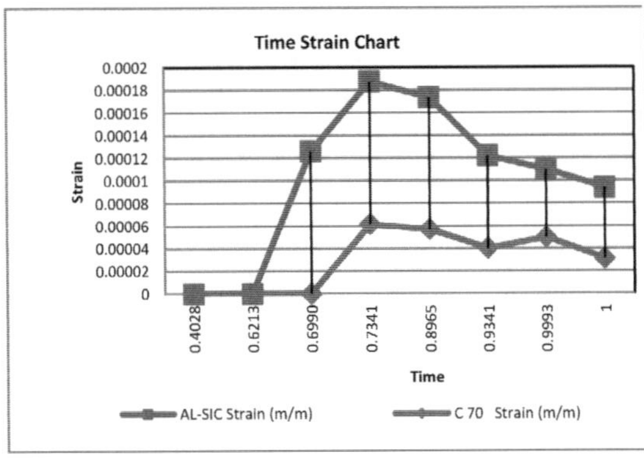

Figure 12. Strain Vs Time on Al/SiC and C70 Steel

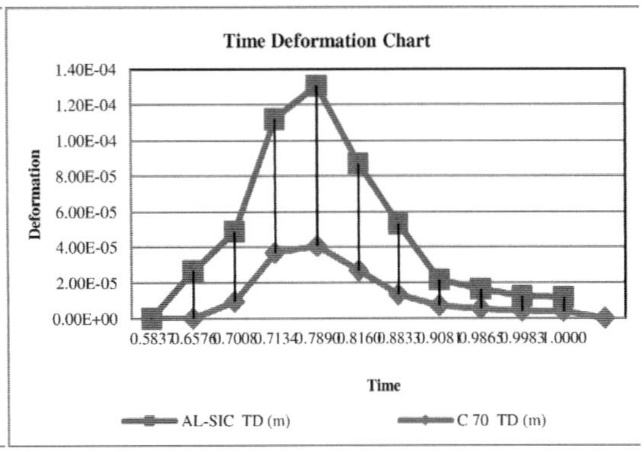

Figure 13. Deformation Vs Time on Al/SiC and C70 Steel

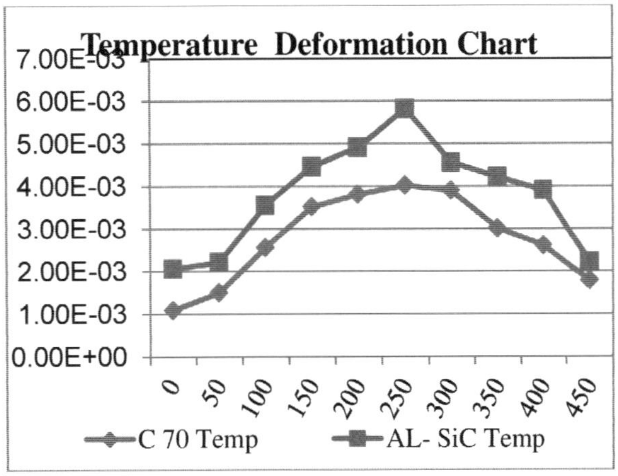

Figure 14. Temperature analysis of Al/SiC and C70 Steel

Table 5. Analytical results of ANSYS

Properties	Level	C70 Steel	Al60/SiC40
Stress	Max	4.16e7	4.16E7
Stress	Min	1.02e4	9.9e3
Strain	Max	2.13e-4	4.36e-4
Strain	Min	5.08e-8	1.02e-7
Deformation	Max	9.78e-6	2.005e-5
Deformation	Min	0	0
Temperature	Max	1.6e-3	1.3e-3
Temperature	Min	1.50e-5	1.19e-5

4. Conclusion

The manufacture of MMC with three different proportions was made. The proportions are Al60/SiC40, Al70/SiC30 and Al75/SiC25 through the Stir casting process. The proper estimation of properties was done with appropriate testing methods. Mechanical Tastings, and required design analysis were discussed. Its based on the requirement Al60/SiC40 was identified. In further, the Al60/SiC40 MMC was subjected to the stress, strain, deformation and Thermal analysis with C70 steel. The results were analyzed.

- Al/SiC has less tensile strength than C70 Steel.
- Al/SiC has hardness nearer when compared to C70 Steel.
- Al/SiC can withstand more temperature than C70 Steel.

5. References

[1] Sudershan Kumar K, Tirupathi Reddy K and Syed AltafHussan, 2012 Modeling and analysis of two Wheeler connecting rod, *International Journal of Modern Engineering Research* **2 (5)** 3367-3371.

[2] Vivek C pathade, BhumeshwarPatle and Ajay N Ingale, 2012 Stress Analysis of I.C. Engine Connecting Rod by FEM, *International Journal of Engineering and Innovative Technology* **1(3)** 12-15.

[3] Pushpendra Kumar Sharma and Borse Rajendra R, 2012 fatigue analysis and optimization of connecting rod using finite element analysis, *International Journal Of advance research in Science and Engineering* **1(1)** 3367-337.

[4] Ram Bansal, 2013 Dynamic simulation of connecting rod made of aluminium alloy using finite element analysis approach, *IOSR Journal of Mechanical and Civil Engineering* **5(2)** 01-05.

[5] Pranav G Charkha and Santosh B Jaju, 2009 Analysis & Optimization of Connecting Rod, Second *International Conference on Emerging Trends in Engineering and Technology* .

[6] Pravardhan S Shenoy and Ali Fatemi, 2005 Connecting Rod Optimization for Weight and Cost Reduction *SAE International.*

[7] Rasekh M, Asadi M R, Jafari A, Kheiralipour K, 2009 Obtaining Maximum Stresses in Different Parts of Tractor (Mf-285) Connecting Rods Using Finite Element Method, *Australian Journal of Basic and Applied Sciences* **3** 1438-1449.

[8] Hassan AM, Alrashdan A, Hayajneh MT, Mayyas AT, 2009 Wear behavior of Al–Mg–Cu based composites containing SiC particles TribolInt **42** 1230–1238.

[9] Ghosh SK and ParthaSaha explains Crack and wear behavior of SiC particulate reinforced aluminium based metal matrix composite fabricated by direct metal laser sintering process Mater Des **32** 139–145.

[10] Subrata Kumar Ghosh, ParthaSaha, 2011 Crack and wear behavior of SiC particulate reinforced aluminium based metal matrix composite fabricated by direct metal laser sintering process, Materials & Design **32(1)** 139-145.

IOP Conf. Series: Materials Science and Engineering **183** (2017) 012010 doi:10.1088/1757-899X/183/1/012010

Thermal analysis on Al7075/Al_2O_3 metal matrix composites fabricated by stir casting process

S Jacob[1] , S Shajin[2] and C Gnanavel[3]
[1]Assistant professor, Department of Automobile Engineering,Vels University Chennai, Tamilnadu, India
[2]Assistant professor, Department of Mechanical Engineering,Narayanaguru Siddhartha College of Engineering, Athencode,Tamilnadu, India
[3]Assistant professor,Department of Mechanical Engineering, Vels University, Chennai, Tamilnadu, India

E-mail: jacobthermal@gmail.com

Abstract. Metal matrix Composites (MMC's) have evoked a keen interest in recent times for various applications in aerospace, renewable energy and automotive industries due to their superior strength , low cost, easy availability and high temperature resistance [1]. The crack and propagation occurs in conventional materials without any appreciable indication in a short span. Hence composite materials are preferred nowadays to overcome this problem [2]. The process of metal matrix composites (MMC's) is to unite the enviable attributes of metals and ceramics. The Stir casting method is used for producing aluminium metal matrix composites (AMC's). A key challenge of the process is to spread the ceramic particles to achieve a defect free microstructure [2]. By carefully selecting stir casting processing specification, such as stirring time, temperature of the melt and blade angle, the desired microstructure can be obtained. The focus of this work is to develop a high strength particulate strengthen aluminium metal matrix composites, and Al7075 was selected which can offer high strength without much disturbing ductility of metal matrix [4]. The composites will be examined using standard metallurgical and mechanical tests. The cast composites are analysed to Laser flash analysis (LFA) to determine Thermal conductivity [5]. Also changes in microstructure are determined by using SEM analysis.

1. Introduction

A composite material is a combination of two or more definite materials, having a identifiable by the interface between them. Composites are used in modern world due to its high strength to weight ratio, less weight and low cost. Modern composite materials are most effectively used to achieve a equivalence of properties for a given applications. Most of the composites are made of just two materials [7]. One is the matrix and it surrounds and binds together fragments of the other material, which is called the reinforcement. The biggest advantage of modern composite materials is that they are light and strong. Composite has the ability to be easily modified because many of them can be molded into different shapes. The mostly preferred matrix material for making a composite is Aluminium because of its cheapest and reduced weight. Aluminum 7075 alloy has higher toughness and are preferred in aircraft and automotive sector [9]. Composites able to meet the design requirements with weight as well as high strength as compared to conventional materials. Composite materials used in aerospace, automotive, trucking, and mass movement will all have fuel consumption. The products manufactured with composites that need less energy to transport or ship than traditional materials [10]. While composites have their worth as weight-saving materials, the current senario is to make them cost effective. The efforts to produce composite components have several manufacturing methods now used in the composites industry. The various feature of materials to be noted when making sport equipment are strength, ability to deform under tensile stress, modulus (damping), and cost [12].

2. Methodology

The preparation of composite is by the Stir casting process. The various aspects of the Stir casting process on the microstructure are the dissemination of the reinforcing particles. It depends on the processing steps involved. During pouring of the melt composite, the content may vary from one form to another or even it can vary in the same process from one region to another. Therefore the spreading of the particles in the melt is a necessary condition for homogeneity in the castings. Hence the study of the distribution of the particles in the laminate is of great significance. Aluminium 7075 Alloy is melted in a crucible by heating it in a muffle furnace up to 950°C for one hour. The grain size of the Aluminium oxide particles used is 150 μm. The Aluminium oxide particles are preheated at 700°C for one to three hours to make their surfaces oxidized. A pink coloured powder (Scum powder) Hexochloroethane is added to the melted Aluminium to remove the Slag from the melt. A blue tablet (Degassing flux) is mixed to remove the bubbles from the melt. Then the melted Aluminium is cooled below the liquid state to keep the slurry in Semi solid state. Automatic stirring was carried out with the help of radial stirrer operated by an electric DC motor for about 3 to 4 minutes at stirring rate of 280 RPM. In order to minimize gas trapping in the molten matrix during stirring, the mixing speed and the stirrer location was placed at a particular location. At this stage, the preheated Aluminium oxide particles were added manually to the vortex. In the final mixing processes the temperature was controlled within 700 ± 10°C. After the process the melt was settled in the sand mould to get desired shape of specimen. The preparation of particle-reinforced MMC was quite a challenge, as the particles are very hard and fragile compare to the matrix materials. This composition of hard and soft materials makes it complex to avoid damages like cracks and broken reinforcement particles, and relief the particle and soft matrix during preparation. The process used is similar for specimens with different compositions (5%, 10% & 15%) of Aluminium oxide.

3. Results and Discussion

3.1 Hardness Analysis Of Specimens

The micro hardness of the three specimens are shown in graph.

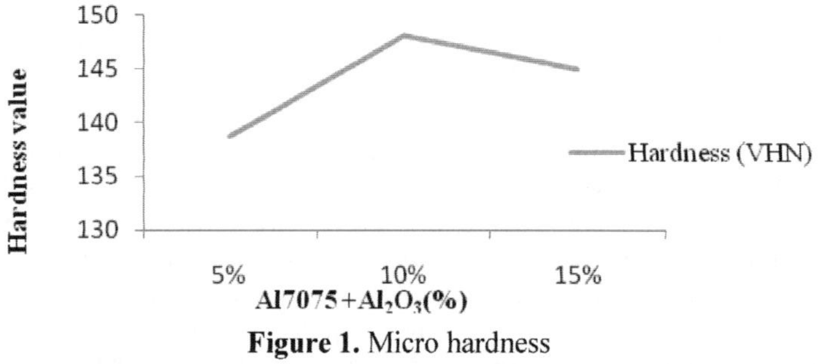

Figure 1. Micro hardness

The hardness values of the specimens were tested by Vickers micro hardness test. From the above graph it reveals that up to a 10 % of reinforcement of the aluminium oxide particles, the micro hardness of the specimen is higher and on further addition of aluminium oxide particles it decreases. The maximum hardness value thus obtained at 10% reinforcement is 148.09 VHN.

3.2 SEM Analysis

The microstructures of the specimens with various wt. % reinforcements of Aluminium oxide with Al7075 are shown in figure

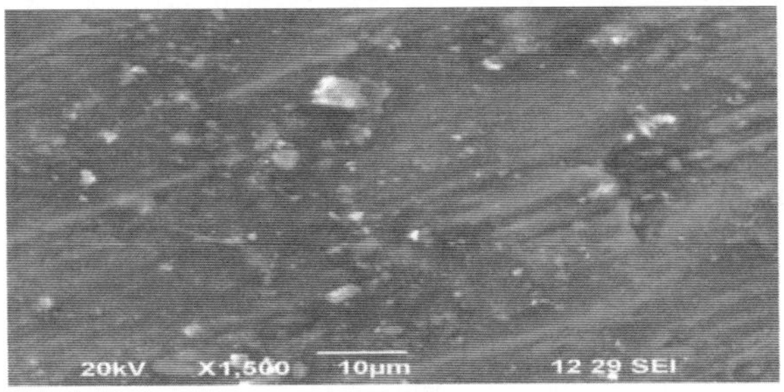

Figure 2(a)

Figure 2(b)

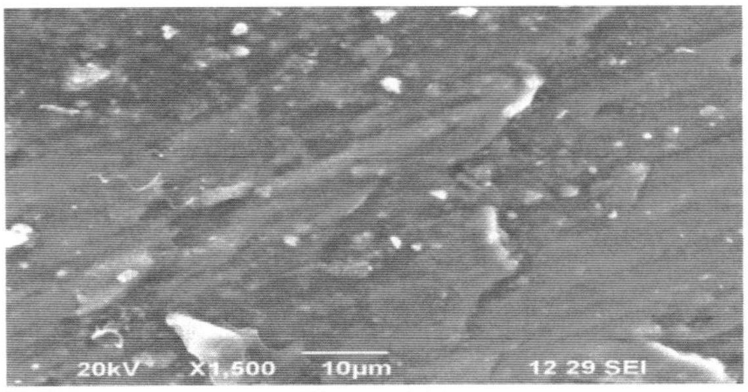

Figure 2(c)

Figure 2. SEM analysis of specimens

The microstructures of Al7075 reinforced with Al_2O_3 particles of various weight percentages obtained are as in Fig. 4.2. The SEM images are obtained at 1500X magnification range. Here dark grey colored particles are Aluminum particles and light white particles are Aluminium oxide particles. The specimen with 10% reinforcement (Fig. 2(b)) is found to have more acquisition of aluminium oxide particles. On the other hand, with the increase of the weight composition of Aluminium oxide particles the uniformity and homogeneity tends to decrease.

3.3. EDAX Analysis

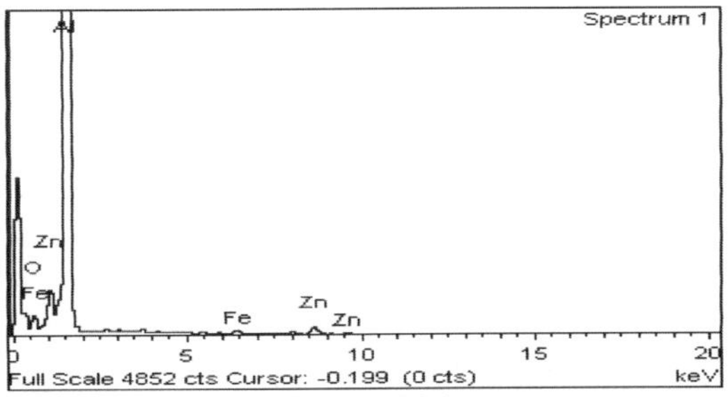

Figure 3(a)

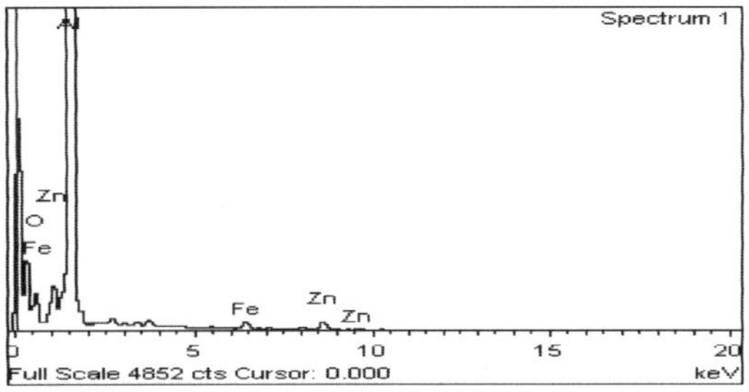

Figure 3(b)

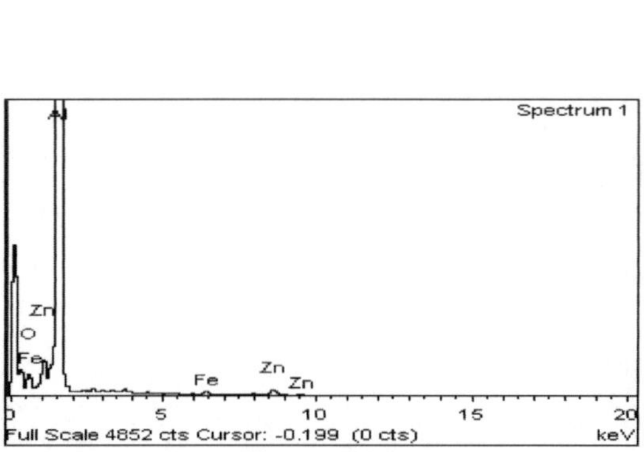

Figure 3(c)

Figure 3. EDAX analysis of specimens

The above charts show the different material composition of the three specimens. For all the three specimens Aluminium is the superior material. The dominant material next to aluminium is Zinc. The other materials like Fe and O are fewer in composition.

3.4. Tensile Testing

The dimensions for the preparation of specimens is shown in the

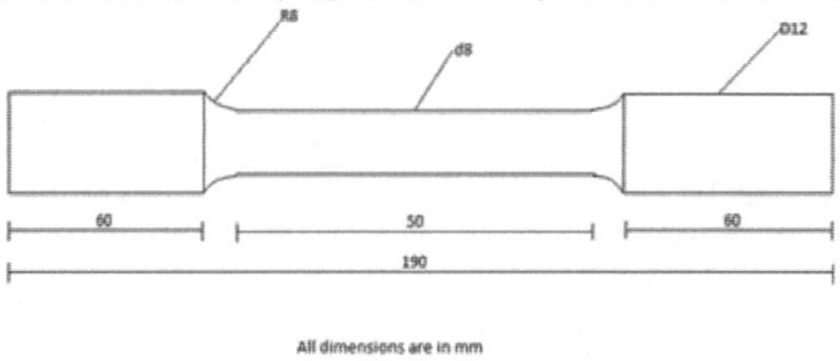

Figure 4. Tensile test specimen

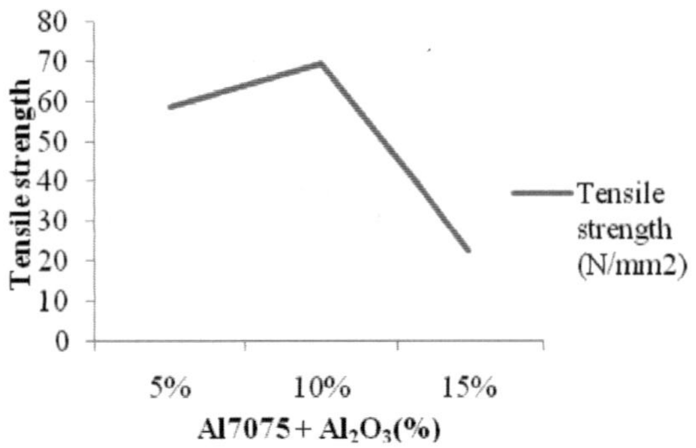

Figure 5. Tensile strength of the specimen

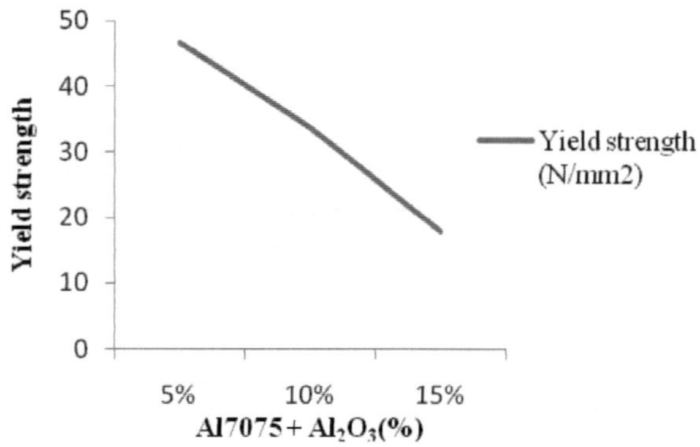

Figure 6. Yield strength

From the above graph it can be seen that the Tensile strength of the material is maximum at 10% reinforcement. The maximum tensile strength at 10% reinforcement is 69.5 N/mm^2 and is obtained at a peak load of 5.46kN. The graph (figure. 6) shows the Yield strength of the specimen with various reinforcement percentages of Aluminium oxide (Al_2O_3).The maximum yield strength is obtained at 5% reinforcement at a peak load of 4.6kN. The higher value of yield strength is 17.825N/mm^2.

3.5. Thermal Conductivity

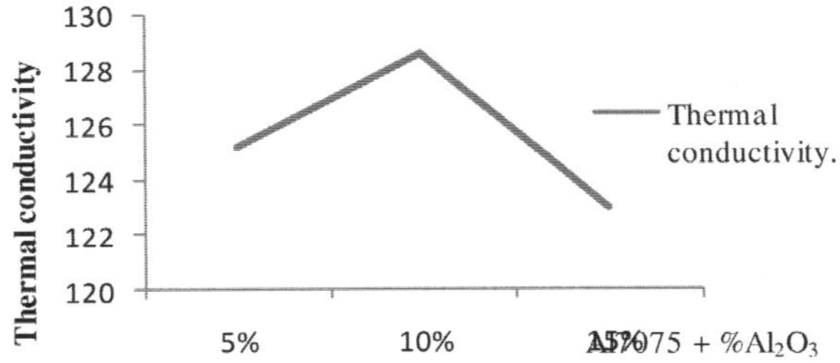

Figure 7. Thermal conductivity

The above figure 7 shows the thermal conductivity of the composite specimens. The heat conductivity of the cast specimens obtained is maximum at 10% reinforcement and is 129.927 W/mK and the lowest value obtained at 15% reinforcement is 125.01W/mK. So with the further addition of Al_2O_3 the value of thermal conductivity decreases.

3.6. Temperature Distribution

Figure. 8(a) shows the Temperature distribution of Al7075/Al_2O_3(5%) composite. From the figure it can be noted that the temperature transfers from the high temperature end to the lower temperature end. The least temperature thus obtained is 606.42 K. Figure. 8(b) shows the temperature distribution of Al7075/Al_2O_3 (10%) composite. Here the lower temperature obtained is 613.407 K. Here the heat transfer increases when compared to Al7075/Al_2O_3 (5%) composite. Figure. 8(c) shows the temperature distribution of Al7075/Al_2O_3 (15%) composite. The heat transferred from one end to the other decreases and the least temperature obtained is 602.623K. From all the three composites, the heat transfer is maximum at 10% reinforcement of Al_2O_3. The transfer of heat to the lower temperature end as obtained is 613.407K.

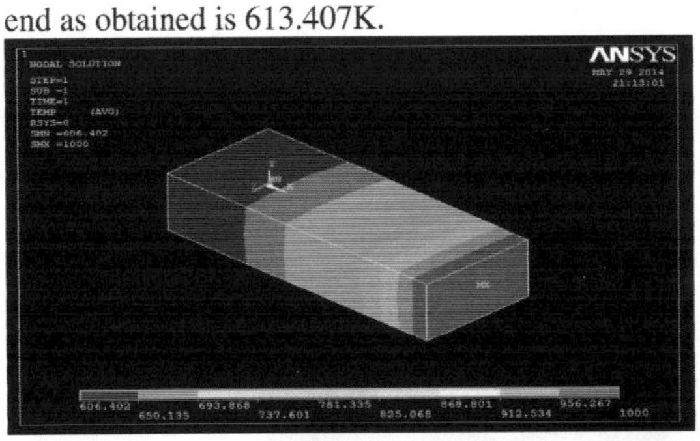

Figure. 8 (a)

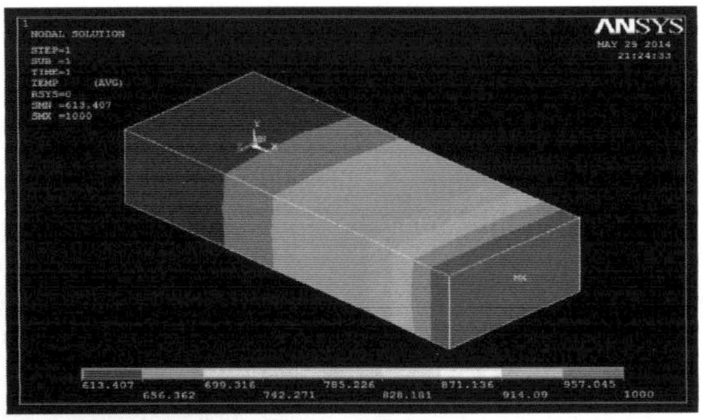

Figure. 8 (b)

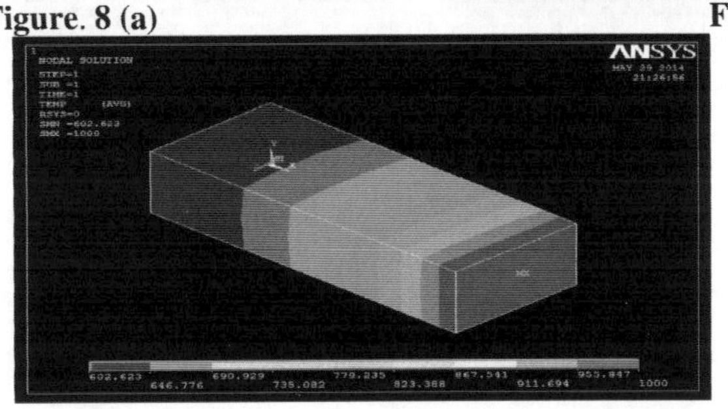

Figure 8 (C). Temperature distribution of specimens

3.7. Thermal Flux Distribution

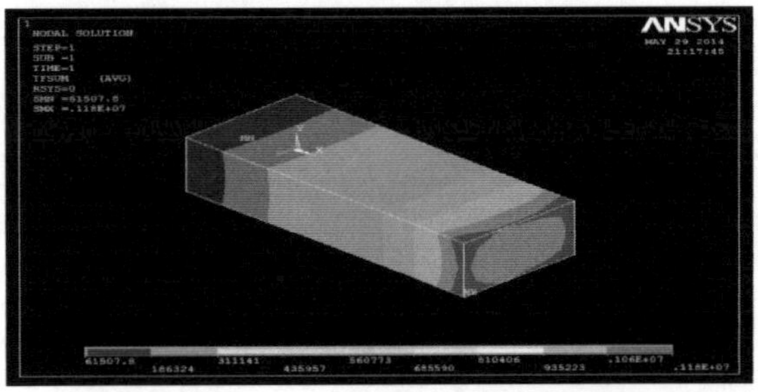

Figure 9 (a) Thermal flux of specimens

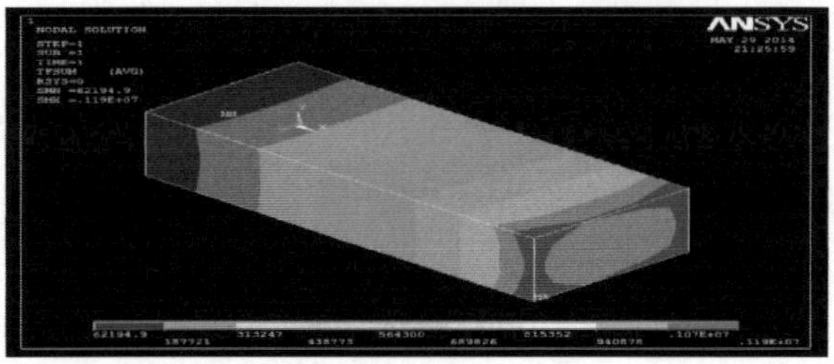

Figure 9 (b) Thermal flux of specimens

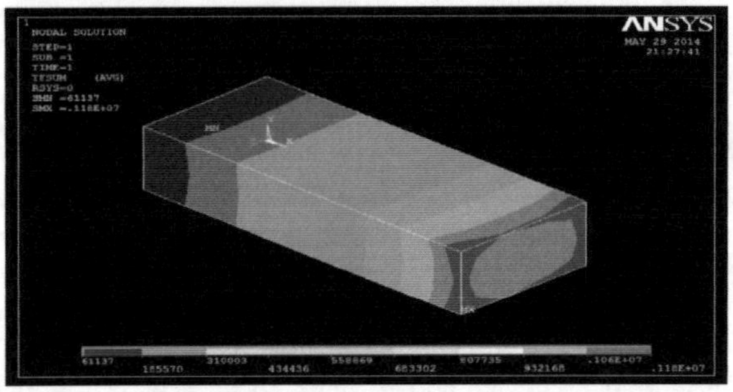

Figure 9 (c) Thermal flux of specimens

Figure. 9(a) shows the thermal flux distribution of Al7075/Al_2O_3 (5%) composite. The maximum thermal flux is at the initial end and is $0.119\times10^7 W/m^2$. In the end the thermal flux is minimum and is $62194.9 W/m^2$. Figure. 9(b) shows the thermal flux distribution of Al7075/Al_2O_3 (10%) composite. Here the maximum and minimum thermal fluxes obtained are $0.119\times10^7 W/m^2$ and $62194.9 W/m^2$.Figure. 9(c) shows the thermal flux distribution of Al7075/Al_2O_3 (15%) composite. The maximum and minimum thermal fluxes obtained are $0.118\times10^7 W/m^2$ and 61137 W/m^2 respectively.From the thermal flux distribution of the composites, the composite with 10% reinforcement of Al_2O_3 is more effective when compared to other composites.The maximum and minimum values obtained for composites of different percentages of reinforcement are tabulated below in Table.

Table 1. Thermal Flux of composites

Al7075 + %Al_2O_3	Minimum Thermal flux (W/m^2)	Maximum Thermal flux (W/m^2)
5%	61507.8	0.118×10^7
10%	62194.9	0.119×10^7
15%	61137	0.118×10^7

3.8. Thermal Gradient

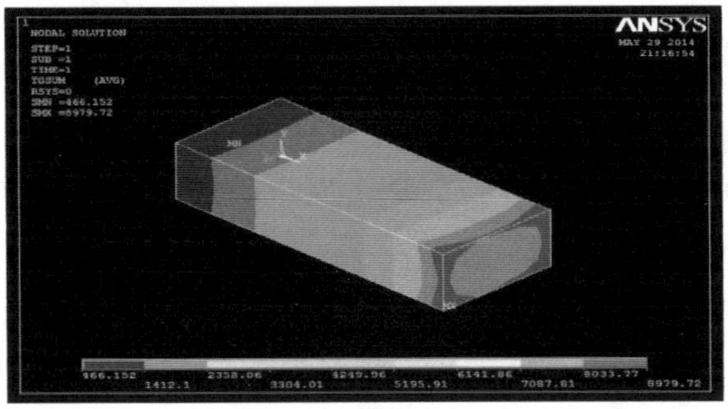

Figure. 10 (a)

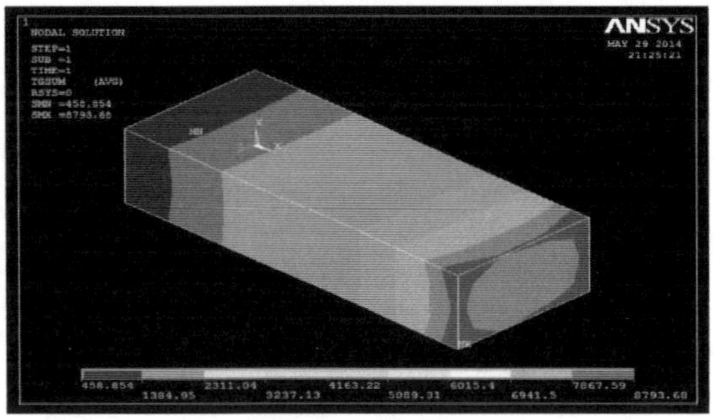

Figure. 10 (b)

Figure. 10(a) shows the thermal gradient of Al7075/Al_2O_3 (5%) composite. The thermal gradient decreases with the increase in the length of the composite material. Minimum and maximum thermal slope obtained are 466.152W/m and 8979.72W/m.

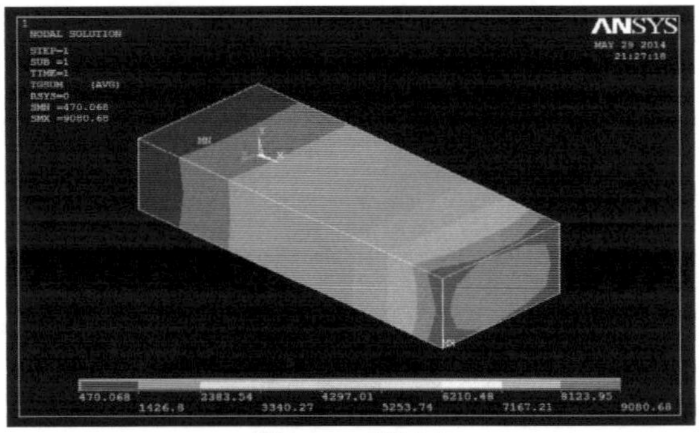

Figure. 10 (c)

Figure 10. Thermal gradient of specimens

Figure. 10(b) shows the thermal flux of the Al7075/Al_2O_3 (10%) composite. Here the heat gradient occurs a maximum value at initial end and the value is 8793.68W/m. The lowest value of thermal gradient is 458.854W/m.Figure. 10(c) shows the thermal gradient of the Al7075/Al_2O_3 (15%) composite. Here the thermal gradient occurs a maximum value at initial end and the value is 9080.68W/m. The minimum value of thermal gradient is 470.068W/m.From the results of the thermal gradient obtained, the composite with 10% reinforcement of Al_2O_3 is the material with less thermal gradient and the material with 15% reinforcement of Al_2O_3 occurs higher.

Table 2. Thermal gradient of specimens

Al7075 + %Al_2O_3	Minimum Thermal gradient (K/m)	Maximum thermal gradient (K/m)
5%	466.152	8979.72
10%	458.854	8793.68
15%	470.068	9080.68

4. Conclusion

The following are concluded from the above analysis

- The higher ultimate tensile strength (69.519N/mm^2) is obtained at 10% reinforcement of Aluminum oxide.
- The maximum Micro hardness value (148.09 VHN) is obtained at 10% reinforcement of Aluminum oxide.
- Thermal conductivity is also maximum at 10% reinforcement with 129.927 W/mK.
- The Temperature distribution and Thermal flux are superior at 10% reinforcement of Al_2O_3 in the Al7075 matrix.

So by all the tests carried out, the specimen with 10% reinforcement is having superior properties.

5. References

[1] AlptekinKisasoz, Kerem AltugGuler and AhmetKaraaslan, 2012 Infiltration of A6063 aluminium alloy into SiC−B4C hybrid performs using vacuum assisted block mould investment casting technique *Transactions of nonferrous metals society of china* **22** 563-1567.

[2] Bayraktar. E, Katundi. D, 2010 Development of a new aluminium matrix composite reinforced with iron oxide (Fe3O4) *Journal of achievements in materials and manufacturing engineering* **38** 7-14.

[3] Krupiński M, Labisz K, Dobrzański. L. A, and Rdzawski A, 2010 Image analysis used for aluminium alloy microstructure investigation *Journal of achievements in materials and manufacturing engineering* **42** 58-65.

[4] Loizaga A, Niklas, Fernandez-Calvo A I and Lacaze J 2009 Thermal analysis applied to estimation of solidification kinetics of Al-Si aluminium alloys *International Journal of Cast Metals Research* **22** 345-352.

[5] MadevaNagaral, Auradi V and Ravishankar M K 2013 Mechanical Behaviour Of Aluminium 6061 Alloy Reinforced With Al_2O_3 and Graphite Particulate Hybrid Metal Matrix Composites *International Journal Of Research In Engineering & Technology* **1** 193-198.

[6] Muhammad Hayat Jokhio, Muhammad Ibrahim Panhwar and Mukhtiar Ali Unar, 2010 Manufacturing of Aluminum Composite Material Using Stir Casting Process *Mehran University Research Journal Of Engineering & Technology* **30** 53-64.

[7] Naher S, Brabazon D and Looney L 2003 Simulation of the stir casting process *Journal of Materials Processing Technology* **24** 567-571.

[8] Narasimalu Srikanth, KhineKhine TUN and Manoj Gupta , 2003 Finite Element Based Energy Dissipation Studies of Al-SiC Composites *Journals of composite materials* **35** 1385-1410.

[9] Neelima Devi, Mahesh V, Selvaraj N 2011 Mechanical Characterization Of Aluminium Silicon Carbide Composite *International Journal Of Applied Engineering Research* **1** 793-799.

[10] Pardeepsharma, Gulshanchauhan, Neerajsharma 2003 Production of AMC by stir casting *International journal of contemporary practices* **2** 23-46.

[11] Phool Kumar, Azada A M, Vinod Kumar 2013 Frictional and Wear Characteristics of Stir-Cast Hybrid Composite Aluminium Al6061Reinforced with SiC Particulates *International Journal of Engineering Research & Technology* **2** 403-410.

[12] Sudindra S and Anil Kumar C 2013 Studies on Al6061/Al_2O_3 and graphite hybrid metal matrix composites *International journal of metallurgical & materials science and engineering* **3** 35-42.

IOP Conf. Series: Materials Science and Engineering **183** (2017) 012011 doi:10.1088/1757-899X/183/1/012011

Comparative Investigation on Modal analysis of LM25 Aluminium alloy with other Aluminim alloys using Finite element analysis software

S Arunkumar [1], V P M Baskaralal [2], and V Muthuraman [3]

[1, 2] Assistant Professor Department of Mechanical Engineering, Vels University, Chennai, India -600117

[3] Associate Professor and Head, Department of Mechanical Engineering, Vels University, Chennai India -600117.

E-mail:gct.arunkmar@gmail.com

Abstract. The rudimentary steps of the modal analysis and simulation are carried out . The modal analysis is carried out on the different Aluminum Alloys cantilever beam. The cantilever beam is designed in the graphical environment of the ANSYS. The cantilever beam was fine-tuned on one end with all degree of liberation on this end were taken, beam cannot move and rotate. Mode shapes and natural frequencies are premeditated in platforms ANSYS with arithmetical formulation of the direct solver including the block Lanczos method. Aluminum alloys are widely utilized in much application due to their estimable weight to vigor property. Many examination works have been distributed out to make developments the mechanical properties of aluminum alloys. The composition of alloying elements plays a consequential role in deciding the properties of an alloy. In this study a numerical analysis implement i.e., finite element analysis (FEA) is utilized. The work obtainable in this paper is aimed at the study of effect of modal analysis of different aluminum alloys. The modeling and analysis is carried out utilizing ANSYS FEA software. A modal analysis is carried out to understand the modes of frequency demeanor of the material considered. The modal analysis play a vital role in the design of components subjected to high vibration.

1. Introduction

Comparative investigation on modal analysis deals with the dynamics comportment of mechanical structures under the dynamics excitation. Utilizing it, we can amend the overall performance of the system in certain operating conditions. A model testing has become to identify the considerate and simulate vibrant demeanor and replication of structure. Frequency replication functions are acclimated to extract such modal parameters as natural frequency and mode shape. Model testing is a non-destructive testing based on pulsation retort of the structures. It helps to vibrations that cause damage of the integrity of system components.. We ken two rudimentary methods of the modal analysis, namely the numerical modal analysis and the experimental modal analysis. In wide range of practical applications the modal parameters are required to evade resonance in structures affected by external periodic dynamic loads. modal analysis over sundry fields of science engineering and technology. In particular, numerous researches cognate to aeronautical engineering, automobile engineering mechanical engineering and dynamic FEM updating of structures. The present investigation reports the dynamic characteristics of mundane structural materials

The two rudimental methods of the modal analysis, namely the numerical modal analysis and the experimental modal analysis. The experimental modal analysis deals with quantification input data from which a mathematical model is derived. However, it has to take different calibers of analysis, from which the model is constructed[2].
Modal analysis provides a potent implement for efficiently simulating the demeanor of deformable objects. Modal analysis has been shown to be a subsidiary implement for interactively engendering authentic simulations of elastic deformation. Both the analytic calculation of modal amplitudes utilizing involute oscillators and the abstraction of high frequency modes have a stabilizing effect on simulations, sanctioning for immensely colossal time steps to be taken[8].
The method to study the dynamic nature of plate is to go for dynamic analysis. It predicts various modes of vibration. Plate because of high strength to weight ratio are in use for many structural applications. Such structures are subjected to dynamic load many times over its life span. It was found that modal frequency is increasing with the addition of stiffeners. Overall stiffness of the plate can be effectively increased by addition of stiffener which can yield better result[5].
In this paper it has been recorded that an increasingly extended movable wing has lower natural frequencies. The experimental statement is reliable with information in structural dynamics that a longer structure has lower natural frequencies. Aluminum and its alloy concrete gravity is approximately 2.7 g/cc, the composites of Aluminum plays a very momentous role in light weight machine components. Silicon, copper, magnesium, iron, manganese, and zinc are paramount elements in Aluminum composites[6].

2. Modeling and analysis of cantilever beam

A rectangular bar element has modeled and analyzed in ansys software. Plane 182 and solid 185 have been used as element type. The mechanical properties such as the young are modulus, poisson's ratio and physical property density are applied in the material property and beam has considered as isotropic. The element edge length of 0.5 is given and meshed on the surface area, meshing is done to discretizes the area and boundary condition is given by arresting all degrees of freedom in one end and other end free. The figure 2 shows the boundary conditions and figure 3 to figure 8 show the six different set of modal frequency of beam.

Table 1. Properties of Aluminum alloys

Property	Aluminum Alloys				
	LM25M	Al2024	Al 5052	Al6061	Al 7075
Density kg/m3	2670	2780	2680	2700	2810
Elastic Modulus N/mm2	71000	73100	70300	68900	71800
Tensile strength (Map)	180	324	193	276	503
Percentage elongation (%)	3	19	12	12	11
Hardness (BHN)	60	120	60	95	150
Poisson's ratio	0.33	0.33	0.33	0.33	0.33

3. Properties of alloys

3.1Mechanical and Physical properties

The physical property density and mechanical properties young's modulus and Poisson's ratio play an important role in modal analysis. These properties are shown for different Al alloys in the table 1.

3.2. Chemical Composition

The table 2 shows the chemical composition of Aluminum alloy. In this LM 25 has considerable amount of silicon, and remaining alloys will have very minimum level of contribution. In AL2024 composition level of copper is more and other alloys are very less weight percentage. In all other aluminum alloys the chemical composition of other alloys weight fraction is very less. This clearly shows that only LM25 and Al 2024 has considerable amount of alloy content in their composition.

Table 2. Chemical Properties of Aluminum alloys

Al Alloy	Cu	Mg	Si	Fe	Mn	Cr	Zn	Ti	Al
LM25	0.1	0.2-0.6	6.5-7.5	0.5 Max	0.3		0.1	0.05	Rem.
Al 2024	3.8-4.9	1.2-1.8	0.5 Max	0.5 Max	0.3-0.9	0.1 Max	0.25 Max	0.15 Max	90.7-94.7
Al 5052	0.15-0.14	2.2-2.8	0.25 Max	0.4 Max	0.1 Max	0.15-0.35	0.1 Max		95.7-97.7
Al 6061	0.15-0.4	0.8-1.2	0.4-0.8	0.7 Max	0.15 Max	0.04-0.35	0.25 Max	0.15 Max	95.8-98.6
Al 7075	1.2-2	2.1-2.9	0.4 Max	0.5 Max	0.3 Max	0.18-0.28	5.1-6.1	0.2 Max	87.1-91.4

4. Result and Discussion

4.1. Modal Analysis

The experimental modal results (Modal frequency and mode shapes) are discussed in the following section. Natural Frequency: Modal analysis gives the value of natural frequency. These natural frequencies were calculated by beam theory. The natural frequencies of an aluminum alloy beam for the six modes are given in below table 3. The comparative analysis of modal analysis has been plotted in the Figure 1. This clearly shows the frequency is for LM25 and Al 2024 are comparatively higher than the other alloys.

Table 3. Modal frequency of different Aluminum alloys

	Frequency					
Al Alloy	1	2	3	4	5	6
LM25	38.04	57.20	227.31	309.24	312.95	410.08
Al 2024	38.36	57.17	228.33	306.37	313.61	409.55
Al 5052	37.85	56.91	226.18	307.71	311.41	408.05
Al 6061	37.33	56.13	223.09	303.42	307.15	402.47
Al 7075	37.33	56.13	223.08	303.48	307.13	402.45

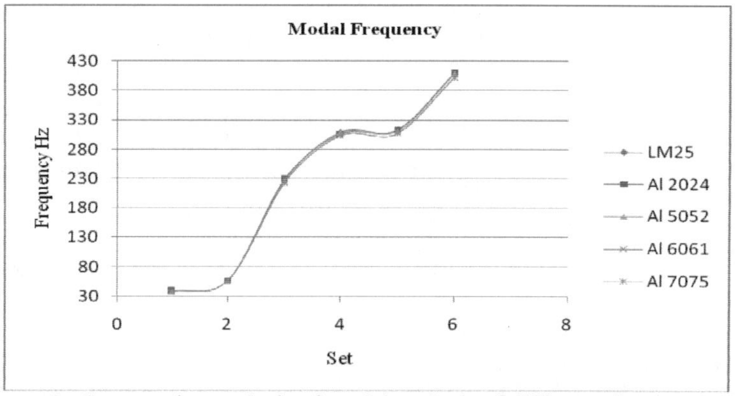

Figure 1. Comparative analysis of modal analysis of different aluminum alloy

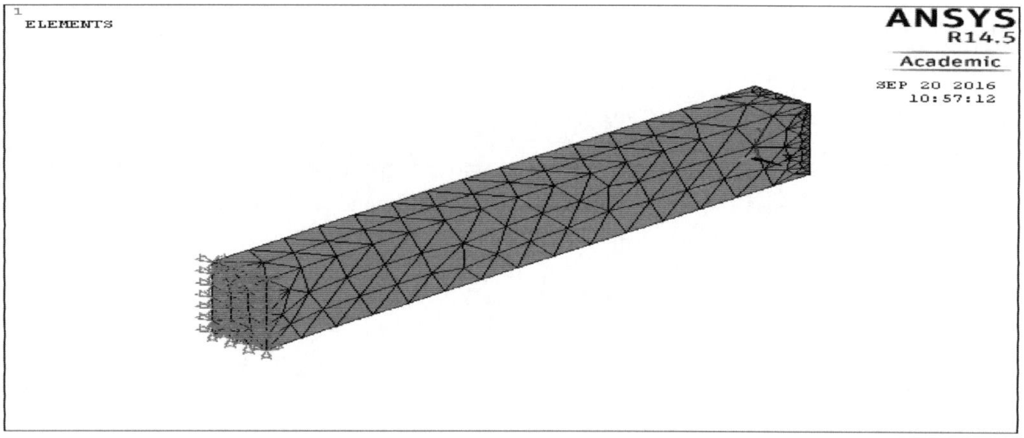

Figure 2. Meshing and Boundary conditions

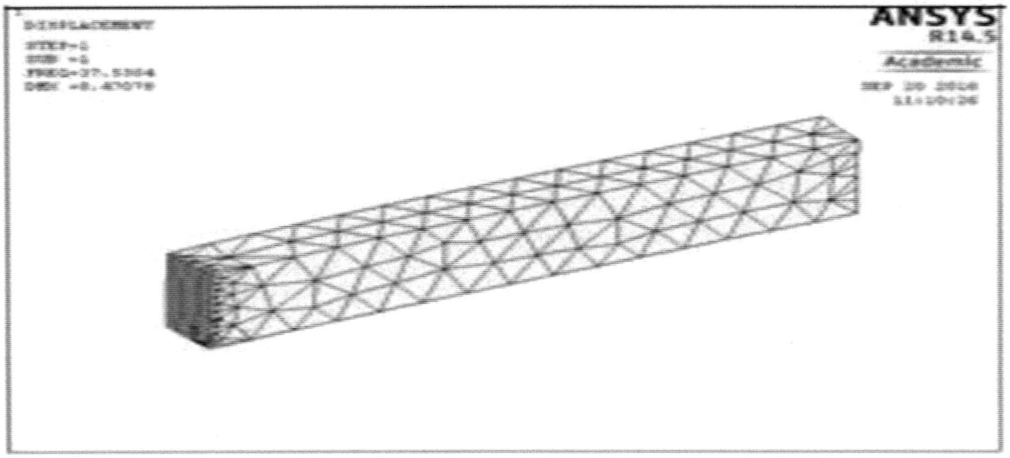

Figure 3. First mode shape at set 1

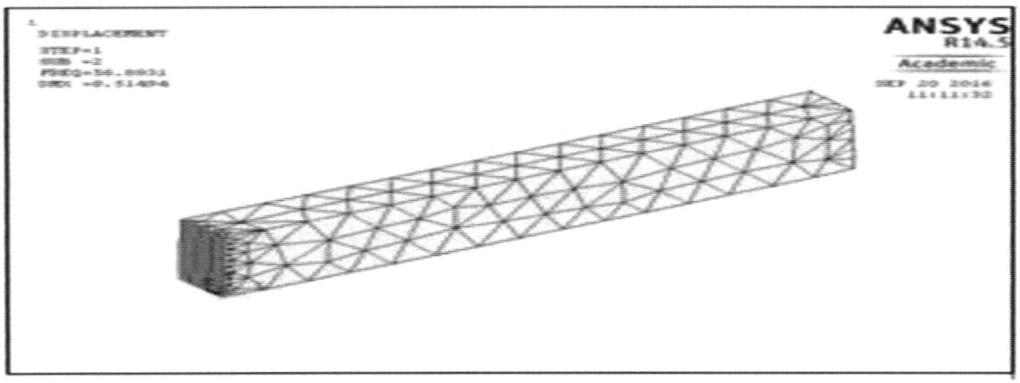

Figure 4. Second mode shape at set 2

IOP Conf. Series: Materials Science and Engineering **183** (2017) 012011 doi:10.1088/1757-899X/183/1/012011

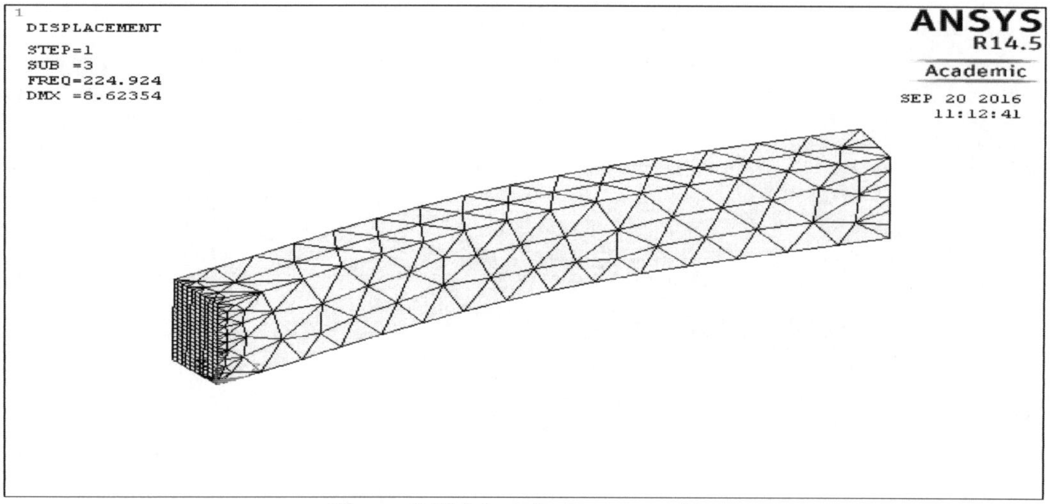

Figure 5. Third mode shape at set 3

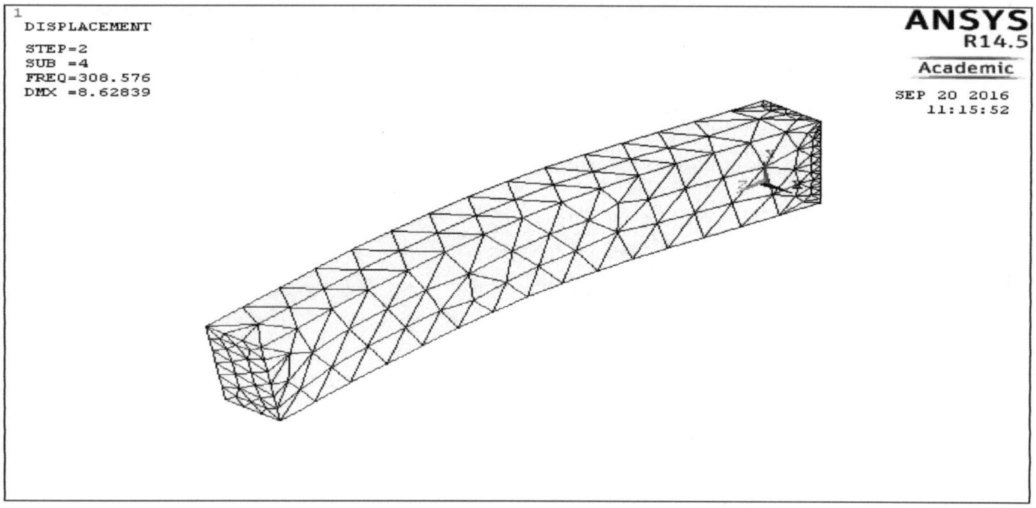

Figure 6. Fourth mode shape at set 4.

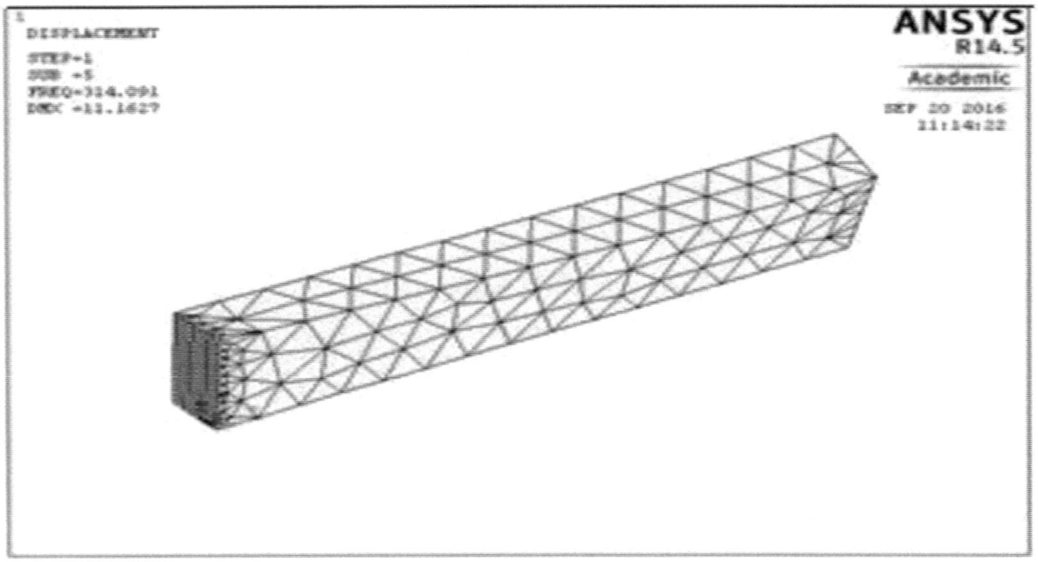

Figure 7. Fifth mode shape at set 5

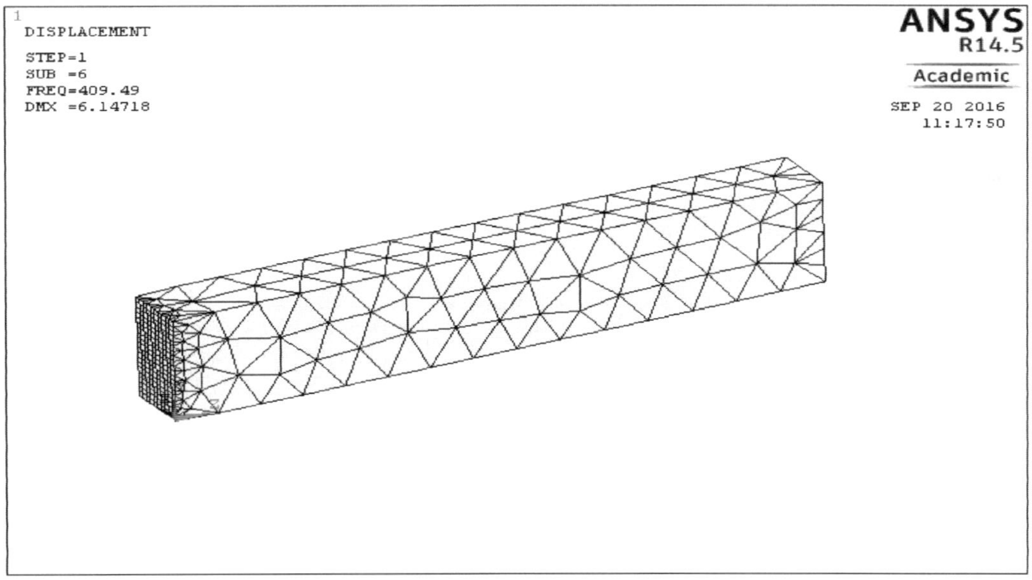

Figure 8. Sixth mode shape at set 6

Given the following equations have the frequencies of the modes and their shapes and have been deduced from

4.2. Euler-Bernoulli Beam Theory.

$\omega_n = \alpha^2_n \sqrt{(E I /mL3)}$; n= 1, 2, 3….

α_n =1.875, 4.694, 7.855; m = ρV = ρxlxhxw; I =wh3/12;

ρ = density in kg/mm^3

From Strength of Materials, the deflection, x, at the tip of a cantilever beam is given by

x = P * L3 / 3 E I

Where x - tip displacement; P - applied load (also referred to as F); L - length of the cantilever beam

E - Young's Modulus of Elasticity; I - bending moment of inertia

The deflection at the end of the cantilever beam can be expressed as

F = k x and therefore, the stiffness of the cantilever beam can be expressed as

k = 3 E I / L3.

Realizing that the natural frequency is

$\omega_n = \sqrt{}$ (k/m) (in rad/Sec) and the natural frequency of the first mode of the cantilever from the continuous solution is

ω_n = (1.875)2 $\sqrt{}$ (E I g/wL4) (where w - weight and

g – gravitational constant) allows the effective mass at the tip of the cantilever beam to be determined. This approximation allows the cantilever beam to be modeled as a single degree of freedom system since the mass and stiffness are known.

4.3. Deformation

The deformation for different Al Alloy is shown in table 4. The values show that for LM25 the deformation is more compared to other aluminum alloy. Al 2024 and Al 7075 have less deformation. The density and young's modulus are more than other Al alloy plays a vital role in deformation of Al alloy.

Table 4. Deformation of different Aluminum alloys

	Deformation					
Al Alloy	1	2	3	4	5	6
LM25	8.47	8.51	8.64	8.58	11.15	6.15
Al 2024	8.24	8.25	7.95	8.12	11.04	6.06
Al 5052	8.47	8.51	8.64	8.58	11.15	6.15
Al 6061	8.44	8.48	8.61	8.55	11.11	6.13
Al 7075	8.27	8.31	8.44	8.38	10.89	6.01

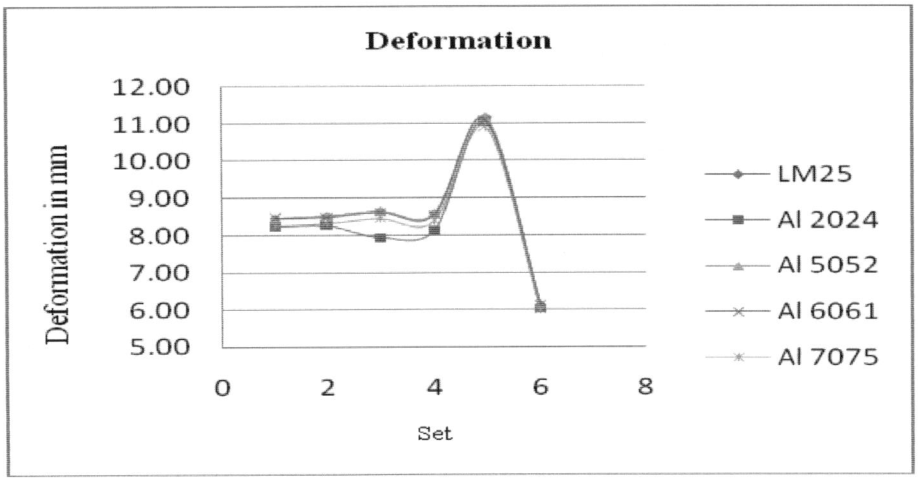

Figure 9. Deformation graph of different aluminum alloy

5. Conclusion

The work obtainable in this paper is intended to investigate the effect of the vibration characteristics and deformation of different Aluminum alloys. The performances of vibration on the alloys are tabulated. From the analysis, it is concluded that the natural frequency of the beam and deformation is related to young's modulus and density. It is observed that the natural frequency is nearly equal for Al 2024 and for LM25 Aluminum Alloy. The deformation for AL2024 and AL7075 is less than other AL Alloys.

This clearly indicates that increase of silicon content in the alloy LM25 increase modal frequency and decrease of young's modulus, density and copper content increases the deflection in LM25.

6. References

[1] Vinay V, VijayKumar N Chalwa, S and N. Kurbet Aravind M 2014 Finite Element analysis of Aluminum alloys for their vibration characteristics *International Journal of Research in Engineering and Technology* **03**(3) 598-600.

[2] Pavol Lengvarský and Martin Hagara 2013 Modal Analysis of Titan Cantilever Beam Using ANSYS and Solid Works *American Journal of Mechanical Engineering* **1**(7) 271-275.

[3] Pragnesh K Chaudhari, and Vipul Patel 2014 Theoretical and Software Based Comparison of Cantilever Beam: Modal Analysis *International Journal of Innovative Research in Advanced Engineering* **1**(5) 75-79.

[4] Azoury.C, Kallassy, A, Combes.B, Moukarzel.I Boudet.R, 2012, Experimental and Analytical Modal Analysis of a Crankshaft, IOSR *Journal of Engineering*, **2**(4), 674-684.

[5] Ash tosh Kumar, Rachayya R. Arakerimath 2015 A Review and modal analysis of stiffened Plate International *Research Journal of Engineering and Technology* **02**(1)581-588

[6] Jia P,Lai S K Zhang, W Lim C W 2014 Experimental and FEM Modal Analysis of a Deployable-Retractable Wing *Modern Mechanical Engineering* **4**(2) 183-197.

[7] Chaphalkar S P Subhash N Khetre and Meshram M 2015 Modal analysis of cantilever beam Structure Using Finite Element analysis and Experimental Analysis *American Journal of Engineering Research* **4** (3) 178-185.

[8] Kris K , Hauser Chen Shen James F. O'Brien 2003 Interactive Deformation Using Modal Analysis with Constraints Graphics *Interface.*

Experimental evaluation of machining parameters in machining of 7075 aluminium alloy with cryogenic liquid nitrogen coolant

V Muthuraman[1], and S Arunkumar [2]
[1] Associate Professor, Department of Mechanical Engineering, Vels University,Chennai,India
[2]Assistant Professor, Department of Mechanical Engineering, Vels University,Chennai,India

E-mail:hodmech@velsuniv.ac.in

Abstract. The experimental results on investigation on the turning of 7075 aluminium alloy, using Cryogenic Liquid Nitrogen (LN_2) as a coolant was analyzed in this paper. The influence of the cryogenic LN_2 coolant compared with that of the conventional coolant on the cuting performance parameters, such as the cutting force, cutting temperature, andbsurface finish was analysed and investigated. The use of the cryogenic liquid nitrogen coolant influenced the cutting temperature and the cutting force by about 17 to 29% and 11 to 20% reduction respectively. The surface finish value of the machined workpiece is about 15 to 23% better than that of the conventional coolant.
Keywords: 7075 aluminium alloy, LN_2 cooling, Tool wear, Cutting temperature

1. Introduction

Al-Zinc alloys are one of the indispensable materials of the present industry, because of their superior mechanical properties over other aluminium alloys. The Aluminium- Zinc 7075 alloy is predominantly used in aerospace and automobile industries due to its excellent mechanical strength.

Machining of Aluminium alloys causes Built Up Edge formation (BUE), which affects the surface roughness [1],[7]. When the cutting velocity is increased, the tendency of BUE vanishes, due to the low cutting force and high temperature [9].

Cutting fluids are used in turning operations primarily to improve the tool life and surface finish of the work-piece. They also help in reducing the temperature and removing the debris during the turning operation. However, the use of cutting fluids has many adverse effects. The cutting fluids used in machining, contain harmful chemical components which is harmful to environment. These conventional cutting fluids are difficult to dispose of and can cause harmful diseases to the operators. In the turning of the 6061 aluminium alloy under Minimum Quantity Lubrication conditions, the amount of cutting fluid influences the material adherance on the tool surface, and tools used for turning experienced nose wear and flank wear [11]. The elimination of cutting fluids in turning causes high friction between the tool and workpiece, which leads to higher temperature in the machining zone. Dimensional inaccuracies and excessive tool wear are also the problems in dry machining [12].

An environmentally safe coolant viz, cryogenic LN_2, is used to overcome these difficulties. Researchers carried out experiments earlier by using cryogenic LN_2 as a coolant for the turning other materials, and obtained good results in reducing the cutting temperature, cutting force, surface roughness and tool wear [4],[6],[10]. It was also observed that when cryogenic LN_2 was used as a coolant in the turning of Ti-6Al-4V alloy and AISI 4045 Steel, the process parameters such as surface roughness, tool wear, cutting force and cutting temperature were reduced [3],[5]. The gases and water vapor also used as cutting coolants by researchers and contributed well to the metal cutting industry [1],[8].

This experimental work aims on investigating performance of cryogenic LN_2 as the cutting fluid in machining 7075 alloy and compare the results with conventional wet machining in terms of cutting temperature, cutting force and surface roughness .

2. Experimental procedure

In this experimental work, Al 7075 alloy was used as the workpiece material. A bar with dimensions of 75 mm diameter and 300 mm length bar was used for the experimental work.The following conditions are considered for the machining work, as mentioned in Table 1.

Table 1 Experimental conditions

Workpiece	7075 Aluminium alloy
Workpiece dimension	Φ 75 mm x 300 mm
Cutting tool insert	Uncoated Tungston carbide insert
Rake angle	0°
Process parameters	
Cutting velocity	51,118 and 181 m/min
Feed rate	0.079,0.159 and 0.205 mm/rev
Depth of cut	1 mm
Machining conditions	Wet and LN_2

The main purpose of this experiment is to investigate the effect of LN_2 as a cutting fluid, and compare it with that of the conventional coolant in the machining of the 7075 Al alloy. The cryogenic LN_2 machining experimental set up used is shown in Figure 1.The various sets of experiments were performed by the turning operation on a High speed automatic lathe (NAGMATI175) shown in Figure 2. The Kistler type 9257B Piezoelectric Dynamometer was used to measure cutting force. The cutting temperature was measured using a non-contact type IR thermometer on the cutting face of the tool insert during the turning process. The Talysurf surface roughness tester was used for measuring surface roughness after completion of turning the workpieces at a constant cutting velocity of 51 m/min and constant feed rate of 0.079 mm/rev. conditions.

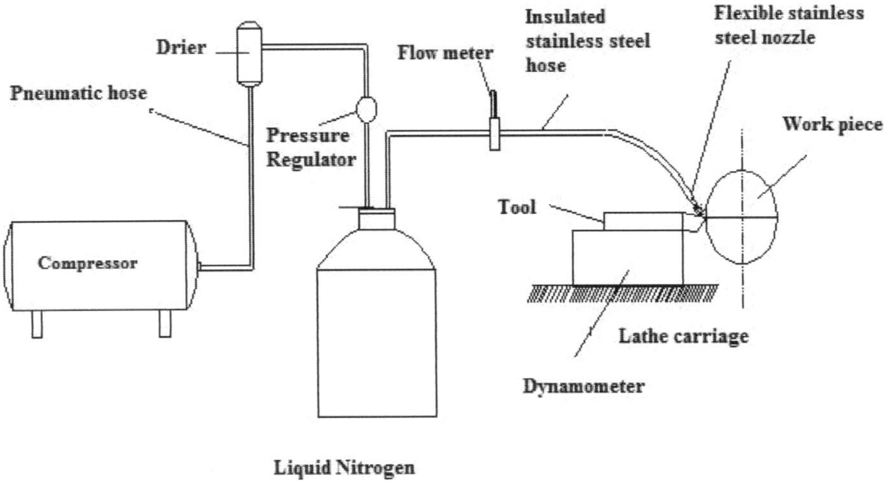

Figure 1. Schematic view of experimental set up for cryogenic LN_2 machining

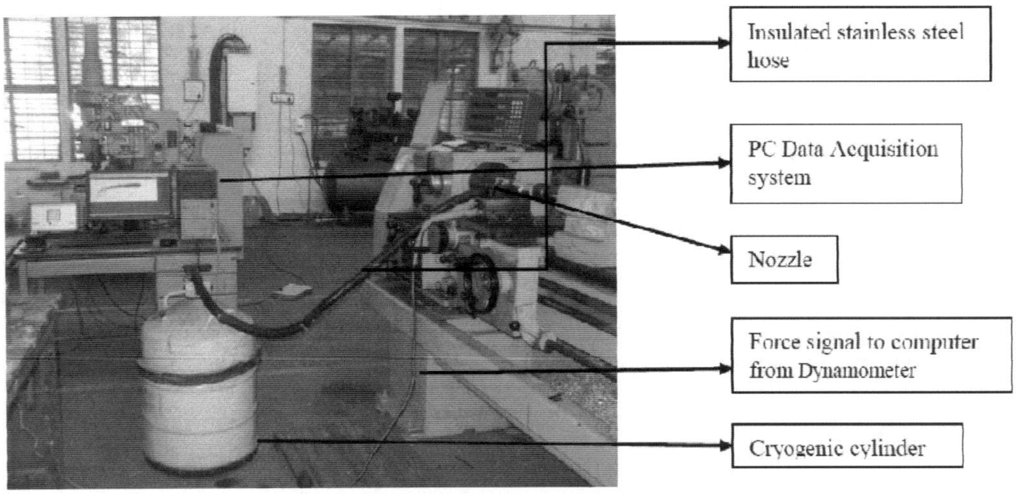

Figure 2. Photographic view of Experimental setup

3. Results and Discussion

3.1. Cutting temperature

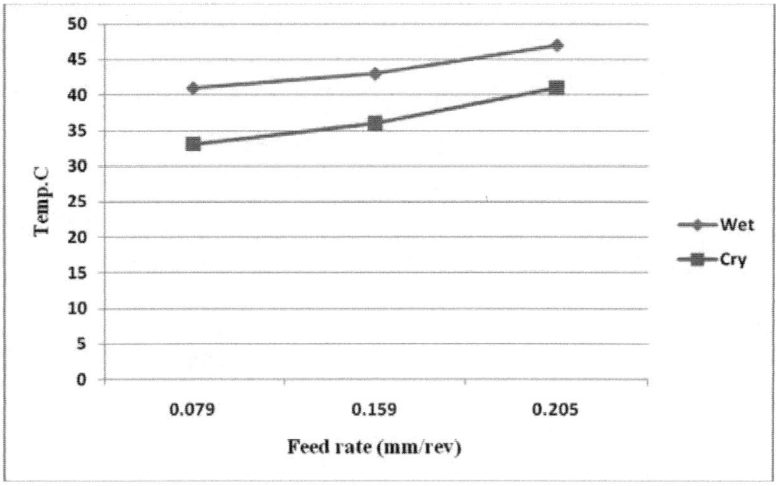

Figure 3a. Cutting temperature at cutting velocity of 181m/min with different feed rate

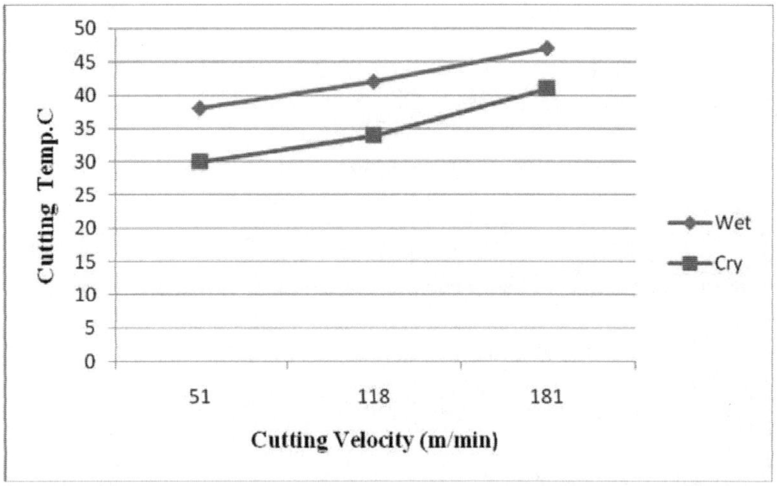

Figure 3b. Cutting temperature at feed rate of 0.205 mm/rev with different cutting velocity

The cutting tool temperature was measured using a non contact type thermometer for different cutting velocity and feed rate conditions. Figure 3a and 3b shows the variation of the cutting temperature under cryogenic and conventional coolant conditions at a constant cutting velocity of 181 m/min.and constant feed rate of 0.205 mm/rev respectively. The result showed that the cutting tool temperature was increased proportionally with the increase in the feed rate and cutting velocity.This was due to more plastic deformation and contact friction between the workpiece and tool. The LN_2 applied between the workpiece and cutting tool absorbs the high heat energy generated in the cutting zone area. Due to its low boiling point (-196°C), the LN_2 gets evaporated by absorbing the high heat energy generated between the workpiece and tool insert used, since the LN_2 was impinging in liquid and gaseous forms between the workpiece and the tool, the friction at the chip tool contact surface and temperature is reduced. This reduction of temperature is about 17-29 % compared to wet machining conditions.

3.2. Cutting force:

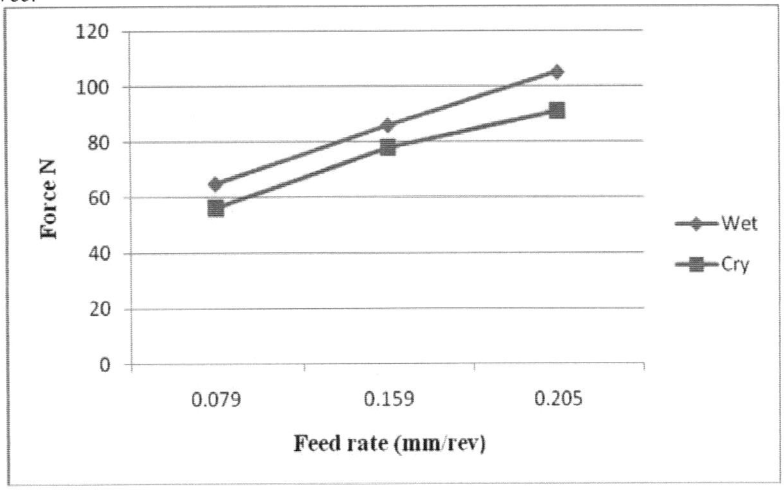

Figure 4a. Cutting force at cutting velocity of 51m/min with different feed rate

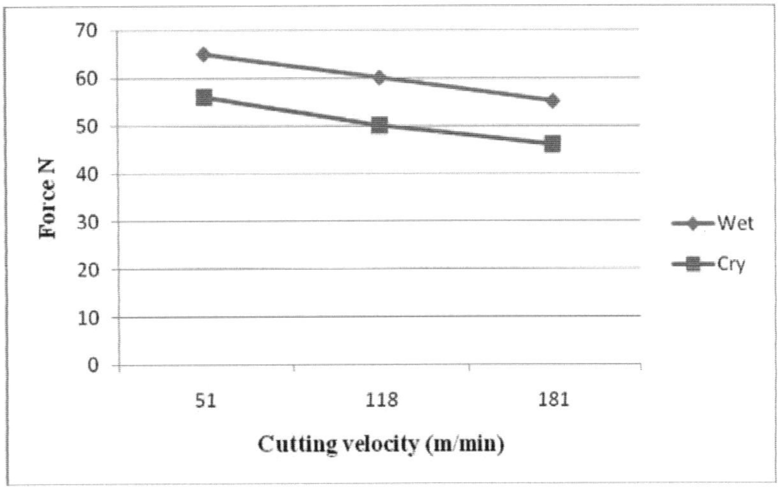

Figure 4b. Cutting force at feed rate of 0.079 mm/rev with different velocity

The main cutting force for various cutting velocity and feed rate conditions was measured.The variation of the cutting force under cryogenic and conventional coolant conditions at a constant velocity of 51 m/min and constant feed rate of 0.079 mm/rev, is shown in Figure 4a. and Figure 4b. respectively.

It is observed from Figure 4a that with the increase in feed rate cutting force is increased. This is due to the increase in the material removal rate and work done per unit time. It is also observed that when cryogenic LN_2 was used as the coolant the cutting force was reduced

The Figure 4b shows the variation of the cutting force at a constant feed rate of 0.079 mm/rev.It is observed that the cutting force decreased with the increase in the cutting velocity. This is due to less duration of the contact between the material and the tool. When the cutting temperature is increased at higher velocity, the material plasticity also increases, which reduces the cutting force. The formation of BUE at high temperature vanished, resulting in a reduction in the cutting force. It can be observed that the application of LN_2 reduces the cutting force more than wet machining conditions. This is due to the adhesion produced by the penetration of LN_2 in liquid and gaseous forms between the workpiece and the tool. The reduction of temperature and reduced friction at the chip tool interface, reduced the cutting force to a great extent. This reduction of the cutting force is about 11-20 % compared to wet machining conditions.

3.3. Surface roughness

The surface roughness value of the machined workpiece was measured at a constant feed rate of 0.079 mm/rev and constant velocity of 51 m/min, by varying the other parameters of cutting velocity and feed rate respectively. The mean surface roughness values (R_a) were taken for both the conditions for evaluation. The surface roughness values at a constant feed rate of 0.079 mm/rev are shown in Figure 5. The surface roughness values at a constant velocity of 51 m/min are shown in Figure 5.

It is clearly seen from figure 6. that the surface roughness decreased with the increase in the cutting velocity. This was due to less duration of contact of the workpiece material tool . The lower material removal rate and reduction of BUE due to increase in temperature, also caused a reduction in the surface roughness value.

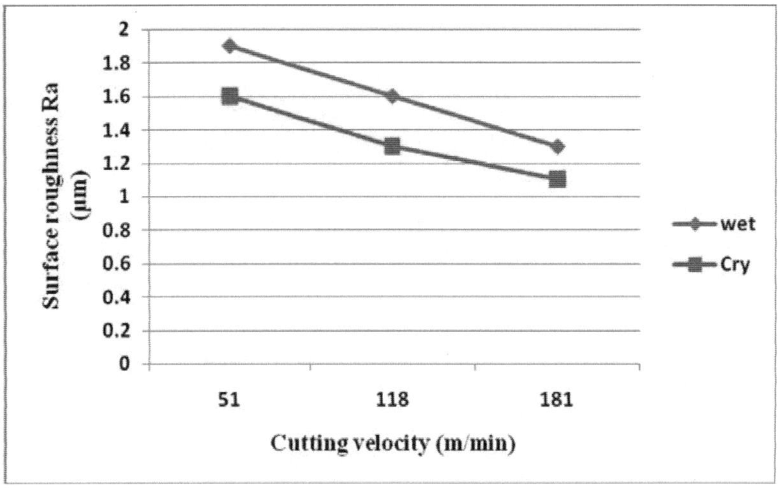

Figure 5.Surface roughness at the feed rate of 0.079 mm/rev with different cutting velocity

It is clear from the above illustration that the use of LN_2 coolant reduced the surface roughness values compared to the conventional wet machining conditions. This is because of the reduction of the cutting temperature and cutting force compared to wet machining conditions. It is also noted that the lesser adhesion of the material over the tool, reduced friction at the chip tool interface due to a better cushioning effect, and higher tool hardness improved the surface finish values.The use of low temperature LN_2 produced better surface finish values by about 15– 23% when compared with conventional wet machining conditions.

4. Conclusion

The following conclusions can be drawn based on turning the Al 7075 alloy by using cryogenic LN_2 as a coolant :

1. The use of cryogenic liquid nitrogen coolant reduced cutting temperature 17–29% less compared to conventional wet machining.
2. The application of cryogenic LN_2 reduced cutting forces at about 11–20% less compared to conventional wet machining..
3. Surface roughness was reduced on use of the cryogenic LN_2 coolants. The application of the LN_2 coolant reduced the surface roughness value by about 15-23 % compared to wet machining conditions.

5. References

[1] Çakir O,Kiyak M and Altan E 2004 Comparison of gases applications to wet and dry cuttings in turning *Journal of Material Processing Technology.* **153** 35–41.

[2] Dae E.K and Dong H H 1998 Experimental investigation of the contact sliding behavior of metals *Journal of Manufacturing Science and Engineering.* **120** 395-400.

[3] Dhananchezian M and Pradeep Kumar M 2011 Cryogenic turning of the Ti–6Al–4V alloy with modified cutting tool inserts *Cryogenics.* **51** 34–40.

[4] Dhar N R,Paul S and Chattopadhyay A B 2002 Machining of AISI 4140 steel under cryogenic cooling– tool wear, surface roughness and dimensional deviation". *Journal of Material Processing Technology.* **123** 483–9.

[5] Dilip Jerold B and Pradeep Kumar M 2012 Experimental comparison of carbon-dioxide and liquid nitrogen cryogenic coolants in turning of AISI 1045 steel *Cryogenics.* 52 : 569–574.

[6] Hong S Y, Markus I and Jeong W 2001 New cooling approach and tool life improvement in cryogenic machining of titanium alloy Ti–6Al–4V *International Journal of Machine Tools Manufacturing,* **41** 2245–2260.

[7] Jeelani S and Musial M 1986 Dependence of Fatigue Life on the Surface Integrity in the Machining of 224-T 351 Aluminium Alloy Unlubricated Conditions *Journal of Materials Science.***21** 155-160.

[8] Liu J, Han R and Sun Y 2005 Research on experiments and action mechanism with water vapor as coolant and lubricant in Green cutting *International Journal of Machine Tools Manufacturing.* **45** 687–694.

[9] Oishi K 1996 Mirror Cutting of Aluminium with Sapphire Tool *Journal of Material Processing Technology* **62 (4)** 331-334.

[10] Paul S and Dhar N R and Chattopadhyay A B 2001 Beneficial effects of cryogenic cooling over dry and wet machining on tool wear and surface finish in turning AISI 1060 steel *Journal of Material Processing Technology.* **116** 44–48.

[11] Sreejith P S 2008 Machining of 6061aluminium alloy with MQL, dry and flooded lubricant conditions *Materials Letters.* **62** 276 –278.

[12] Sreejith P S and Ngoi B K A 2000 Dry machining — machining of the future *Journal of Material Processing Technology,* **101 (1-3)** 289–293.

Experimental Investigation of Turning Parameters on AA 6061-T6 Material

A Parthiban [1], R Pugazhenthi [2], R Ravikumar[3] and P Vivek[4]
[1,4]Assistant professor, Department of Mechanical Engineering, School of Engineering, Vels university, Chennai-117
[2]Associate professor, Department of Mechanical Engineering, School of Engineering, Vels university, Chennai-117
[3]Professor, Department of Mechanical Engineering, Ariyalur Engineering College, Melakaruppur-621707

E-mail: [1]parthibana@yahoo.co.in

Abstract. Turning is a broadly used material removal process to manufacture cylindrical products. The effects of various process parameters on turning like spindle speed, feed rate and depth of cut have been investigated to Impact on Material Removal Rate (MRR) and surface roughness (Ra) by using Response Surface Methodology. Experimental plan is performed by a Box-behenken design. The main purpose of this work is to study the effect of process parameters on Aluminium alloy AA6061-T6 surface, and to develop the mathematical model for Material removal rate and surface roughness on milling process. The quadratic model is best agreement with experimental data; finally the numerical optimization technique has been used to find out best optimum milling parameters. The optimal set of process parameters has also been predicted to maximize the MRR and minimize the surface roughness.

1. Introduction

Turing is one of the most popular machining processes and AA6061-T6 is playing vital role in modern industries. Now a day's aluminium alloy has significant role in aeronautical industries. However AA6061 is one of the important materials due to its inherent properties such as high strength to low weight ratio, good corrosion resistance, high thermal conductivity and easy machinabilty and formability. Still machining of aluminium alloy with tuning is a difficult task due to its high speed and depth of cut. Consequently Material removal rate and Surface roughness play the most significant role in determining the productivity and the quality of a product produced with AA6061. So this work takes AA6061 aluminium alloy as a work piece material for turning [1]. With regard, to the Spindle speed, feed rate and depth of cut are considered as a predominant principal parameter in turning. Lot of researcher and investigations has been done in analyzing the cost and quality of turning. However with regard to the reduction of wastage that is caused in material removal rate has not been taken up for serious study [2]. Hence this study considers the principal parameters on material removal rate and surface roughness. Consequently, tuning is one of the extensively used to make cylindrical product based contact type turning process in the order of all type of material [3]. However the MRR and surface roughness are important during turning, but the MRR are sensitive with Depth of cut, speed and the work piece material. Although, MRR is usually affected with Spindle speed and Depth of cut, the turning quality of the material depends mainly on feed rate and Speed [4, 5]. The previous research on MRR and Surface roughness had less attention for best parameter selection on aluminium alloys during All geared lathe turning. But few works were employed in the RSM, Artificial Neural Network and Fuzzy logic for predicting the parameters in turning. Thus the main aim of this work is to develop a mathematical model using RSM, to optimize turning operation using numerical optimization.

2. Experimental procedures

The experimental procedure is planned based on Box – Behnken design from the design of experiment concepts for turning operation

2.1. Machine tool

The experiments are conducted on All geared lathe machine to use of turning operation as shown in Figure.1. Based on the machine tool specification the experimental factor's range and levels are selected. The selected ranges and levels are given as an input to machine [7].

Figure 1. Turning operation

2.2. Work piece material

The work piece material considered in this work is AA6061 – T6 aluminium alloy. AA6061 – T6 is the newer material utilized in modern manufacturing industries such as Aeronautical and ship building industries due to its inherent properties. And the specimen size was round rod before machining 25mm diameter and length was 75mm

2.3. Experimental plan

The experimental work conducted based on Box-Behnken design and it requires three levels for each factor [8]. So the ranges and levels of each factor considered for this work is tabulated in table 1. The factors are Feed rate Spindle speed, and depth of cut. Turning operation carried out on work piece with 50mm length and single pass were obtained in each experimental run

2.4. Measurements of responses

The material removal rate to find out by using following relation $MRR = A_p \times v_c \times v_f$

and where A_p are Depth of cut v_c Are cutting speed and V_f are feed rate, and to measure surface roughness are to use measuring mittayo surface roughness meter.

3. Result and discussion

3.1.Response surface methodology

Response surface methodology Montgomery is a collection of mathematical and statistical techniques and to optimize this response. From the previous research the effect by Material removal rate and surface roughness are considered factors in this work were Spindle speed, feed rate and depth of cut on Material removal rate and surface roughness. In order to matamatical model the relations between these variables, the response surface methodology was assumed [9].

$$R = b_0 + \sum_{i=1}^{2} b_i X_i + \sum_{i=1}^{2} b_{ii} X_{ii}^{2} + \sum_{i=1}^{2} \sum_{j=1,(i<j)}^{2} b_{ij} X_i X_j + \varepsilon \tag{1}$$

IOP Conf. Series: Materials Science and Engineering **183** (2017) 012013 doi:10.1088/1757-899X/183/1/012013

Where X1, X2 and X3 represented the spindle speed, feed rate and depth of cut respectively, Xi2 and Xi Xj, the squares and the interaction terms of these factors, the constants b, the regression coefficients of these parameters and ε, the experimental error.

Table 1. Experimental plan

S.No	Speed (rpm)	Feed rate (mm/min)	Depth of cut (mm)	MRR (mm^3/ min)	Surface roughness (μm)
1	945	1.25	2.5	0.187	1.657
2	945	1.25	2.5	0.198	1.786
3	945	1.5	3	0.187	1.654
4	840	1.25	2	0.165	1.354
5	945	1.25	2.5	0.197	1.765
6	945	1.25	2.5	0.1765	1.897
7	1050	1.25	2	0.259	0.987
8	945	1	3	0.1909	1.987
9	840	1.5	2.5	0.249	1.678
10	1050	1.5	2.5	0.1892	0.879
11	840	1	2.5	0.1987	1.789
12	945	1.25	2.5	0.21	1.987
13	945	1	2	0.11	1.786
14	840	1.25	3	0.166	1.987
15	1050	1.25	3	0.259	0.9087
16	1050	1	2.5	0.173	0.9078
17	945	1.5	2	0.231	1.23

3.2. Analysis of variance (ANOVA)

ANOVA is the statistical method it is used to calculate the size of the variation in data set [10]. The main elements of ANOVA table are sum of squares, mean square, source of variance, DOF, F ratio and the probability associated with the F ratio. The source of variance deals with parameters that are called factors (Spindle speed, Feed rate and Depth of cut).

The ANOVA Table 2 shows that the models are significant and the F-value are 11.50 implies. And low signal noise. The probabilty values are less than 0.0500 indicate model terms are significant. In this case Speed and feed are significant model terms. And the R^2 and Adj R^2 values are 0.9718 and 0.8891. so that model are find the design space.

The ANOVA Table 3 shows that the models are significant and the F-value are 13.86 implies. And low signal noise. The probabilty values are less than 0.0500 indicate model terms are significant. In this case Speed and feed are significant model terms. And the R^2 and Adj R^2 values are 0.946 and 0.8786. so that model are find the design space.

Table 2. ANOVA table for Material Removal Rate

Source	Sum of Squares	df	Mean Square	F Value	p-value
Model	0.0219	12	0.0018	11.4989	0.0152
A	0.0092	1	0.0092	57.6116	0.0016
B	0.0021	1	0.0021	13.0778	0.0224
C	0.0002	1	0.0002	1.2631	0.3240
AB	0.0006	1	0.0006	3.7354	0.1254
AC	0.0001	1	0.0001	0.7338	0.4399
BC	0.0039	1	0.0039	24.5360	0.0077
A^2	0.0018	1	0.0018	11.2958	0.0283
B^2	0.0060	1	0.0060	37.9791	0.0035
C^2	0.0000	1	0.0000	0.1168	0.7497
A^2B	0.0003	1	0.0003	2.0135	0.2289
A^2C	0.0002	1	0.0002	1.0135	0.3710
AB^2	0.0093	1	0.0093	58.3959	0.0016
Error	0.0006	4	0.0002		
Total	0.0226	16			

Table 3. ANOVA table for Surface roughness

Source	Sum of Squares	Df	Mean Square	F Value	p-value
Model	2.5355	9	0.2817	13.8648	0.0011
A	1.8347	1	1.8347	90.2965	< 0.0001
B	0.0717	1	0.0717	3.5299	0.1023
C	0.0152	1	0.0152	0.7500	0.4152
AB	0.0017	1	0.0017	0.0831	0.7814
AC	0.1265	1	0.1265	6.2251	0.0413
BC	0.0124	1	0.0124	0.6119	0.4597
A^2	0.7786	1	0.7786	38.3175	0.0004
B^2	0.0236	1	0.0236	1.1637	0.3165
C^2	0.0264	1	0.0264	1.3002	0.2917
Residual	0.1422	7	0.0203		
Lack of Fit	0.0777	3	0.0259	1.6044	0.3216
Error	0.0646	4	0.0161		
Total	2.6777	16			

3.3.Response surface models

Response Surface model, are to predict the Material removal rate and Surface roughness values, is developed using RSM. The cubic mathematical models have been developed to predict the Material removal rate and surface roughness. The developed mathematical models for AA 6061-T6 are showed equations 2-3

$$MRR = +23.81 - 0.0374 \times A - 27.1 \times B - 1.07 \times C + 0.034 \times A \times B + 3.07e^{-3} \times A \times C - 0.249 \times B \times C$$
$$+1.1e^{-5} \times A^2 + 9.6 \times B^2 - 8.4e^{-3} \times C^2 - 4.58e^{-6} \times A^2 \times B - 1.63e^{-6} \times A^2 \times C - 0.01 \times A \times B^2 \quad (2)$$

$$Ra = -39.128 + 0.077 \times A + 0.628 \times B + 4.522 \times C + 7.8e^{-4} \times A \times B - 3.38e^{-3} \times A \times C + 0.446 \times B \times C - 3.9e^{-5} \times A^2 - 1.1 \times B^2 - 0.3168 \times C^2 \quad (3)$$

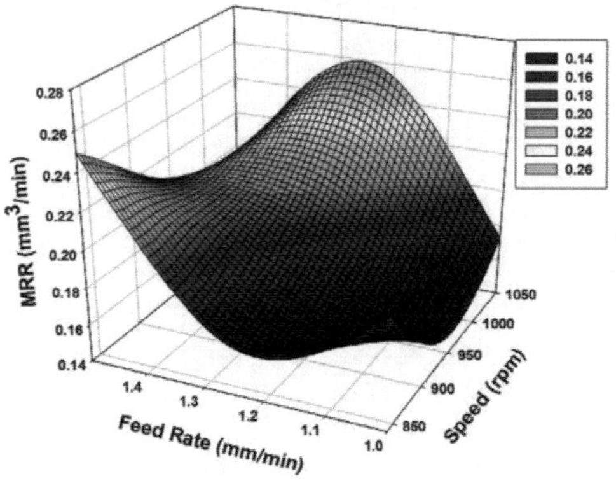

Figure 2. Effect Speed and Feed rate on MRR

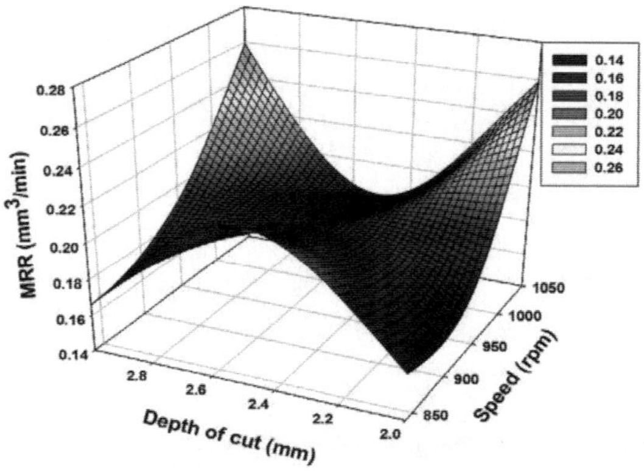

Figure 3. Effect Speed and Depth of cut on MRR

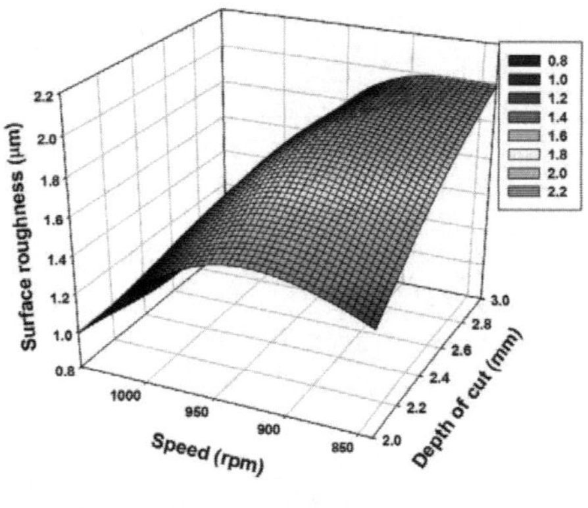

Figure 4. Effect Speed and Depth of cut on Surface roughness

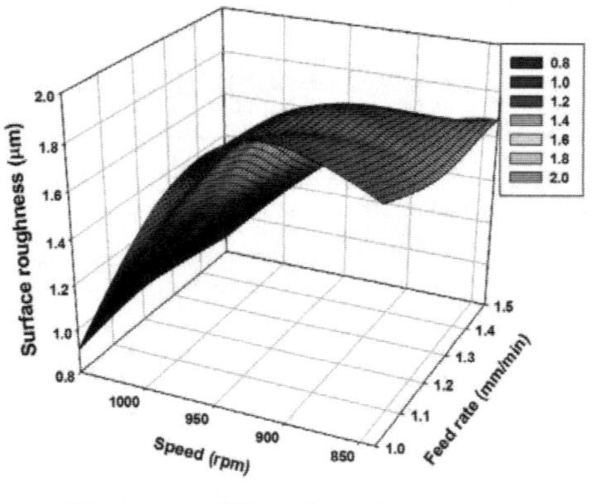

Figure 5. Effect Speed and Feed rate on Surface roughness

The figure.2 express, firstly the Material removal rate is low (range from 0.14 to 0.16 mm3/min) at mid level of the Feed rate (1.2-1.25 mm/min) and Spindle speed at 950 rpm. MRR is gradually increased with respect to increase the feed rate and low speed. The MRR is maximum (0.26 mm3/min) at the high level of spindle speed (1000-1060 rpm). The fig.3 suggests, at constant spindle speed of 850 rpm for increasing depth of cut the MRR increases towards the mid-point and then starts decreases towards the end point. At constant depth of cut 2mm for increasing spindle speed the MRR increases. At maximum spindle speed and Depth of cut the MRR are maximum.

The Figure.4 shows, initially the surface roughness is low (1µm) at high level of the spindle speed (1000-1050 rpm) while low level of depth of cut 2mm. Surface roughness is gradually increased with respect to increase the depth of cut and decrease the spindle speed. Surface roughness is maximum

(2.1 μm) at the high level of Depth of cut (2.6-3mm).From the Figure 5 shows that all level of feed rate at high spindle speed to achieve low surface roughness. Minimum spindle speeds at all level of feed rate (1-1.5 mm/min) the surface roughness's are high.

3.4 . Validation of experimental results

To predict and validate the improvement in Material removal rate and surface roughness for turning of Aluminium alloy AA6061-T6 with respect to the chosen initial parameter setting, confirmation tests are used.

Figure. 6 –7 shows the validation of experimental results for the Material removal rate and surface roughness respectively for AA 6061-T6. Thus, the response equations for Material removal rate and surface roughness to develop through response surface methodology can be used to predict the MRR and surface roughness for any combination of turning parameters within the experimentation range.

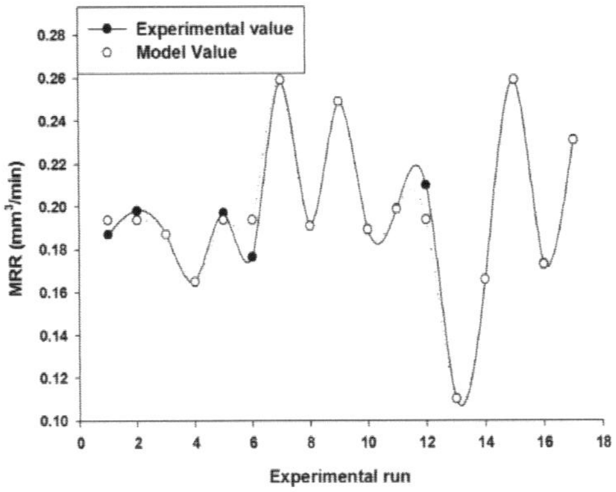

Figure 6. MRR on Experimental vs model value

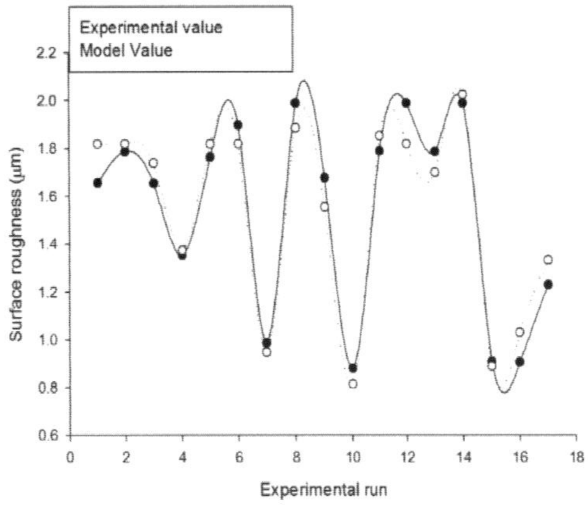

Figure 7. Surface roughness on Experimental vs model value

4.Optimization

Optimization can be defined as the process of finding the situations that give the maximum or minimum value of a function. Optimization, in its broadest sense, can be applied to solve any engineering problem [11].In this work optimization of turning parameters on MRR and Surface roughness, The MRR and Surface roughness is a quality, but this work is tried to achieve same turning parameters at AA 6061-T6. The numerical optimization was used; the multi response optimization is solved desirability approach by using design expert software. The numerical optimization is set the goal as shown in Table 4 to minimize the response to achieve high quality. Table 5 shows the optimized value for MRR and Surface roughness of input parameters to the corresponding response value. the finally to predicted the response value Material removal rate 0.2926 mm^3/min and Surface roughness 0.3879 μm has been achieved. And the Figure 8 shows overlay plot drawn between the Spindle speed and feed rate the Depth of cut at 2mm. These ranges of graphical optimization result shows on the figure 8.the shaded area on the overlay plot regions are to be achieved for proposed criteria and select the optimal turning parameters.

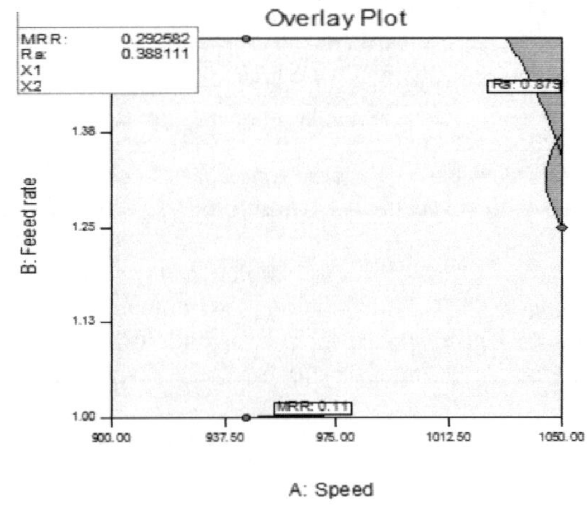

Figure 8 . Over lay plot shows the region of optimal cutting condition

Table 4 . Criteria for Numerical optimization

Name	Goal	Lower Limit	Upper Limit
Speed (rpm)	maximize	900	1100
Feed rate (mm/min)	is in range	1	1.5
Depth of cut (mm)	minimize	2	3
MRR (mm3/min)	maximize	0.11	0.259
Ra (μm)	minimize	0.879	1.987

Table 5 . optimization result for cutting conditions

Si. No	Speed (rpm)	Feed rate (mm/min)	Depth of cut (mm)	MRR (mm^3/min)	Ra (μm)
1	1100	1.16	2	0.2926	0.3879
2	1100	1.17	2	0.2995	0.3807
3	1100	1.13	2	0.2817	0.3964
4	1100	1.2	2	0.3080	0.3678
5	1100	1.41	2	0.2842	0.2130
6	1100	1.31	2	0.3147	0.2989
7	1100	1.37	2	0.2996	0.2477
8	1100	1.34	2	0.3092	0.2758
9	1100	1.11	2	0.2702	0.4030
10	1100	1.19	2	0.3049	0.3731

5. Conclusion

This work to findings the experimental Investigation of the effect of Spindle speed, Feed rate and Depth of cut on the MRR and Surface roughness in Turning for AA 6061-T6 the following conclusions are made

- The ANOVA tables of the MRR and Surface roughness shows the models are significant the probability value is < 0.0001
- The numerical optimization is carried out the combinations of process parameters are identified to achieve the maximum MRR and minimum surface roughness.
- The Depth of cut plays a dominant role in the turning conditions of AA 6061-T6.This indicates that the high Spindle speed and minimum Depth of cut to achieve small surface roughness and Maximum MRR.
- Response graph can be used graphically for selecting the turning parameters and providing the preferred MRR and surface roughness values.
- The minimum Surface roughness and maximum MRR is obtained from the study were 0.3879 µm and 0.2926 mm, when the process parameters such as spindle speed, feed rate and depth of cut were maintained at 1100 rpm, 1.16 mm/min, and 2mm. It is possible to obtain the minimum Surface roughness and MRR using the above values of process parameters.

6. References

[1] Puneet Saini, Shanti Parkash and Devender Choudhary 2014 Experimental Investigation of Machining Parameters For Surface Roughness In High Speed CNC Turning of EN-24 Alloy Steel Using Response Surface Methodology *Int Journal of Engineering Research and Applications* **4** 153-160

[2] Pankaj Sharma and Kamaljeet Bhambri 2013 Experimental Investigation of Parameters of CNC Turning by Taguchi based Grey Relational Analysis *International Journal of Engineering Research and Applications* **3** 429-436

[3] Philip Selvaraj D and Chandramohan P 2010 Optimization Of Surface Roughness Of AISI 304 Austenitic Stainless Steel In Dry Turning Operation Using Taguchi Design Method *Journal of Engineering Science and Technology* **5** 293 – 301

[4] Neeraj Saraswat Ashok Yadav Anil Kumar and Bhanu Prakesh Srivastava 2014 Optimization of Cutting Parameters in Turning Operation of Mild Steel *International Review of Applied Engineering Research* **4** 251-256

[5] Rahul Davis Vikrant Singh and Shaluza Priyanka 2014 Optimization of Process Parameters of Turning Operation of EN 24 Steel using Taguchi Design of Experiment Method *Proceedings of the World Congress on Engineering* **2** 2-6

[6] Sahoo P 2011 Optimization of turning parameters for surface roughness using rsm and GA *Advances in Production Engineering & Management* **6** 197-208

[7] Hascalik A and Caydas U 2008 Optimization of turning parameters for surface roughness and tool life based on the Taguchi method *International Journal of Advanced Manufacturing Technology* **38** 896-903

[8] Jayant A and Kumar V 2008 Prediction of surface roughness in CNC turning operation Using Taguchi design of experiments *Journal of the Institution of Engineers (India)* **88** 19-25

[9] Kandananond K 2009 Characterization of FDB sleeve surface roughness using the Taguchi Approach *European Journal of Scientific Research* **33** 330-337

[10] Javidi A Riegger U and Eichlseder W 2008 The Effect of Machining on the Surface Integrity and Fatigue Life, *International Journal of Fatigue* **30** 2050-2055

[11] Mahendra Korat and Neeraj Agarwal 2012 Optimization of Different Machining Parameters of En 24 Alloy Steel in CNC Turning by Use of Taguchi Method *International Journal of Engineering Research and Application* **2** 2248-9622

Investigation of Machine-ability of Inconel 800 in EDM with Coated Electrode

K.Karunakaran[1], Assistant Professor and research Scholar,
Department of Mechanical Engineering, Vels University, Chennai, TN, India.
ksivasakthi1966@gmail.com
M.Chandrasekaran2 , Professor and Director,
Department of Mechanical Engineering, Vels University, Chennai, TN, India.
E-mail : chandrasekar2007@gmail.com

Abstract. The Inconel 800 is a high temperature application alloy which is classified as a nickel based super alloy. It has wide scope in aerospace engineering, gas Turbine etc. The machine-ability studies were found limited on this material. Hence This research focuses on machine-ability studies on EDM of Inconel 800 with Silver Coated Electrolyte Copper Electrode. The purpose of coating on electrode is to reduce tool wear. The factors pulse on Time, Pulse off Time and Peck Current were considered to observe the responses of surface roughness, material removal rate, tool wear rate. Taguchi Full Factorial Design is employed for Design the experiment. Some specific findings were reported and the percentage of contribution of each parameter was furnished

1. Introduction

Based on the review it was found that, some research works were reported on EDM in machining performance improvements on such difficult-to-cut materials. Kuppan et al[1] investigation was on EDM deep hole drilling on Inconel 718. The process variables like peak current, pulse on-time, duty factor and electrode speed were considered for their investigation as factors. In which the authors developed mathematical models for the responses of MRR and depth averaged surface roughness (DASR) and used response surface methodology. Rajyalakshmi and Venkata Ramaia [2], investigations were for the performance measures like MRR, SF, and SG in Wire EDM on Inconel 825 material they used Taguchi grey relational analysis for optimizing process parameters. MuthuKumar et al[3] investigation was in WEDM process to optimize parameters by Gray–Taguchi Method for machining Incoloy800 super alloy with multiple performance characteristics such as MRR, surface roughness and Kerf. Lin et al[4] analyzed parameter optimization on micro milling EDM for the machining of Inconel 718. The another attempt was made by Muthu Kumar et al [5] on WEDM process to optimize parameters for improving responses of MRR, SR and Ker width for machining Titanium alloy. The authors recently developed a Mathematical Model for Radial Overcut on Electrical Discharge Machining for Incoloy 800. And they used Response Surface Methodology[6]. Rathi and Mane[7] studied the Effect of Powder Mixed dielectric in EDM of Inconel 718 machining with powders like Silicon carbide, Aluminum oxide, Graphite. Karunakaran and Chandrasekaran[8] work was on nPMEDM to study the influence of Al, Si and MWCNTs nonopowders. This work analyse the machine-ability performance on Inconel 800 with Silver coated electrolyte copper Electrode as well as plain electrode[9].

IOP Conf. Series: Materials Science and Engineering **183** (2017) 012014 doi:10.1088/1757-899X/183/1/012014

2. Experimentation

The factors peak current varied at three levels as 5a, 10A and 15A, similarly the pulse on time and Pulse off times were considered at three levels as 6μs, 7μs , 8μs and 3 μs, 4 μs, 5μs respectively. Taguchi Full factorial Design is shown in table 1. The MINITAB release 16 software was used for obtaining such design. The gap voltage 240V is maintained. In this experiment, the work piece (ϕ22×20mm) was Inconel 800. Additionally, the tool material (ϕ25×23mm) was Silver Coated electrolytic copper. The Coating Thickness is neglected. Polarity used was straight. The observations like SR, MRR and TWR were recorded in table 2. The average of three trails was considered for recording to avoid observational error. The machining time per run is 5 minutes. The MRR and TWR are computed by using the equation (1) and (2). The Electronica Machine Tools make Xpert 1model die sinking type CNC EDM machine was employed in this research. Kerosene is dielectric fluid in this machine because of its property of very low viscosity and it gets flushed away easily. The Taylor Hobson make, Surtronic3+ branded contact type profile-meter was used to measure surface roughness with 0.8mm cut of length. MRR is mathematically expressed equation 1, where Wj1 and Wj2 are weights of the job/ Sample before and after the machining respectively. The machining time is t minutes.

Table 1 Taguchi Full Factorial Design

Experimental. Run	*Current (A)*	*Pulse On Time (μs)*	*Pulse Off Time (μs)*
1	5	6	3
2	5	6	4
3	5	6	5
4	10	6	3
5	10	6	4
6	10	6	5
7	15	6	3
8	15	6	4
9	15	6	5
10	5	7	3
11	5	7	4
12	5	7	5
13	10	7	3
14	10	7	4
15	10	7	5
16	15	7	3
17	15	7	4
18	15	7	5
19	5	8	3
20	5	8	4
21	5	8	5
22	10	8	3
23	10	8	4
24	10	8	5
25	15	8	3
26	15	8	4
27	15	8	5

Table 2 Experimental Results

Experimental. Run	***MRR (g/min)***	***TWR (g/min)***	***SR (μm)***
1	0.07148	0.00042	1.34
2	0.09725	0.00059	1.37
3	0.10321	0.00068	1.67
4	0.15427	0.00085	1.4
5	0.1763	0.00114	1.51
6	0.20078	0.00127	1.71
7	0.22187	0.00158	1.43
8	0.2665	0.00256	1.55
9	0.31891	0.00147	1.82
10	0.23619	0.00166	1.43
11	0.27624	0.00171	1.65
12	0.29222	0.00197	1.93
13	0.31491	0.00225	1.45
14	0.33568	0.00277	1.73
15	0.37154	0.00294	2.05
16	0.4653	0.00379	1.55
17	0.50848	0.00415	1.86
18	0.53285	0.00272	2.28
19	0.36433	0.00384	1.58
20	0.41234	0.00407	1.91
21	0.45822	0.0045	2.24
22	0.48716	0.00483	1.63
23	0.52833	0.0056	2.03
24	0.57562	0.00583	2.41
25	0.59838	0.00654	1.92
26	0.67215	0.00668	2.07
27	0.74133	0.00702	2.58

MRR = (Wj1 –Wj2) / t (1)

TWR is expressed mathematically in equation 2.

TWR= (Wt1 - Wt2) /t (2)

Where Wt1 and Wt2 are weights of the job/ sample before and after the machining respectively.

3. Results and Discussion

A. The Material Removal Rate (MRR)

The results were graphically analyzed and depicted the machinabilty characteristics. The MRR performances when maintained the constant Pulse on Times of 6 μs, 7 μs and 8 μs are depicted in figure 1 to figure 3.

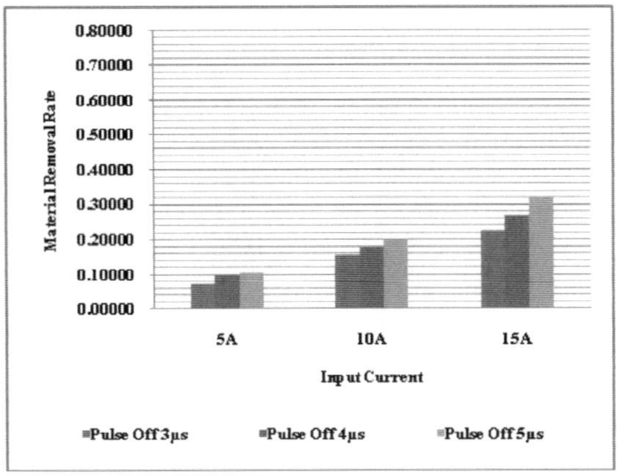

Figure 1. MRR at Pulse on Time 6 µs

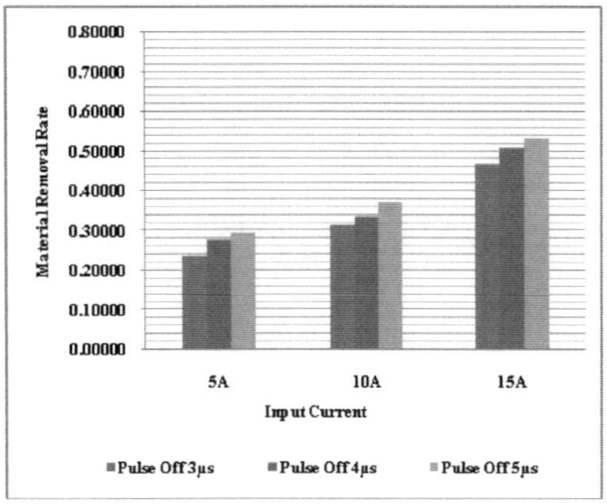

Figure 2. MRR at Pulse on Time 7 µs

The results show that the MRR linearly increases with increase of peak Current. The MRR with respect to pulse off time is also significant. The parameters influence with respect to maximum MRR were numerically estimated and tabulated in the Table 3. The Pulse on Time increment also increases the MRR. Hence controlling all three parameters will help to get desired machine-ability quality for meeting the job requirements.

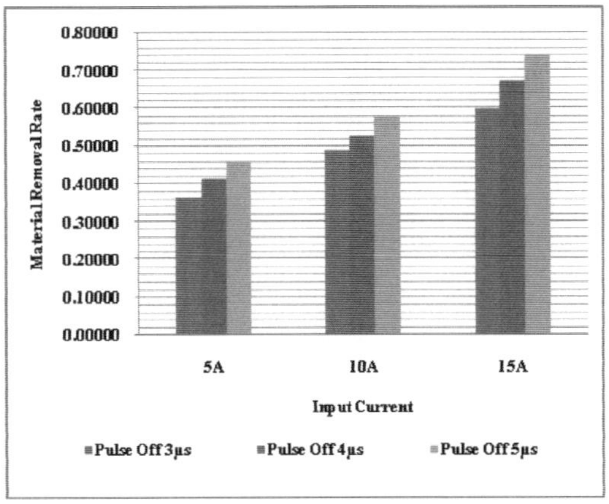

Figure 3. MRR at Pulse on Time 8 µs

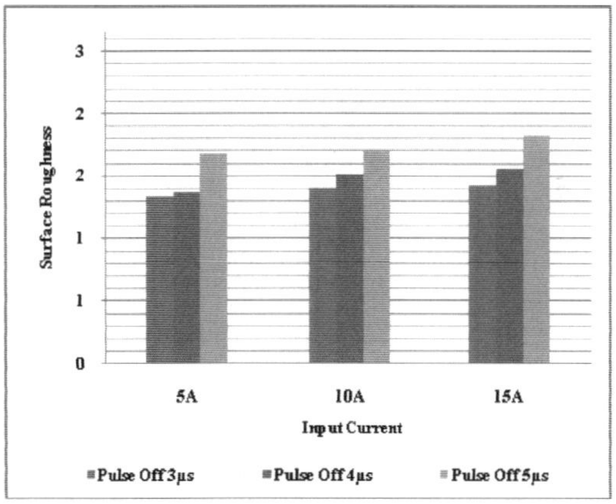

Figure 4. SR at Pulse on Time 6 µs

The Surface Roughness (SR)

The surface roughness on machined is a measure of surface finish.

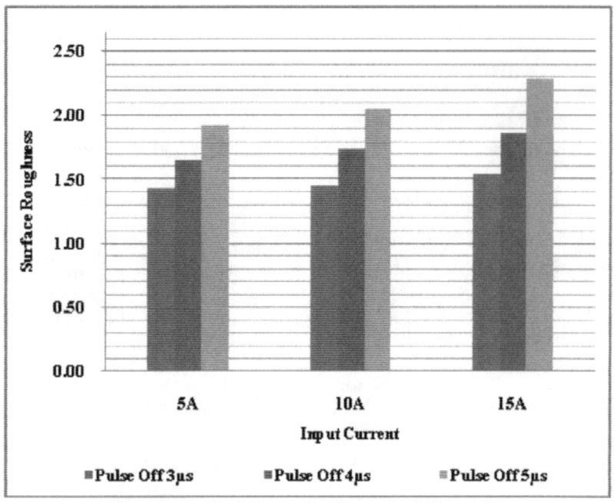

Figure 5. SR at Pulse on Time 7 μs

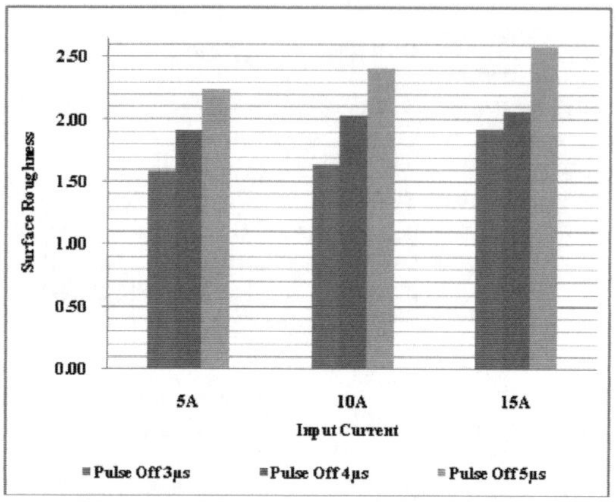

Figure 6. SR at Pulse on Time 8 μs

The SR performances when maintained the constant Pulse on Times of 6 μs, 7 μs and 8 μs are depicted in Figure 4 to Figure 6 respectively. The good surface quality usually obtained at minimum settings of operating parameters. But their individual contribution was analyzed with respect to maximum tool wear. The results were furnished in Table 4.

3. Tool Wear Rate (TWR)

The TWR was monitored and graphically depicted in Figure 7 to Figure 8 for the constant pulse on Time 6 μs, 7 μs and 8 μs respectively. TWR observed were minimum values at low pulse on Time, low pulse off time and low Peak Current settings. The TWR is comparatively lower than MRR. The TWR gradually increases with increase of the operating parameters values. Apart from general studies on Tool Wear behavior the in particular study also done. The contribution of each operating parameter with respect to maximum value of TWR was analyzed and furnished in Table 4.

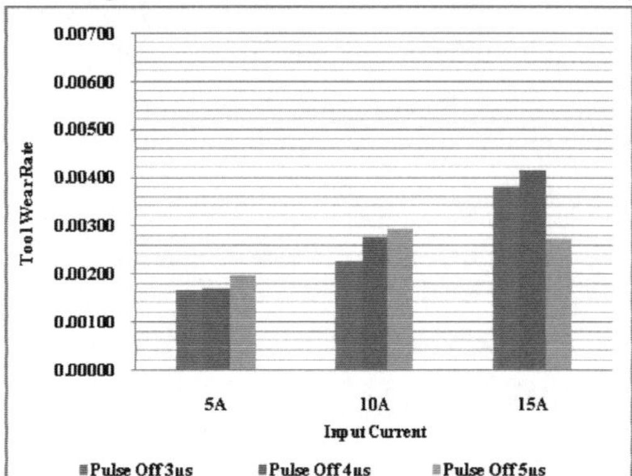

Figure 8. TWR at Pulse on Time 7μs

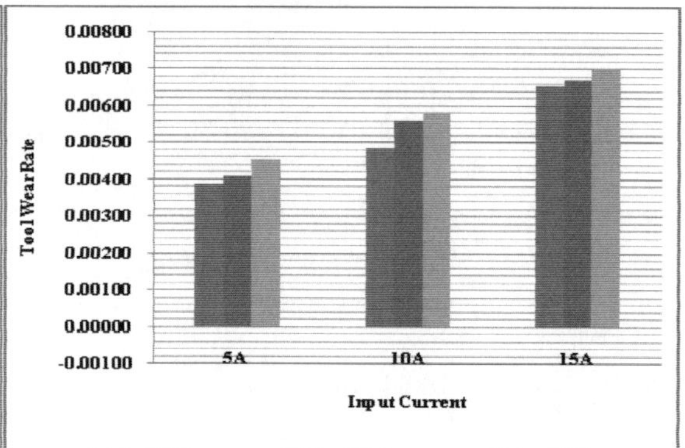

Figure 9. TWR at Pulse on Time 8 μs

Table 4. Contribution of Process Parameters

Experimental. Run	*MRR g/min*	*TWR g/min*	*SR (μm)*
1	9.60%	6.00%	51.90%
2	13.10%	8.40%	53.10%
3	13.90%	9.70%	64.70%
4	20.80%	12.10%	54.20%
5	23.80%	16.20%	58.50%
6	27.10%	18.10%	66.30%
7	29.90%	22.50%	55.30%
8	35.90%	36.50%	60.10%
9	43.00%	20.90%	70.50%
10	31.90%	23.70%	55.40%
11	37.30%	24.30%	64.00%
12	39.40%	28.00%	74.80%
13	42.50%	32.00%	56.20%
14	45.30%	39.40%	67.10%
15	50.10%	41.90%	79.50%
16	62.80%	54.00%	59.90%
17	68.60%	59.10%	72.10%
18	71.90%	38.70%	88.40%
19	49.10%	54.70%	61.10%
20	55.60%	57.90%	74.00%
21	61.80%	64.10%	86.80%
22	65.70%	68.80%	63.20%
23	71.30%	79.80%	78.60%
24	77.60%	83.00%	93.40%
25	80.70%	93.20%	74.50%
26	90.70%	95.20%	80.20%
27	100.00%	100.00%	100.00%

4. Conclusion

The machine-ability study on Inconel 800 is successfully carried out in EDM. The results are analyzed graphically. The Taguchi full factorial Design gives detailed results for this study. In this study the maximum MRR at higher setting was achieved as 0.74133 g/min. The SR obtained in the range of 1.34 μm to 2.58 μm. The tool wear was in the range on of 0.004 g/min to 0.007 g/min. The wear ratio (WR) is expressed as the ratio between TWR and MRR. The minimum of this ratio (0.0046) is economy which was obtained for the setting of 15A of Peak Current, Pulse on Time 6 μs and Pulse off Time 5μs and the yielded results are 0.31891g/min of MRR, 0.00147g/min of TWR and 1.82μm SR.

References

[1] Kuppan P, Rajadurai A and Narayanan S, 2008 Influence of EDM process parameters in deep hole drilling of Inconel 718 *Int J Adv Manuf Technol* **38** 74–84.

[2] Rajyalakshmi G and Venkata Ramaiah P, 2013 Multiple process parameter optimization of wire electrical discharge machining on Inconel 825 using Taguchi grey relational analysis *Int J Adv Manuf Technol* **69** 1249–1262.

[3] MuthuKumar V, SureshBabu A, Venkatasamy R and Raajenthiren M, 2010 Optimization of the WEDM parameters on machining Incoloy 800 Super alloy with multiple quality characteristics, *International Journal of Engineering Science and Technology* **2** 1538-1547.

[4] Lin M Y, Tsao C C, Hsu C Y, Chiou A H, Huang P C, Lin Y C, 2013 Optimization of micro milling electrical discharge machining of Inconel 718 by Grey-Taguchi method *Trans. Nonferrous Met. Soc* **23** 661−666.

[5] MuthuKumar V, SureshBabu A, Venkatasamy R, and Raajenthiren M, 2010 Multi response optimization of machining parameters of Wire-EDM using grey relational analysis in the Taguchi Method *International Journal of Applied Engineering Research* **5** 2325-2338.

[6] Muthukumar V, Rajesh N, Venkatasamy R, Sureshbabu A and Senthilkumar N, 2014 Mathematical Modeling for Radial Overcut on Electrical Discharge Machining of Incoloy 800 by Response Surface Methodology *Procedia Materials Science* **6** 1674 – 1682

[7] Mahendra G Rathi and Deepak V Mane 2014 Study on Effect of Powder Mixed dielectric in EDM of Inconel 718 *International Journal of Scientific and Research Publications* **11(4)** 1-7

[8] Karunakaran K and Chandrasekaran M, 2016 Experimental Investigation of process parameters influence on machining Inconel 800 in the Electrical Spark Eroding Machine, Second *International Conference on Mechanical Engineering and Automation Science* 13-15.

[9] Karunakaran K and Chandrasekaran M, 2016 Experimental investigation Nano particles influence in NPMEDM to Machine Inconel 800 with Electrolyte Copper Electrode *2nd International Conference on Frontiers in Automobile and Mechanical Engineering* 7-9 .

Restructured review on Electrical Discharge Machining - A state of the art

C Gnanavel[1], R Saravanan[2], M Chandrasekaran[3] and R Pugazhenthi[4]
[1]AssistantProfessor,Department of Mechanical Engineering, Vels University, Chennai, India
[2]Professor,Department of Mechanical Engineering, Ellenki College of Engineering and Technology, Hyderabad, Telangana, India
[3]Professor, Department of Mechanical Engineering, Vels University, Chennai, India
[4]AssociateProfessor, Department of Mechanical Engineering, Vels University, Chennai, India.

E-mail: gnanavelmech1986@gmail.com

Abstract. In this literature survey, an attempt made to review the Electrical Discharge machining and its related machineoperating parameters and machining parameters were studied the various research works. This work is unique and innovatively constructed which helps to be aware of each parameter in machining collectively and individually. This review furnishes both indispensable and restructured information EDM, PMEDM, WEDM, MEDM, NDEDM etc. The choice of electrodes, optimization methods and influencing parameters and their influences are precisely presented and concluded with research gaps.

1. Introduction

The modern materials possess superior properties and attract for produce highly precise components. Machining of such extremely hard and brittle materials by the conventional manufacturing processes is economically difficult. These high strength materials can be processed by using higher strength tools like polycrystalline diamond (PCD) and carbide cutting tools. On the other hand, it offers increasing of the cost of thetool and high tool wear rates and thus the manufacturing process of arduous materials like MMCs by standard means it increases the manufacturing cost. Therefore the unconventional manufacturing processes can be adapted for economic manufacturing of the arduous materials.

2. Electrical Discharge Machining(EDM)

In the EDM process the heat or thermal force is used as a driving force formachining the conductive materials and non-conductive materials. The basic construction and working of EDM is shown in figure 1. In EDM high-frequency of electric discharge supplied to theworkpiece which conducts electricity in controlled manner. This resultsin the removal of small amount of particles from the workpiece. The melting and evaporation of the material is processed.Complex shapes and geometry can be shaped with fine and high precision work can be carried out. It has some advantage in the manufacture of mould cavities, die casting, automobile, spaceresearchand surgical equipments.Generally, the EDM involves the incident of spark of initialization, dielectric breakdown and thermo-mechanical erosion of metals [1]. There are several non-conventional machining process are availableamong them the Powder mixed Electric Discharge machining (PMEDM) has found to be effective processing of hard materials like composites materials [1,2].

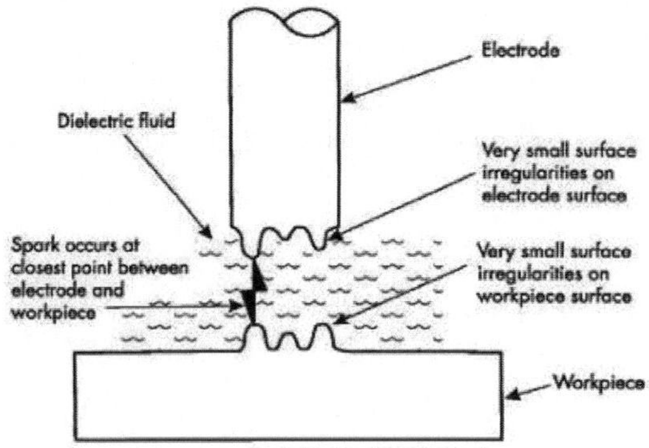

Figure 1. Basic workingprinciple of EDM

3. Powder Mix Electrical Discharge Machining (PMEDM)

The principle of PMEDM is shown in Figure 2. It is used to improve the surface finishing of the manufacturing surfaces by improve the MRR and reduced TWR by increasing the gap between the electrodeand workpiece [36]. In PMEDM the suspended particles are presence in the dielectric fluid and significantly which increases the surface finish and machining efficiency.The recent review of DivyaRana et al[8] reported that PMEDM is excellent research shows the potential area.

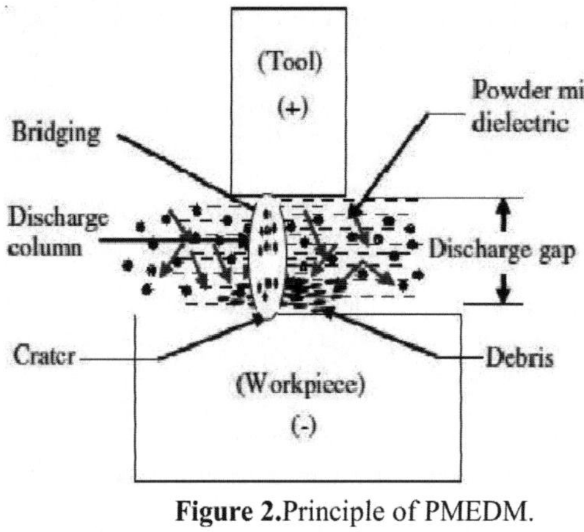

Figure 2. Principle of PMEDM.

4. Micro – Electrical Discharge Machining (M-EDM)

The M-EDM is another modified form of the EDM process;it has the ability to machine the complex3D- shapes on any electric power conductible materialsin spite of its hardness, toughness, and strength.

5. Wire Electrical Discharge Machining (WEDM)

The WEDM is especially used in aerospace, nuclear and automotive industrial components manufacturing industries, in the figure 3 the basic principles operation of Wire EDM is shown in Figure 3. It is more popular to machining various hardest materials like molybdenum, titanium, nimonics, zirconium, etc,withthehighaccurate surface finish and intricate shapes. It is an economical process to cut hardest materials with complex shapes withexact contours in low volume and larger variety. Discharge current, discharge capacitance, pulse duration, pulse frequency, wire speed, wire tension, average working voltage and dielectric flushing condition are the most important control parameters of the WEDM.

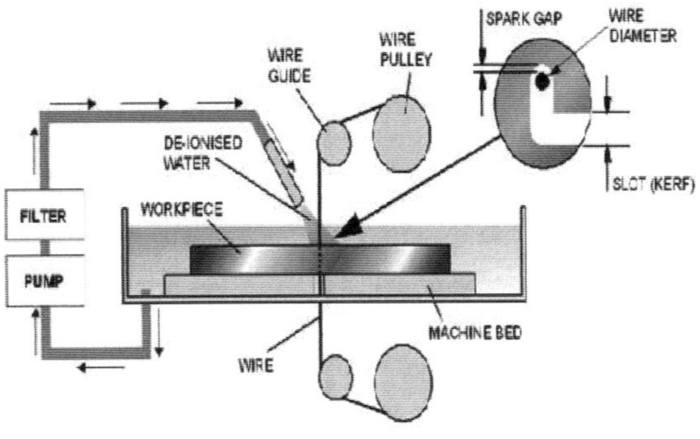

Figure 3 Principles of Wire EDM

6. Parameters and their Influences

The parameters commonly involved in EDM, PMDEM WEDM were shown as Fish Bone Diagram in Figure 4.

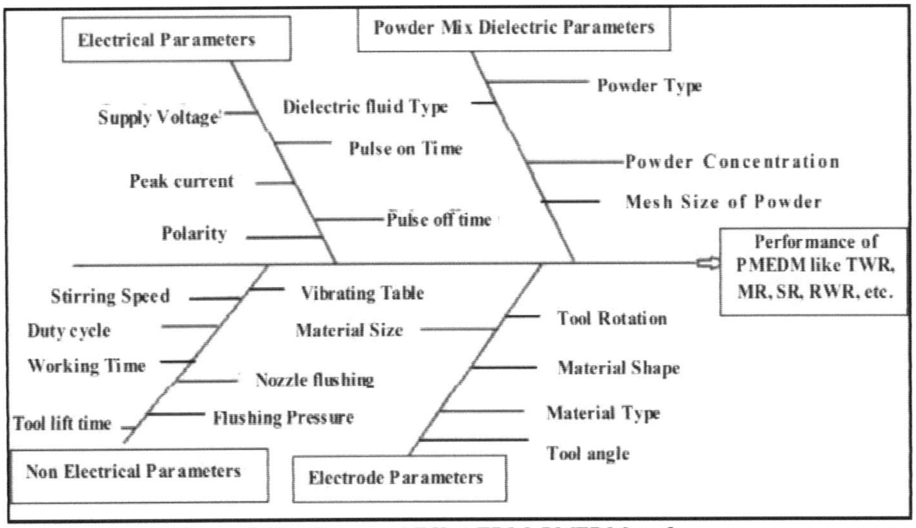

Figure 4. The Cause and Effect EDM, PMEDM performance

7. Electrical Parameters

7.1. Polarity

- Straight or reverse depending on the application [1].
- In reverse polarity, tool/ electrode and workpiece connect to the positive and negative terminal of the D.C. source respectively [1].

- Positive polarity improves MRR and negative polarity improves surface finish [1].
- Material removal rate with reversed polarity was considerably lower than that of straight polarity [2].
- The machining time was about eight times longer than straight polarity [2].
- The surface obtained by reversing polarity was found to be smoother than that using straight polarity [2].

7.2. Peak Current

- MRR will be improved by applying high current, but it results in tool wear and some surface roughness [3].

7.3. Pulse duration

- Increase in pulse durationup to 750μs has improved MRR for Inconel 718 [8].
- Higher peak current and pulse duration capable of increasing the performance and productivity of EDM Inconel 718.

7.4. Discharge Voltage

- The flushing condition is improved by increasing the gap and cut operation is stabilized by setting it to higher voltage [3].
- By increasing the open circuit voltage the electric field strength is increased which results in increase in surface roughness, TWR and MRR [3].

7.5. Pulse ON Time and Pulse OFF Time

- Metal removal is directly proportional to the amount of energy applied during the pulse on time [1].
- When longer pulse is given to the work piece material, a substantial amount of metal will be melted extra than the short pulse duration, results the crater to be broader and deeper. This develops surface roughness, heat affected zone will be deeper and has a larger recast layer. [4].
- There will be counter productiveness when the pulse duration is on the extreme side [3].
- If the interval and duration are not set properly many pulses will have failure by which the machining efficiency will be lost [1].
- The material removal rate starts decreasing when the optimum pulse duration for each tool and the workpiece combination is exceeded [4].
- The output of electrode can be restricted by applying long pulse for the machining process[2].
- More is pulse off time greater will be the machining time [5].
- Pulse off time also governs the stability of the process[4].

7.6. Spark frequency

- An increase in spark frequency results in an improved SR[5].
- The material removalperiod of time is shared by a large number of sparks,the size of the crater is reduced to increase in frequency, and thus SR decreases [5].

7.7. Duty factor

- The flushing time is very less when the high duty factor and this might lead to the short circuit condition [6].
- The high pulse off time arrases for small duty factor of the low machining rate [6].

8. Powder Parameters

In EDM the suitable powdermaterial is to be mixedinto the dielectric fluid of EDM for enhancing the machining performance called PMEDM [29]. PMEDM is a different machining of mechanism as compared to the conventional EDM. Hence a detailed review of recent literature is presented here below to illustrate the effects of powder type, powder size (mesh size) and powder mixing ratio with adielectric (powder concentration).

8.1. Type of Powder

- The powder added into the dielectric fluid could increase the material removal rate and it decrease the (TWR) and which improves the surface quality of the work quite clearly [5].
- The different powders have a different impact on the manufacturing process output of the PMEDM process [8].
- Some of the inorganic oxide powders notdissolve uniformly and persistently in kerosene, concentrate and precipitate quickly, so they do not works a good role in improving the MRR, decreasing the SR and TWR [14].

The requirements of properties of Powder which can be suspended into thedielectric fluid of EDM must have thefollowing [14].

- It should be electrically conductive in nature.
- It must have good suspension capabilities.
- It should be in toxic and odorless.
- It should have good thermal conductivities
- It must be non-magnetic in nature.

8.2. Concentration of added powder

- In addition to that the appropriationquantity of powder into dielectric fluid plays a very important role on MRR, SR, and TWR. The finding of various powder concentrations is presented in the article for better understanding [15].
- The MR depth reached the maximum value of appropriate concentration. Sometimes increases the TWR and SR. So it is to be optimized based on machining requirements [16-19].
- Increase or decrease in the concentration with the powder would decrease the MRR [20].

8.3. Mesh size of powder

The size of the powder particles affects the PMEDM performance. A large diameter of the powder particle (mesh size) increases the gap, but simultaneously decreases the MRR and then increases the SR [10].

8.4. Electrical properties of powders

The electrical conductivity of the added Powder directly affects EDM performance [8].
The added powder increases the conductivity of the dielectric fluid and results in the extension of the gap distance

9. Non- Electrical Parameters

The electrical conductivity of the added Powder directly affects EDM performance [8].
The added powder increases the conductivity of the dielectric fluid and results in the extension of the gap distance.

9.1. Dielectric Flushing

- Lonardo and Bruzzone[13]exposed that the MRR and TWR are in fluencies the flushing roughing operation, while in the finishing operation; it influenced the SR.

- The influence of flushing rate also plays a vital role in the crack density and recast layer; it is also minimize the optimal flushing rate [12].
- Poor flushing ends up with the stagnation of dielectric fluid and builds up of machining residues, which apart from low MRR also leads to short circuits and arcing [8,21, 22].

9.2. Electrode Gap

The presence of backlash is particularly undesirable.

- Gap width is not measured directly, but can be inferred from the average gap voltage [9, 11].

9.3. Rotating the electrode

- As the tool rotation serves as an effective gap flushing technique, the MRR and SR can be significantly improved by introducing the rotary motion to the electrode [24 - 27].

9.4. Electrode Parameters

The electrodes don't only differ by materials, but also its shape, size, and coating over it. The electrode material for specific applications is shown in table 1. The certain clearance usually provided in the electrode design based on the materials of theelectrode and the workpiece, and MRR. The details are furnished in Table 2 are side clearance inatool with respect to finish [29].

Table 1. Electrode Material and Applications

Sl. No	Material	Wear Ratio	MRR	Fabrication	Cost	Application
1	Copper	Low	Rough range is high	Easy	High	Metals
2	Brass	High	High at finishing	Easy	Low	Metals
3	Tungsten	Lowest	Low	Difficult	High	Drilling the small holes
4	Tungsten copper alloy	Low	Low	Difficult	High	Accuracy is high
5	Cast iron	Low	Low	Easy	Low	Only on few materials
6	Steel	High	Low	Easy	Low	Finishing work
7	Zinc based alloy	High	High on rough range	Easily die-casted	High	Metals
8	Copper graphite	Low	High	Difficult	High	Metals

Table 2. The operating Condition vs side clearance

Sl. No.	Rate of Cutting	Finish	Side Clearance (mm)
1	Rapid	Coarse	0.6 – 0.7
2	Medium	Medium	0.3 – 0.4
3	Very Low	Fine	0.04 – 0.07

9.5. Conventional and Non-Conventional Optimization Methods

The conventional and non- conventional methods of optimization were consolidated and presented in the chart in Figure 5.

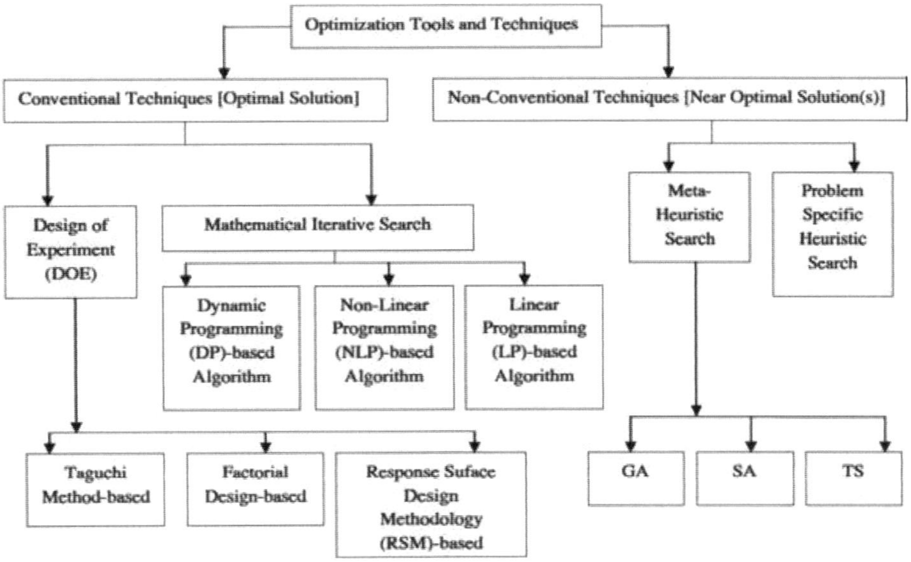

Figure 5. Optimization Method

9.6. Recent Investigations on Parameters Influence

Amandeep Singh and Ranjot Singh[48] comprehensively reviewed the time of past works, the mechanism ofthePMEDMprocedure, and study the research literature in this area. But this review is unique to consolidate the recent research upto date to reveal the innovative methods of research like theuse of new materials, methods, powders, concentration ratios etc. Razak et al.[10] investigation was on the EDM of premium stainless mold steel (Stavax) material. SiC powder mixed with various concentrations (5, 10 and 15 g/l) kerosene dielectric fluid, the mesh size also varied at three levels (10, 20 and 30 μm) and concluded that PMEDM improves MRR and reduces the SR, TWR, Machining Time and Machining Cost. Muniuet.al.[15]investigated the effects of suspending diatomite powder in distilled water dielectric fluid in the machining of mild steel workpiece withagraphite electrode. The increase of powder concentration in the distilled water dielectric fluid improves the MRR. XueBai et al.[16]studied a unique case of the influence of various combinations of workpiece and tool electrodes powder mixed near dry electric discharge machining (PMND-EDM) process with the parameter variations in pulse off time, peak current, and pulseON time, air pressure, tool rotation and powder concentration on the MRR. They reported that brass is the best tool for machining W18Cr4V on the basis of maximum MRR. While work with same brass tools for machining C45 carbon steel workpiece exhibited superior MRR when there was improper heat dissipation.

Nimo Singh Khundrakpam et al.[17] reported that powder concentration and peak current were significant parameters to maximizing the MRR after they studied the effect of mixing Zinc powder in kerosene dielectric fluid forthe EDM of EN-8 steel. They investigated the effects of process parameters like tool electrode diameter, peak current, flushing pressure, pulse off time and powder concentration. Taguchi's L 27 orthogonal array was employed in research design. Satpal Singh and Kalra[18] investigated the influence of mixing of tungsten powder in the dielectric fluid for the EDM of EN 24 steel alloy. The authors used Taguchi's L 27 orthogonal array to analyze the effects of process parameters like peak current, tungsten powder concentration, duty cycle, and pulse on time on TWR as well as in MRR. They used powder concentration in the range of 0 to 4g/l. Khalid et al.[20]studied kerosene dielectric with aluminium PMEDM of Al-SiC metal matrix composite by using copper electrodes. Their variable process parameters were Peak current and aluminium powder concentration.

They found that the powder concentration is indirectly proportional to surface roughness. At higher powder concentration, the lowest peak current value of 2A setting gives least surface roughness values.

The investigation of the effect of both electrical and non-electrical machining parameters of WEDM, such as current, pulse on-time, off-time, servo reference voltage, maximum feed speed, wire speed, flushing pressure and wire tension along with workpieces of metal matrix composites (Al/Al2O3) for the responses of cutting speed, surface finish, and kerf width. They reported that Reinforcement percentage, current, and on-time was found to have a noteworthy effect on the responses. Wire breakages were found to pretense limitations on the cutting speed in machining of Al/Al2O3p composite materials. Wire shifting was bringing into being to get worse the machined surfaces [30]. Anuraag and Vijay Kumar Reddy[31] conducted research by introducing material as one of the variable parameters. That isthe various proportions of reinforced aluminium matrix with silicon carbide weight percentages like 3%, 9% & 15% were considered. The other parameters are various concentrations of Aluminium powder in Kerosene dielectric, Pulse on time (discharge time) and peak current. Polarity used to beNegative, The tool was acopper electrode. The reinforcement by weight and pulse on time increases the rate decrement of MRR increases. The Peak current increases the MRR significantly increased. At constant peak current the workpiece variation influence is very less.

Jahan et al. [32]investigated the effects of various operating parameters such as capacitance, resistance, supply voltage, electrode rotational speed and gap control parameters on the micro-EDM of Duralumin by engraving micro-grooves on the surface of it. The responses set wherethe material removal rate (MRR), tool wear ratio (TWR), thesurface roughness (SR) and machining depth of the microgrooves. The authors reported that increase of capacitance and voltage results in anincrease of MRR, TWR, and SR. Lal et al.[33] conducted a study of WEDM of different reinforced AMC (Al7075/7.5 % SiC/7.5 % Al2O3) work material along with the other variable parameters like the wire drum speed on the kerf width, discharge duration, discharge current and pulse interval time and reported that the discharge current influenced more significantly on kerf than discharge duration.

Satishkumar et al.[34]investigated WEDM of Al6063/SiCp composites to observe the effects of parameters such as an experimental study of wire electrical discharge machining characteristics of Al6063/SiCp composites. In this work, the effect of WEDM parameters such as gap voltage, pulse off time, wire feed and pulse on time, on MRR and SR. Prabu et al.[35]investigated the EDM parameter effects on TiB2 reinforced aluminium AMCs (Al-TiB_2) work materials TiB2 with thelow-frequency vibrating tool. And found that when setting of the current increased, the MRR and surface cracks of the workpiece also increased. The increase in reinforcement ratio decreases the MRR and surface cracks.Zain et al.[37, 8]studied the effect of Tantalum Carbide (TaC) powders,the stainless steel workpiece machining in PMEDM thedifferent levels of powder concentration and peak current (5, 10 and 15g/l.)were studied. The result was current proportionate to MRR and surface roughness (Ra). However, the response of powder concentration variation was not uniform.

Similarly,Rizi et al.[38]StudiedSiC PMEDM of Titanium alloy Ti-6Al-4V workpiece with thecopper electrode and included mesh size of particles at different levels. They achieved littleamount of MRR, surface finishare increased and alessamount TWR is reduced. Ali et al.[39] carried out their study on SiC PMEDM of Ti-6Al-4V alloy workpiece with tungsten carbide electrode. They investigated the effect of discharge energy and powder concentration in micro-EDM machining. And reported that powder concentration factor has influenced lesser than the discharge current. Ajay and Anirban[40] used different PMEDM setup with powders of Graphite, aluminium, copper and tungsten materials and used different dielectric fluids such as refined mineral oil, EDM oil and kerosene for machining Die Steel H11 and H13 grade. The results were that even at the lower current setting, the micro-hardness was improved in tungsten powder mixed EDM and with thetungsten-copper electrode.

The relative comparison of micro-hardness between deposited and non-deposited regions exhibited a boost of 56% for H13 and 37% for H11 than EDM.

Pichai and Apiwa[41] explored a unique case of surface improvement by a titanium coating layer in the tungsten carbide surface in electrical discharge coating (EDC) in b PMEDM (titanium powder mixed with oil). The oil used was employed as dielectric fluid as well as produce carbide material for coating. The current and duty cycles were varied for observing the titanium coating layer thickness. Kansal et al.[42] conducted the FEA,it was infered that the temperature profiles and material transformations occur in the workplace are resulted by high temperature, large deformations and transient operation.Gopalakannan et al.[43] developed mathematical modeling by using response surface methodology for estimating the optimal process parameters for the EDM of Al 7075-B_4C MMCs. The variable factors were Pulse Gap voltage, current, Pulse off time, Pulse on time for the targeted responses of MRR, TWR, and SR. Kumar et al.[44] developed mathematical modeling to fix optimal Wire speed, Pulse on/off time for WEDM of Al-SiC(20%) AMC to obtain desired MRR and SR.

Gopalakannan et al.[45] statistically optimized EDM process parameters such as current, Pulse on time, Pulse off time and Gap voltage for the EDM of aluminium Hybrid metal matrix composite for the responses of MRR, TWR and SR. Grey. Suresh Kumar et al.[46] investigated machinability of new material of Al (6351)- 5% SiC-10 % B4C Hybrid Composite in EDM. The variable process parameters such as current, Pulse on time, Duty cycle, and Gap voltage were considered and studied the TWR and SR performances. Laxman et al.[47] considered the factors such as peak current, tool lift Time, pulse off time and pulse on time, In EDM of Titanium Super Alloy for investigating MRR and TWR performances. Ndaliman et al.[49] reported an interesting case of influence of tool material that is enhancing microhardness of titanium alloy (Ti-6Al-4V) through modification of electrical discharge machining process parameters. They considered Cu–TaC composite electrode, urea mixed distilled water asadielectric fluid, process variable parameters, peak current, the pulse duration, the duty factor and Urea concentration. The result revealed that micro-hardness of the electrical discharge machined surfaces with Cu–TaC electrode/urea dielectric fluid was generally higher than that of those with Cu electrode/distilled water dielectric fluid. The highest microhardness of 1795 Hv was attained with 10 g/L of urea concentration.

10. Conclusion

The contribution of EDM research to industries formachining the newlyinvented hard materials to machining is makes the EDM technology to remains as indispensable. This review of the research trends in EDM and some of its types like PMEDM, NDEDM, Micro EDM and WEDM from the recent contributions was presented. Some innovative approaches were found in the review All most the available research works on PMEDM have studied for enhancing the cutting speed (MRR), reducing the tool wear, and improving surface quality (low SR). Some common results are listed below

- The MRR improved by mixing powders in the EDM's dielectric Fluid. Which namely PMEDM.
- The MRR increased by the increase of Peak Current.
- Higher Peak Current leads to higher MRR, TWR, and SR.
- The MRR decreased or less responded by the increase of Pulse off time.
- The TWR comparatively lesser in PMEDM than EDM
- The use of optimum use of powder concentration and particle size (mesh size) leads to reach the maximum efficiency of EDM process. Either completely experimental or partly experimental to establish the relationship among parameters to find optimal parameters. The statistical methods like GRA, RSM, ANOVAetc.are used to it.

11. Future Directions

- The simulation and FEA need to be addressed more.
- Most of the research work has been reported with use of Al, Si, and graphite powders and some literature available on the uses of powder like Cr, Ni, Mo, tungsten, SiC etc., but only a few literature was found in using Nanopowders.
- More studies need to address in Role of powder concentration and mesh size of powders in the surface properties like microhardness, corrosion resistance, microstructure, wear resistance and fatigue resistance etc.
- Higher peak currents produce rougher surfaces in EDM process. The applicability of PMEDM process for industrial purposes is constrained by several key technical issues that need to be further investigated.
- The investigation on the influence of process parameters such as tool vibration, tool rotation, and workpiece rotation has been taken up by very few researchers. More investigations need to be addressed to gain a better understanding of the process.

12. References

[1] Narender Singh P, Raghukandan K, Rathinasabapathi M and Pai B C2004Electric Dischargemachining of Al-10% SiCp as – cast metal matrix composites*Elsevier Journal of materials processing technology* 155-156.

[2] Sathish Kumar D, Kanthababu M, Vajjiravelu V, Anburaj R, ThirumalaiSundararajan N and Arul H2011Investigation of wire electrical discharge machining characteristics of Al6063/SiCpcomposites*Int J AdvManufTechnol*975-986.

[3] Dilshad Ahmad Khan and Mohammad Hameedullah 2011 Effect Of Tool Polarity on The Machining Characteristics In Electric Discharge Machining of Silver Steel And Statistical Modelling of The Process *International Journal of Engineering Science and Technology***6(3)**5001-5010.

[4] Kuppan P, Rajadurai A and Narayanan S 2008 Influence of EDM process parameters in deep hole drilling of Inconel 718 *International Journal Advance Manufacturing Technology***38** 74-84.

[5] Kumar A, Maheshwari S, Sharma C and Beri N 2011 Analysis of machining characteristics in additive mixed electric discharge machining of nickel-based super alloy Inconel 718 Materials and Manufacturing Processes **26(8)** 1011-18.

[6] Kansal HK, Singh, S , Kumar P2005 Application of Taguchi method for optimization of powder mixed electric discharge machining *International Journal of Manufacturing Technology and Management***3(7)**29–341.

[7] Kansal HK, Singh S, and Kumar P2005Parametric optimization of powder mixed electrical Discharge machining by response surface methodology*Journal of Materials Processing Technology***3(169)** 427–436.

[8] DivyaRana, Ajay Kr. Pal and PoojaTiwari 2015 Study of Powder Mixed Dielectric in EDM-A Review *Intl J EnggSciAdv Research***2(1)**, 69-74.

[9] Crookall J R , Heuvelman C J 1971Electro-discharge machining—the state of the art *Annals of the CIRP***1(20)**113–120.

[10] RazakM A, Abdul-RaniA M, and Nanimina A M, 2015, Improving EDM Efficiency with Silicon Carbide Powder-Mixed Dielectric Fluid *International Journal of Materials, Mechanics and Manufacturing***1(3)**40-43.

[11] Bayramoglu M and DuffillA W 1995 Manufacturing linear and circular contours using CNC EDM and frame type tools*International Journal of Machine Tools and Manufacture***8(35)** 1125–1136.

[12] Wong YS, Lim LC and Lee, LC 1995Effect of flushing on electro-discharge machined surfaces *Journal of Materials Processing Technology***48** 299–305.

[13] Lonardo PM and BruzzoneAA 1999 Effect of flushing and electrode material on die sinking EDM *Ann. CIRP***1(48)**123– 126.

[14] SharanjitSingh and ArvindBhardwaj 2011 Review to EDM by Using Water and Powder-Mixed Dielectric Fluid *Journal of Minerals & Materials Characterization & Engineering***2(10)**199-230.

[15] MuniuJ M, IkuaBW, NyaangaD M and GicharuS N 2013 Applicability of Diatomite Powder-Mixed Dielectric Fluid in Electrical Discharge Machining Processes in *International Journal of Pure and Applied Sciences and Technology***19(1)**19-36.

[16] XueBai, Qinhe Zhang, JianhuaZhang,Dezheng Kong and Tingyi Yang, 2013 Machining efficiency of powder mixed near dry electrical discharge machining based on different material combinations of tool electrode and workpiece electrode in *Journal of Manufacturing Processes* **15** 474–482.

[17] Nimo Singh Khundrakpam, Som Kumar, AmandeepSingh and Gurinder Singh Brar, 2014 Study and Analysis of Zinc PMEDM Process Parameters on MRR in *International Journal of Emerging Technology and Advanced Engineering* **3(4)**.

[18] Satpal Singh and KalraC S 2014 Experimental Study of PMEDM on EN 24 Steel with Tungsten Powder in Dielectric in *International Journal on Emerging Technologies***5(1)** 153-160.

[19] Mahammadumar M Jamadar and KavadeMV2014 Effect of Aluminium Powder Mixed Edm on Machining Characteristics of Die Steel (AISI D3) in Proceedings of 10th IRF International Conference.

[20] Khalid Hussain, AnuraagG P, GangarapuHemanth and Shaik Abdul Subahan 2014 Studies on Surface Roughness of Al-SiC Composites machined by Powder-Mixed EDM in *International Journal of Applied Engineering Research***21(9)**11067-11077.

[21] Guu YH and Hocheng H 2001 Effects of workpiece rotation on machinability during electrical discharge machining *J. Mater. Manuf. Processes***1(16)**91–101.

[22] Kunieda M and Masuzawa T 1988 A fundamental study on a horizontal EDM. *Ann. CIRP* **1(37)**187–190.

[23] Masuzawa T, Okajima K and Taguchi T, 2002 EDM-lathe for micromachining *Ann. CIRP***1(51)** 355–358.

[24] Mohri N, Tsukamoto J and Fujino M 1989Drilling of deep microholes by EDM *Ann. CIRP***1(38)** 195–198.

[25] Yan BH,Wang CC, Liu WD and Huang FY 2000Machining characteristics of Al2O3/6061Al composite using rotary EDM with a disklikeelectrode*Int. J. Adv. Manuf. Technol***5(16)**322–333.

[26] Kagaya K, Oishi Y and Yada K 1986 Micro-electro discharge machining using water as a working fluid—I: micro-hole drilling *Precision Engineering***3(8)**157– 162.

[27] Sato T, Mizutani T, Yonemochiand KandKawata K 1986 The development of an electro discharge machine for micro-hole boring*Precision Engineering***3(8)**163– 168.

[28] Soni JS 1994 Microanalysis of debris formed during rotary EDM of titanium alloy (Ti 6Al 4V) and die steel (T 215 Cr12) Wear **177** 71–79.

[29] Pandey PC and Shan HS1999 Modern Machining Process*Tata McGraw- Hill Publishing Company Ltd* 84-113.

[30] Nilesh G Patil and BrahmankarP K 2010 some studies into wire electro-discharge machining of alumina particulate-reinforced aluminum matrix composites *International Journal of Advanced Manufacturing Technology***5(48)** 537-555.

[31] AnuraagG P and Vijay Kumar Reddy M 2016 Studies on the Material Removal Rate of Al-SiC Composites Machined by Powder- Mixed EDM Technique*International Journal of Engineering and Technology***2(8)** 829-836.

[32] Jahan M P , Pegah Kakavand, Kwang E L M, Rahman M and Wong Y S 2015 An experimental investigation into the micro-electro-discharge machining behaviour of aluminium alloy (AA 2024)*The International Journal of Advanced Manufacturing Technology* **5(78)** 1127–1139.

[33] ShyamLal, Sudhir Kumar Z A, Khan, and Siddiquee A N 2014 Wire electrical discharge machining of AA7075/SiC/Al2O3 hybrid composite fabricated by inert gas-assisted electromagnetic stir-casting process*Journal of the Brazilian Society of Mechanical Sciences and Engineering***36** 335-346.

[34] Satishkumar D, Kanthababu M, Vajjiravelu V, Anburaj R, ThirumalaiSundarrajan N and Arul H 2012Investigation of wire electrical discharge machining characteristics of Al6063/ SiCpcomposites*The International Journal of Advanced Manufacturing Technology***63**1191-1202.

[35] Prabu M, Ramadoss G, Narendersingh P, Christy T V and VedhagiriEswaran V2013Electrical discharge machining of Al-TiB2 with low-frequency vibrating tool *Science and Engineering of Composite Materials***1**1–8.

[36] UmakanthMathapathi 2015 Material removal rate with powder mixed Dielectric on Electric Discharge Machining*International Journal on Emerging Technologies* **6(2)**274-283.

[37] Zain V, Ndaliman M B, Khan A A and Ali M Y 2012 Electro-Discharge Machining of SUS 304 Stainless Steel with TaC Powder-Mixed Dielectric*Advanced Materials Research***576**, 72-75.

[38] RiziM S, RazaviG R, OstadmohamadiM and HavaieA R 2012 Optimization electro discharge machining of Ti-6Al-4V alloy with silicon carbide powder mixed *Advanced Materials Research***566** 466-469.

[39] Ali M Y, RahmanN A and ArisE M 2012 Powder mixed micro electro discharge milling of titanium alloy: investigation of material removal rate*Advanced Materials Research***383** 1759-1763.

[40] Batish and Bhattacharya A 2012 Mechanism of material deposition from powder, electrode and dielectric for surface modification of H11 die steel in EDM process*Materials Science Forum***701** 61-75.

[41] Janmanee P and Muttamara A 2012 Surface modification of tungsten carbide by electrical discharge coating (EDC) using a titanium powder suspension*Applied Surface Science***258** 7255-7265.

[42] Kansal H K, SinghS, and Kumar P 2008 Numerical simulation of powder mixed electric discharge machining (PMEDM) using finite element method *Mathematical and Computer Modelling***47** 1217-1237.

[43] Gopalkannan S, Senthilvelan T and Ranganathan S 2012Modeling and Optimization of EDM Process Parameters on Machining of Al7075-B4C MMC using RSM *International Conference on ModelingOptimization and Computing***38**685-690.

[44] Kumar K and Ravukumar R 2013Modeling and Optimization of Wire EDM Process*International Journal of Modern Engineering Research***3(3)**1645-1648.

[45] Gopalkannan S, Senthilvelan T and Ranganathan S 2013 Statistical optimization of EDM parameters on machining of aluminum Hybrid metal matrix composite by applying Taguchi based Grey Rational analysis*Journal of Scientific & Industrial Research* 358-365.

[46] Sureshkumar S, Uthayakuma M, ThirumalaiKumaran S, Parameswaran P and Mohandas E 2014 Electrical Discharge machining of Al (6351)- 5% SiC-10 % B4C Hybrid Composite : *A Grey Relational ApproachHindawi Publishing Corporation Modeling and Simulation in Engineering* 1-7.

[47] Laxman J and DrGuru Raj K2014 Optimization of EDM process parameters on Titanium Super Alloy Based on Grey Relational analysis *International Journal of Engineering Research* 344-348.

[48] Amandeep Singh and Ranjot Singh2015 Effect of Powder Mixed Electric Discharge Machining (PMEDM) on Various Materials with Different Powders: A Review*International Journal for Innovative Research in Science & Technology***3(2)** 164-169.

[49] Ndaliman M B, Khan A A and Ali M Y 2013 Influence of electrical discharge machining process parameters on surface micro-hardness of titanium alloy*Journal of Engineering Manufacture***227** 1310.

IOP Conf. Series: Materials Science and Engineering **183** (2017) 012016 doi:10.1088/1757-899X/183/1/012016

Cast Off expansion plan by rapid improvement through Optimization tool design, Tool Parameters and using Six Sigma's ECRS Technique

T Gopalakrishnan[1] **and R Saravanan**[2]
[1]Assistant Professor, Department of Mechanical Engineering,
VELS University, Pallavaram, Chennai, Tamil nadu, India
[2] Professor & Principal, Ellenki Institute of Engineering and Technology
Hyderabad, Andra Pradesh, India

E-mail: [1]gopalakrish185@gmail.com

Abstract. Powerful management concepts step-up the quality of the product, time saving in producing the product thereby increase the production rate, improves tools and techniques, work culture, work place and employee motivation and morale. In this paper discussed about the case study of optimizing the tool design, tool parameters to cast off expansion plan according ECRS technique. The proposed designs and optimal tool parameters yielded best results and meet the customer demand without expansion plan. Hence the work yielded huge savings of money (direct and indirect cost), time and improved the motivation and more of employees significantly.

1. Introduction

Nowadays, manufacturing and service industries strive to realize the worth of six sigma and design for six sigma with a wider acceptance, growth and investment for the same [1]. The Byrne et al [2] reported that lean Six Sigma initiatives led to product innovations, redesigned processes, and streamlined supplier chain and increase revenue growth by 80%. Jiju Antony [3] conducted opinion survey and concluded that Six Sigma is uniformly applicable for large, medium and small companies. In fact, the responses are results are more rapidly and much more visible in smaller companies than in larger corporations. The literature evident that six sima approach can be used for improving quality in injection moulding process Vikas Dwivedi et al. [4] , Vikas Tayal and Jitendra Kumar, [7], Rishi Pareek and Jaiprakash Bhamniya, [13] and Bharati et al [5] Explains in small scale jobbing industries Darshak and Desai[9] in metal casting foundry industries (Vijayaram et al (2006), Darshak A and Desai [9] and Gijo et al [4], in rolling mill by Kunal Ganguly [12]. in a manufacturing industry (Mehdiuz Zalan et al. [14]). Simulation based modelling and analysis were used to reduce the manufacturing time (Saravanan and Raju [9] and in specific case of solution to engine-overheating problem by using six sigma approach (Antony et al [6]). Jagdeep Singh and Harwinder Singh [9] discussed important concepts, surveys and case studies in continuous improvement methodology and highlighted future implications. The authors noticed that literature on this classification is limited.

This work is unique by improving productivity through the reduction of machining time by optimizing the tool parameters and tool designs. In this work the six sigma ECRS method was employed.

2. The problem statement

The firm is a single source to the world leading automobile manufacturer. The existing capacity of the supplier is 4800 units per month. But the manufacturer increases his demand to 6000 units per month as increases of their customer demand. Hence there is 41% gap to meet the demand.

3. Action Plan

It is difficult to move to expansion plan due to the constraints of space, investments, skilled labours and time span. So the theme launched as increase the capacity to 6000 units per month by innovation through leading management techniques and technology. The twelve week schedule prepared to execute the improvement plan. The schedule starts with defining the exact problem (2 weeks), make necessary observation (2 weeks), data analysis must be made within 2 weeks, for action taken 4 weeks, check for consistency for two weeks and standardization to be done within two weeks. The Takt time calculated and presented for the existing and aimed (Table 1).

4. Observation

The area of problem identified as flywheel housing line. The product image is shown in Figure 1. The process flow (layout shown in figure 2) with machine code and machining details are as follows. The gravity die casted housings is arrived to machining centre than they are processed in the sequence of: VTL-1 (Facing & Boring), VTL-2 (Facing & Boring), Vmc-1 (Milling, Facing, Drilling, Tapping, Boring, Reaming), Vmc-2 (Milling, Boring, Drilling, Tapping), Vmc-3 (Milling, Drilling, Tapping), Hmc (Milling, Drilling, Tapping), Vmc-4 (Milling, Drilling, Tapping), De-burring, Washing, Drying and Packing. The major operations are seven. The average man machine time for ISX fly wheel housing were observed at various operations and shown in Figure 3 a (Existing). The high cycle time were found at Vmc stations. The tool wise cycle time also observed and presented such cycle time at Vmc 1 in Table 2. By brainstorming the cause and effects for the high cycle time at VMC stations were recorded and shown in Figure 4. And validation such diagram was done with management was presented in Figure 5. The value added and non value added times in seconds at Vmc were observed and for sample such times details at Vmc -1 are furnished in Figure 6.

5. Solution procedure

The productivity can be increased by reducing man timer and/ or machining time. This paper and scope of this work is to reduce machining time (Figure 7). The following wastes in process were identified.

1. Over production due to high over all rejection – 8%
2. More time for Rework & salvaging rejected parts
3. Manual time high due to De-burring
4. Unwanted axis movement for Tool changing
5. Tool change time high
6. More number of tools used for operations
7. Roughing & Finishing operation doing separately
8. Low cutting parameter used.

ECRS Methodology followed in machine to eliminate Waste
E - Eliminate non value adding time (Tool changing time, axes movement, indexing)
C - Combine operation (Combination cutting tools)
R - Reduce value adding time (Optimize parameters)
S - Simplify operation (Programming – Multiple to single pass)
According ECRS the following steps were executed.

Step 1: Air cutting time optimization
Step 2: Tool change time optimization
Step 3: Process parameter optimization with existing tool
Step 4: Process parameter improvement with New tool
Step 5: Balancing of machining cycle time
Step 6: Loading and unloading time optimization
Step 7: Combine tools wherever possible

5.1 *Cycle Time Reduction at Vmc -1*

1. Action Taken – 1: Increasing face Milling Cutter diameter
   For quick machining operation the face milling cutter size is increased from Ø80 mm to Ø100 (Figure 8) and reduced machining time from 92 seconds to 66 seconds.
2. Action Taken – 2: Increase number of tooth
   The feed per tooth is increased by increasing the number of tooth (from two to four). In the new cutter the diametrical and depth variance were introduced(Figure 9). The machining time was reduced from 42 seconds to 17 seconds in that operation.
3. Action Taken – 3: Introducing KAIZER Finish Boring Bar
   By use of KAIZER Finish Boring Bar the 59 seconds cycle time was saved.(Figure 10)
4. Action Taken – 4: Combined Ø7.3 Drill and Inspection Hole Boring Bar
   The Ø7.3 Drill and Inspection Hole Boring Bar were combined (Figure 11) to reduce the tool change time. by this improvement the cycle time saved only seconds.
   The computerized quality checking were carried out (Figure 12) and found the process in control and capable sustainably.

5.2 *Cycle Time Reduction at Vmc -2*

1. Action Taken – 5: Increasing Bore Cutter diameter
   For quick machining operation the Bore Cutter diameter is increased from Ø20 mm to Ø80 (Figure 13) and reduced machining time from 69 seconds to 53 seconds.
2. Action Taken – 6: Combining Facing And Chamfer Cutter
   In the existing operations separate cutters were used for facing and chamfering operations. Now the facing and chamfer cutters were combined with 16mm XEKT insert (Figure 14) to reduce machining time. Accordingly the combined tool reduced from 25 second to 9 seconds.
3. Action Taken – 7: Tool Modification
   The Ø25 End Mill with Interpolation to Ø197 Boring Bar (Figure 15) and hence the time reduced from 35 second to 9 seconds.
   The computerized quality checking were carried out (Figure 16) and found the process in control and capable sustainably.

Figure 1. Fly wheel Housing

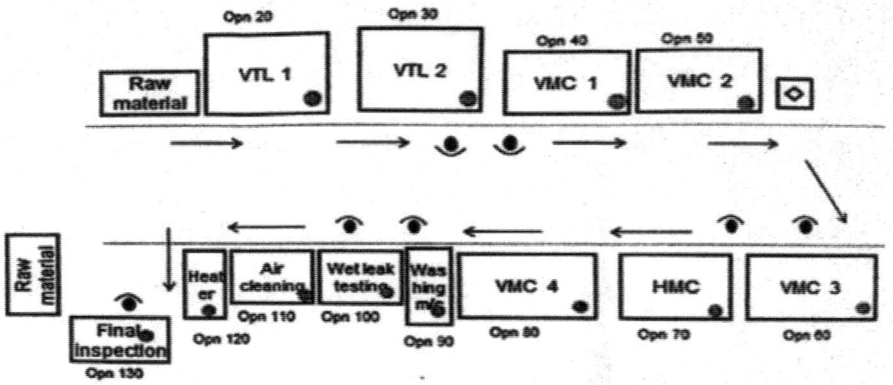

Figure 2. Machining Cell layout

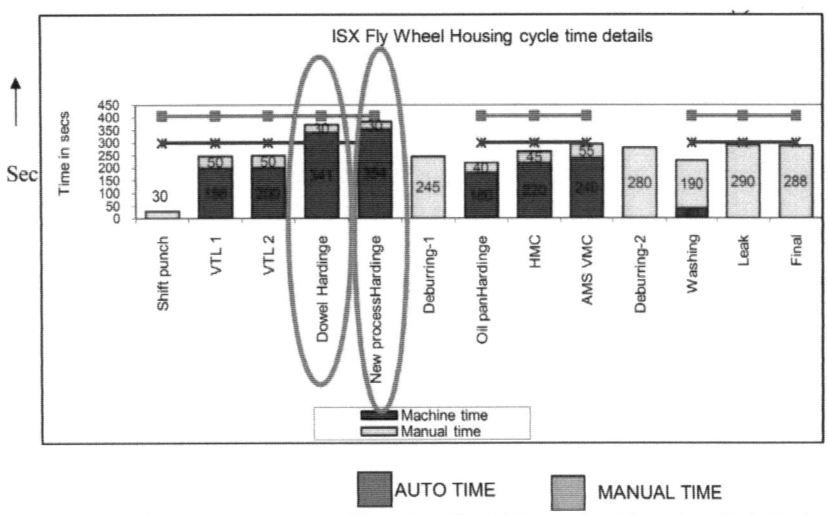

Figure 3a. The average man machine time for ISX fly wheel housing (Existing)

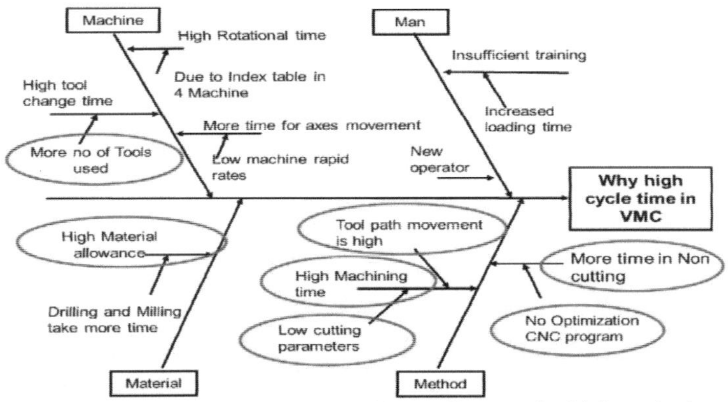

Figure 4. Fishbone diagram or Ishikawa diagram for high cycle time

IOP Conf. Series: Materials Science and Engineering **183** (2017) 012016 doi:10.1088/1757-899X/183/1/012016

Sl. No.	OPTIONS	Productivity	Quality	Cost	Delivery	Morale	Safety
1	Establish Sub Contract machining for Isx flywheel housing			✗	✗		
2	Add additional machines for de-bottlenecking			✗			
3	Work 24 X 7 to compensate for the Quantity	✗	✗	✗		✗	
4	Revisit the cell for the capacity enhancement	✓	✓	✓	✓	✓	✓

✓ Scope available ✗ No scope to improvement

Figure 5. Validation of Ishikawa diagram for high cycle time

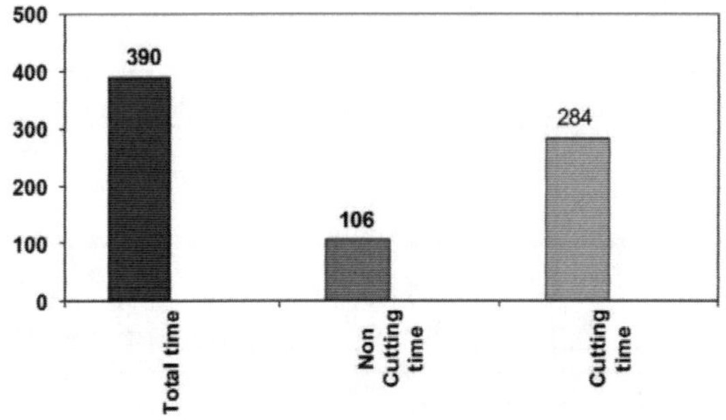

Figure 6. The value added and non value added at Vmc -1

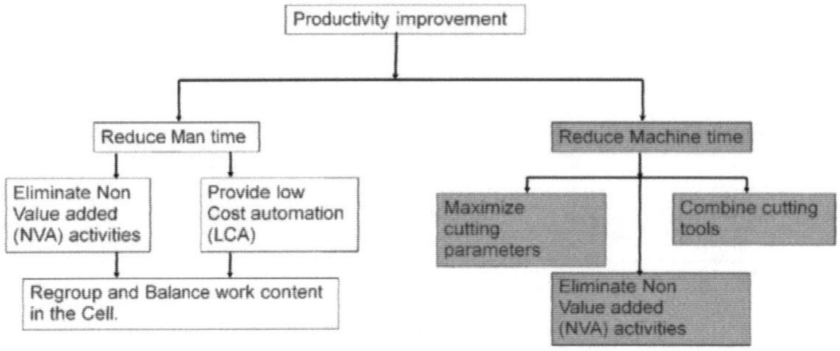

Figure 7. Methodology

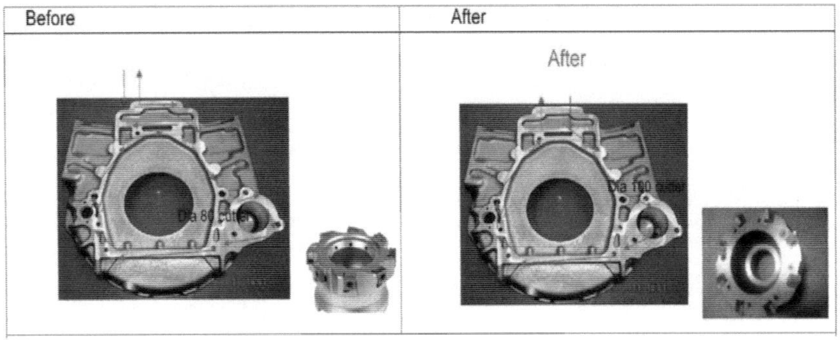

Figure 8. Action Taken – 1

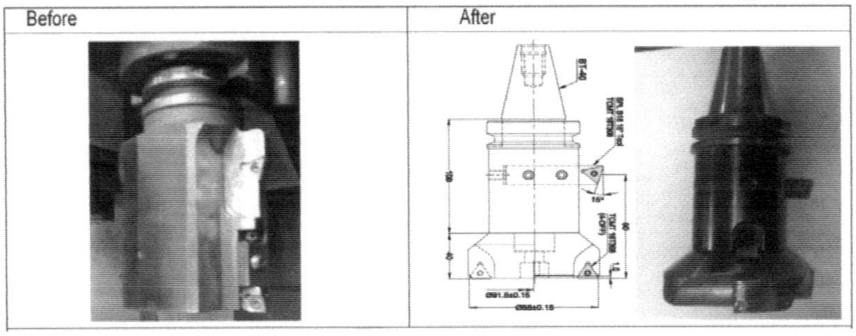

Figure 9. Action Taken – 2

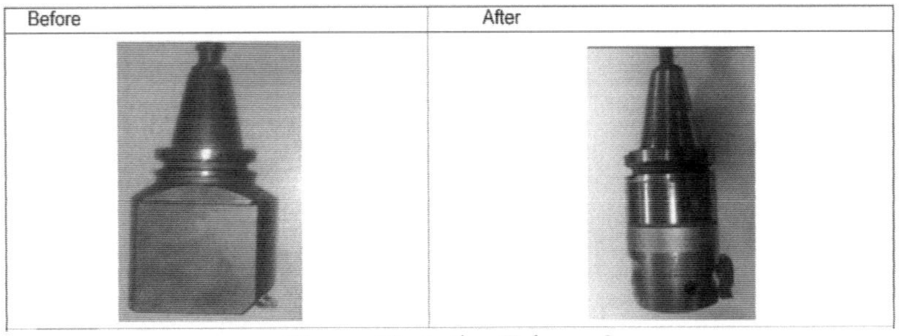

Figure 10. Action Taken – 3

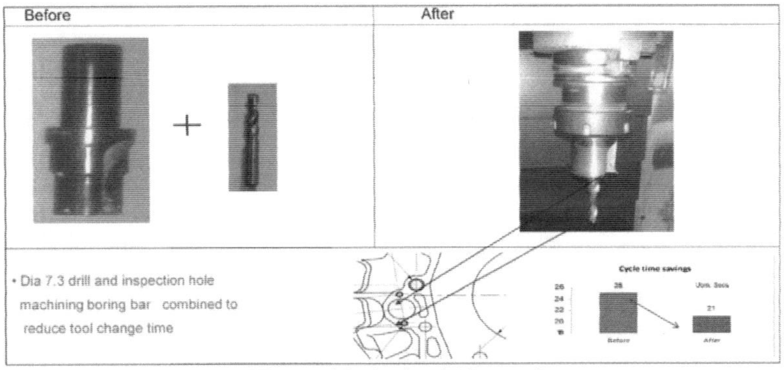

Figure 11. Action Taken – 4

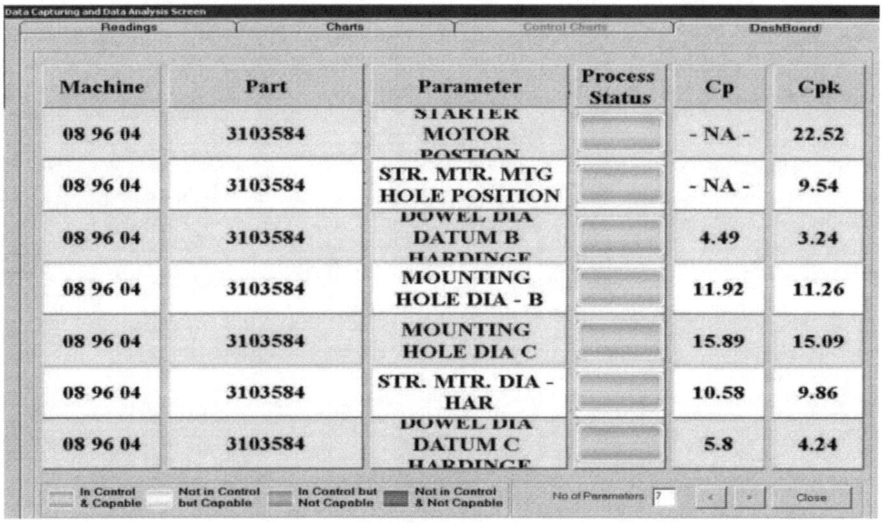

Machine	Part	Parameter	Process Status	Cp	Cpk
08 96 04	3103584	STARTER MOTOR POSTION		- NA -	22.52
08 96 04	3103584	STR. MTR. MTG HOLE POSITION		- NA -	9.54
08 96 04	3103584	DOWEL DIA DATUM B HARDINGE		4.49	3.24
08 96 04	3103584	MOUNTING HOLE DIA - B		11.92	11.26
08 96 04	3103584	MOUNTING HOLE DIA C		15.89	15.09
08 96 04	3103584	STR. MTR. DIA - HAR		10.58	9.86
08 96 04	3103584	DOWEL DIA DATUM C HARDINGE		5.8	4.24

Figure 12. The online report of computerized quality checking at Vmc-1

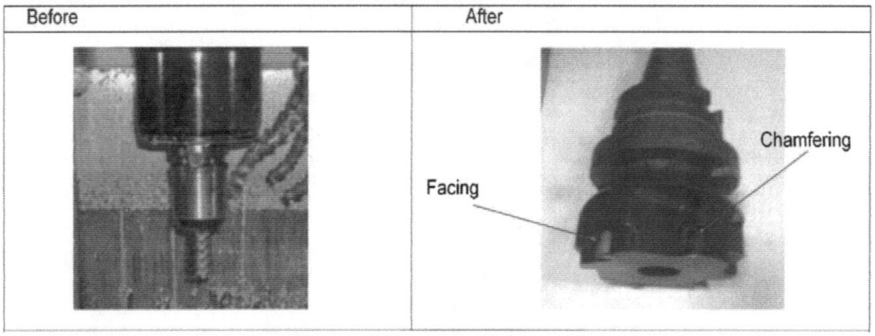

Figure 13. Action Taken – 5

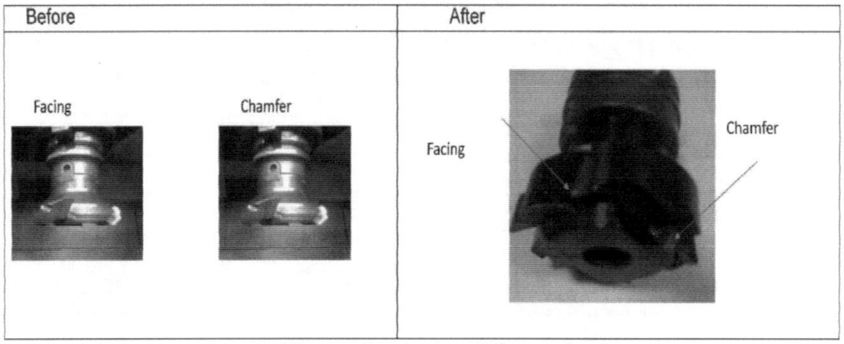

Figure 14. Action Taken – 6

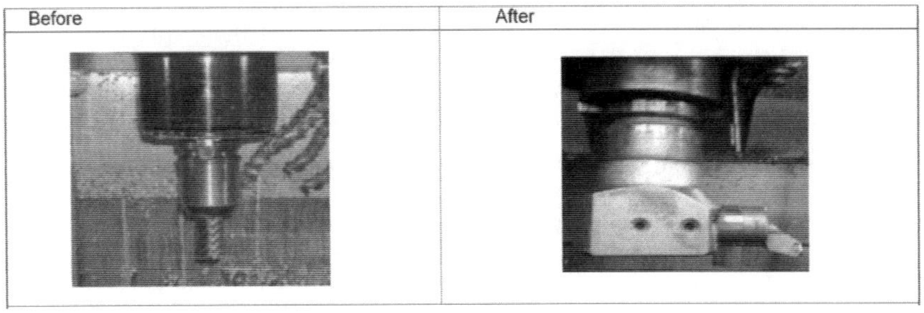

Figure 15. Action Taken – 7

Table 1. Existing and Target.

Description	Existing	Target
Available time in a month(Sec)	1944000	1944000
Average monthly demand(nos.)	4000	6000
Takt time(sec)	486	320
Hourly output (nos.)	10	12
Part price (Rs.)	3500	3500
No. of Operators/day	24	21
Sales/Employee/year (Rs.L)	70	120
Improvement (%)		71

Table 2. Tool wise cycle time VMC-1

Tool No.	Description of Operation	Tool indexing	Auto time	Total time
		Time in Seconds		
1	Face milling		92	92
2	13.8 x 3 drilling	4	13	17
3	13.8 x 3 ball bose drilling	4	5	9
4	5/8 x 11 Unc tap x 3	4	14	18
5	Starter motor rough	4	38	42
6	Starter motor finish	4	23	27
7	Dia 7.3 drill	4	7	11
8	Inspection hole machining	4	6	10
9	Dia 18.25 drill		60	60
10	Dowel finishing x 2	4	11	15
11	R38 end mill	4	36	40
	Loading/Unloading			30
	Cycle time		371	

6. Results and Discussion

The improvement project was carried out realized the improvement. They are as follows:

- The average man machine time for ISX fly wheel housing were observed at various operations and shown in Figure 3 b (Final).
- The Takt time was reduced from 486 seconds to 320 seconds,
- The units per hour improved from 10 units to 12 units.
- Scrap reduced from 0.5 tons per day to 0.26 tons per day.
- The dispatch from 2606 units to 5184 units (Figure 17)

The Tangible Savings includes

- The Average sales increase per month is 6000 Units,
- The Average Sales increase per month is 0.23 Cr ,
- The Average sales increased as 2.7 Cr per year
- Incremental contribution (38%) - 1.026 Cr. and
- Expansion plan not executed. That is Cost Saving for Creating another cell for Fuel filter housing in particularly minimum investment of 2 Vmc machines with the cost of 136 lacs apart from this labour, operation and maintenance expenditures etc.

The intangible savings includes

- The Machine utilization improved,
- The quality ensured by improved Process capability for critical characteristics.
- The Power Consumption reduced.
- The employee morale and motivation improved.

7. Conclusion

The drastic improvement through engineering approach with latest management technique was discussed with case study. The objective of this publication is, one can get clear idea and make a try to make an attempt for drastic improvement before suggest the expansion plan. In this case study the target was achieved without expansion strategy. Hence huge amount of cost savings, profit were realized. The business stability ensured. The scrap rate reduced nearly 50%.
Here the takt time target achieved by introducing Modified tools and optimized the cutting process parameter. The Improved the process flow (line balancing) in the cell was ensured. The Quality and its reliability were significantly improved. Customer satisfaction as well as employee morale were improved.

8. References

[1] Ricardo Bañuelas and Jiju Antony, 2004 Six sigma or design for six sigma *The TQM Magazine* 4(**16**) 250 – 263.
[2] George Byrne, Dave Lubowe and Amy Blitz, 2007 Using a Lean Six Sigma approach to drive innovation *Strategy & Leadership* **2(35)** 5 – 10.
[3] Jiju Antony 2008 b Can Six Sigma be effectively implemented in SMEs *International Journal of Productivity and Performance Management* **5(57)** 420 – 423.
[4] Vikas Dwivedi, Mohd.Anas and Mohd. Siraj, 2014 Six Sigma: As applied in Quality Improvement for injection Molding Process *International Review of Applied Engineering Research* 4(**4**) 317-324.
[5] Bharati P K and M. I. Khan, Harvinder Singh, 2011 Six Sigma Approach for Quality Management in Plastic Injection Moldig Process: A Case Study and Review *International Journal of Applied Engineering Research* **3(6)** 303-314.
[6] Jiju Antony, Kumar M and Tiwari M K 2005 An application of Six Sigma methodology to reduce the engine-overheating problem in an alternative company Proceeding of the Institution of Mechanical engineers *Part B Journal of Engineering Manufacture* **219** 633 – 646.

[7] Vikas Tayal and Jitendra Kumar, 2012 Improvement in production rate by reducing the defects of injection moulding International *Journal of computer science and comm. Engg IJCSCE.*
[8] Saravanan Rathinasamy and Raju R, 2010 Sequencing and scheduling of non-uniform flow pattern in parallel hybrid flow shop *International Journal of Advanced Manufacturing Technology* **49(1)** 213–225.
[9] Darshak A Desai 2012 a Increasing Bottom-line through Six Sigma quality improvement drive: case of small scale foundry Industry *International Journal of Lean Six Sigma* **2(36).**
[10] Darshak A. Desai, 2012b Quality and Productivity improvement through Six Sigma in foundary industry *International Journal of Productivity and Quality Management* **2(9).**
[11] Pugazhenthi R, , Anthony Xavier M and Saravanan R 2016 A Case Study on Effect of Grouping Technique in a Multi-Stage Hybrid Flow Shop *Int. J. Computing Science and Mathematics*. **7(1)** 42-53.
[12] Kunal Ganguly, 2012 Improvement process for rolling mill through the DMAIC six sigma approach *International Journal for quality research* **3(6)** 221-231.
[13] Rishi Pareek and Jaiprakash Bhamniya, 2013 Optimization of Injection Moulding Process using TAGUCHI and ANOVA *International Journal of Scientific & Engineering Research* **4(1)** 2229 – 5518.
[14] Mehdiuz Zalan, Sujitkumar Pattanayak, Arun Chandra Paul, 2013 Study of feasibility of Six Sigma implementation in a manufacturing industry: A case study *International Journal of Mechanical and Industrial Engineering (IJMIE)* **1(3)** 2231-6477.

Comprehensive Study on Wastages of Supply Chain Information Sharing in Automotive Industries

R Sendhil Kumar [1], **S Pugazhendhi** [2], **C Muralidharan** [3] and **S Murali** [4]

[1] Department of Industrial Engineering , Dymos Lear automotive India Pvt Ltd, Hyundai Motor Group, Chennai, India
[2] Directorate of Distance Education, Annamalai University, Annmalainagar, India.
[3] Department of Manufacturing Engineering, Annamalai University, Annmalainagar, India.
[4] Department of Mechanical Engineering, Vels University, Chennai, India.

E-mail: sastha2004@yahoo.com

Abstract. A supply chain is a very extensive concept, which encompasses many problems and features when it comes to controlling. Now a day's lean concept is a very common method to several areas, such as service sectors and manufacturing. Applying the lean concept to supply chain management is a very popular study part, it has attracted many industrial practices and researchers with different applications. Information sharing and technology remain one of the key factors of integrating the supply chain members. Current scenario the competition is no longer between the competing companies, but it happens between the supply chains. So the efficiency of the supply chain is very important. And the effective sharing of information can enhance the supply chain efficiency through minimizing the inventories and Information sharing can increase supply chain efficiency by reducing inventories and stabilizing the production. This paper describes and discusses about the seven deadly wastes of supply chain information with the comparative principle of Toyota production system (TPS) principle approach. How the TPS can be applied to supply chain information sharing And lean tool of 5S concept possibility improve the information sharing.

1. Introduction

The structure of a supply chain management encompasses several elements, facilities, administrations, sections, and so on, giving cooperative efforts to achieve the requirements of the customer. The supply chain coordination is the important element, which line up the different elements working consistently to achieve the common objective. To increase performance of the supply chain, the coordination of the supply chain should be enhanced. The expected supply chain performance of effectiveness and efficiency increase through the effective coordination [1]

Supply chain activities are associated with information through the flow of goods and services from the stages of raw materials to the final end user [1]. The materials are flowing in forward direction and the flow of information in both directions. The flow of material and information move concurrently in the supply chain, could hardly detach one another [2]. The supply chain configuration starts with central business which might be operational service or manufactured; then spreads to the next level of partners. These partners are stated as supply chain actors of central business. Encompassing from the central business, the suppliers are the upstream partners, which could be suppliers of components, materials, tools, technique, resources of manpower, transport facility and so on. In other side the downstream are customers could be mass supply of distribution, retailers ,wholesalers and so on [3] [4].

2. Research methodology

This objective of this study was carried out the Toyota production system seven wastages comparison and same of in supply chain information sharing through literature. And ways to improve the information sharing of 5S concept of information sharing management system.

2.1. Lean Thinking Toyota Production system (TPS)

Toyota Production system (TPS) was initiated by Womack, Jones and Roos in the concept of lean. They observed that an approach of continuous improvement can eliminate the muda from the production process. Muda is the Japanese word meaning of waste. It was identified firs by Taiichi Ohno in the production process. He defined the waste as, which one does not create value in the activities for the end customer. As per the Taiichi Ohno, in the manufacturing environment, there are seven types of wastes available. There are overproduction, inventory, transportation, motion, waiting, extra processing and defects. This seven wastes will describe in detail in the below paragraphs.

2.1.1. Overproduction

Overproduction is the activities of continue the service or product after obtain the required level. It means the manufacture of product quantity or services more than the customer requirement. This excess of production is mentioned as excess as resources are utilized, values are not added from the view of end customer [5]. This overproduction doesn't give the additional value to the service or product because it is not required by the customer. Therefore, it is observed that there is having no value addition from the view of end customer's. This overproduction considered as inventory, which also is defined as waste.

2.1.2. Transport

Movement of material is the unnecessary in the production process. The transportation of material from one production cell to another is considered as the working progress (WIP) and adding the time value the process, without value addition to the service or product from the view of end customers and should be defined as waste. It is also increasing the possibility of damage to the materials [5].

2.1.3. Waiting

If there are no activities performed with the product or service, it is called waiting. Due to waiting of materials, product or service, no values are added to the process. So from the requirement of the end customer, it doesn't add the values to the end product or service and should be considered as waste. The situation of waiting will happen, when the downstream supply chain process is slowed down due to lack of on time activities of upstream [5].

2.1.4. Over Processing

Extra operation in the product or service is defined as over processing. It is defined as waste as the activities, doesn't add the values to the product or service as the view of the customer. The over processing are reprocessing, more handling, additional storage and rework which happen to more inventory, over production and defects.

2.1.5. Inventory

Storing of finished product and work in progress (WIP) is defined as inventory. It doesn't add the values to the product from the customer perspective, and which do not satisfy the customer demand [5].So it should be considered as waste as no added value to the product or service.

2.1.6. Motion

The unwanted movement of materials, employees, equipment, extra production and excess inventory, which are not adding the value to the product or service is called the waste motion.

2.1.7. Defects

The finished product or service doesn't meet the customer required specification is called the defect. Defect required the extra processing, time, resource and inspection to rectify. The correction of defect doesn't value addition to the service or end product. And customer not ready to pay for this correction. Therefore defect is defined as the waste [5].

2.2. Management of information sharing

Production and manufacturing wastages are generally understood and visible. In the area of information sharing and management the concern of waste is tangible and less clear. The key barriers of the information management are the lack of clear view and definition [7]. Only very few previous research publications are analysing about the information sharing and management issue. They define the wastages to the section of lean information sharing. In the next section will describe the seven types of wastages in information sharing and management. A comparison of the definitions of the seven wastages is shown in table-1

Table 1. Seven wastages of information sharing

Seven Waste (TPS)	Waste in information sharing	Description
Over production	Excess information	Unnecessary information, redundant, too much data, over dissemination, pushing data
Transportation	Information transfer. Electronic transfer.	Incomplete information, communication failure, security threats, multiple sources
Waiting	Delay for information Waiting for Information	Unavailable information, late communication, quality of data suspect, waiting for information
Over processing	Preprocessing/ more iteration. Lack of information. Extra steps to provide information.	Too much iteration, unnecessary serial communication and data conversions, format changing, too many verification, unclear criteria.
Inventory	Partial information Storing the information	Poor configuration management, too much information, complicated retrieval. Providing information too early will require storage
Motion	Task Switching / wavering Searching for information	Indirect access, pushed to wrong sources, reformatting, manual intervention
Defect	Wrong information	Wrong information, conversion errors, lack of data quality, incomplete data, ambiguous, more verification required.

2.2.1. Excess Information (Overproduction)

Generating and communicating the excess information is over production as the comparison of TPS concept. Unnecessary additional information leads the process of lack of finding the correct information and less efficient. The excess of information consumed the excess time and resource to find the appropriate information. [7]. The excess of information doesn't create the values to the customer requirement [8].

Excess information is simply producing too much sharing of information, where the value added information data is involved in the overload of information and has to be identified [7]. The waste is associated with the resources and process used to find the correct information from the over loaded information process. The creation and transfer of the excess information are described as the root cause of the waste. The resources used for identifying the required information in the overloaded information condition defined as waste, which utilized the resources without addition of value added to the information and customers' requirements. So the excess information is defined as the creation and maintenance of additional information and resource utilized for finding the valuable information.

2.2.2. Transfer of information (Transport)

In a general manufacturing process transportation increases the processing time and non-value added to the product. It is defined as waste. Same concept in information sharing an electronic transfer doesn't add the value to the information data. This brings into line with the explanation of the manufacturing situation. Transportation of information not defined as a waste because it does not consume the resources. And if the transportation and transfer of information occurs immediately, it doesn't add the time to the process. The waste is not linked with the transfer of the information itself, as it occurs instantaneously, but the information receiver has to decide the received information provide the addition of value or not [7].

The waste of information transfer is associated with the information which is directly communicating with customers or supplier through the communication system like mail. In the case of excess information sharing, the information is not necessarily forced to send between communicator. Waste is characterized by communication data which is received and sent which no one uses. The communication of information sends to the wrong information user. In this wrong information transfer is not a waste of transportation, due to the immediate transfer through electronic. But if the data is not utilized by the receiver, then it is considered as the waste of information. And the delay of information transfer to the right person at the right time is the waste and increase the communication transfer time. The information transfer waste may be linked to the fact of how the information management and organized to communication flow might be wrong.

2.2.3. Waiting of Information (Waiting)

Waiting time of information as the time required to identify the right information to transfer through appropriate resources. The reason of the waste of information waiting happens due to discontinue processing or very simply because the information cannot be found [7]. The information user waiting for information because this ideal time happening due to unavailability or lack of right information [8]. Thus the information needs to flow, but the information not yet read or unavailability. This information waiting of information waste doesn't give any value added to the information and information user.

2.2.4. Lack of Information (Over processing)

The over processing is the state of processing linked with lack of information [6]. The waste of pre-processing or after processing is associated with action undergone to overcome the information unavailability. These actions may contain creating of right new information or non-value added information [6]. The additional information doesn't create the values from the customer's view [5,8]. Hicks et al. [6] Description could be argued to be somewhat incorrect. Additional processing and steps of creating the information to overcome the lack of information and may create the additional support to the information user. As the information is new and irrelevant to the information, then it is considered as waste. At the same time, this is a process which initially should not happen, as valuable information should be provided to the information user in the first place. Providing the additional steps of information may indicate the missing of the process or broken communication in the information chain. So, additional processing is defined as actions which generate no value for the information user and resource due to absence of information.

2.2.5. Storing Of Information (Inventory)
Data of information is normally stored in the platform of digital storage. It doesn't have the digital platform, which does not represent a major financial cost [7].And the digital storage doesn't take the physical space, when comparing with the manufacturing sector. So the inventory storage of information data is not considered by the cost .Hicks et al. [6] Defined storage of information as a waste by as well as the information user. And the waste is linked with the process of information, data retrieval from the electronic storage [7]. The inventory of information is described that in the situation of excess information than customer required at the specified time of instant [8]. So the collected information is stored, but not utilized by information user. It is similar like the excess information of over production, that waste can be defined as inventory. The stored information does not value or required at a given moment in time [8]. So the stored information may be relevant at a later point in time. The information created and available in advance also defined as the waste. Communicating the information in advance require the storage, which is not considered as waste as the cost is considerably less. Information is generated in advance and stored which increase the available information, which creates the process of finding value added information less efficient.

2.2.6. Searching of Information (Motion)
The unwanted motion is defined to be connected to the organizations assumption of "gatekeeper functions". A gatekeeper job is formed when one individual is trained to use a specific software application [7]. Only the gatekeepers are skilled to use the specific software or applications. Thus, high dependability created to access the specific software or application. It creates the unwanted movements between the computers, which described as unwanted motion. In manufacturing environment, motion of people or material may have unwanted movements. But in the information sharing, motion describes as additional processes required to attain the information like the searching of files in the server, additional button clicks on the computer [8]. So the motion is described as unnecessary movements and procedures followed to attain the information.

2.2.7. Wrong Information (Defects)
The key cause of defects is wrong and inaccurate information [6]. The action of modifying and verify the information consider as the waste. Creating results on incorrect/wrong information can also outcome in wrong actions. This type of activities might not be shown if the information provided was right. Incorrect activities taken based on incorrect information are also defending as waste [6]. The defects are defined as error on work, incorrect information, and partial information [8].Waste is also defined by unsuitable activities taken based on incorrect information. Incorrect information will not deliver value to the information user and can outcome in inadequate actions based on the incorrect information.

3.5 S (House keeping)
The Five 5S's are some guidelines for maintaining the workplace organization which intends to establish each operator's workspace for maximum efficiency (figure 1). A comparison of the definitions of 5S approach in manufacturing floor and information area are shown in table-2.

3.1. Sort

Categorized things that what is required and what is not required.so that the things that are repeatedly used are available nearby as easy to find as possible. And things which are less frequently used or not needed should be discarded or relocated.

3.2. Set in order

The required things should be arranged in systematic order for easy use. The aim is to reduce the unnecessary movement of works in his job. For example, files are stored in the rack as per the numerical label order for easy access. Using the tools in the shop floor can be used the separate tool kit for production and maintenance persons.

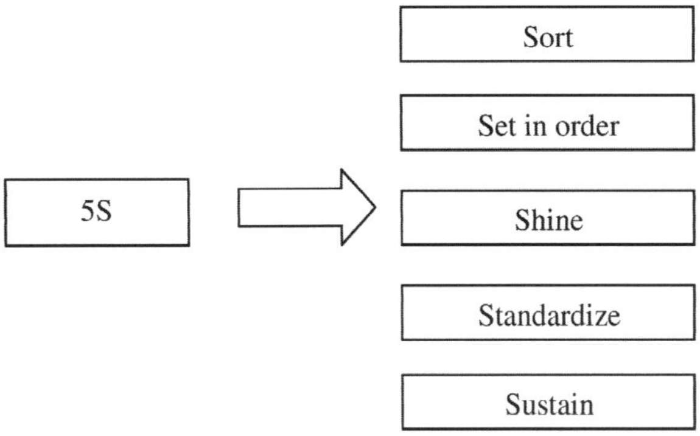

Figure 1. 5S for the workplace

3.3. Shine

Maintain the workplaces, tools and machineries clean so as to avoid the issues related to poor work condition. This will confirm the presence of any non-conformity; such as split of oils in the floor and dust contamination on the work place.

The coating industries faced the issue of poor quality due to dust. Some companies are applying the light colours in their work place and floor to control and easy to identify.

3.4. Standardize

Makes the work place standardization through continuous follow-up and practice the first 3S of sorting, arranging and cleaning. Make the first 3 S's a routine practice by implementing clear procedures for sorting, set in order and shine.

3.5. Sustain

Encourage, proper training and communicate the 5 S's to confirm part of the organization culture. This should include assigning the 5S team to ensure the status of 5S practices and level in the organization to sustain the 5S advantages.

4. Information sharing waste reduction

If the organization is not up to the standard in the level of excellence in effective information sharing, there is a chance to reduce information waste through the help of 5S approach which is depicted in the table 3.

Table 2. 5S approach in the area of manufacturing and information sharing

5S	Manufacturing work place	Information sharing
Sort	Segregation of needed & Not needed material (tools/part)	Check and segregation of files, folders, images, software and unwanted application in the system.
Set in order	Arrangement of required Tools, Parts in systematic order	Organize files and optimize the use of file folders. Create the specific shortcut icons for mostly using files, folder, softwares. Make the flow and Master file for guiding to search the files.Follow the change history to understand and avoid the duplication. Visual identification, control for easy identification of Hard copy folders.
Shine	Clean the workplace & equipment on a regular basis	Eliminate the duplication of files. Update the master file & Delete the duplicated file frequently. Stored the file in the master file path & avoid storing in a common storage location
Standardizes	Make the first 3 S's a routine practice by implementing clear procedures for sorting, set in order and shine.	Establish the procedure to maintain the 1st 3 S
Sustain	Keep to the rules to maintain the standard & continue to improve every day. Encourage, proper training and communicate the 5 S's to confirm part of the organization culture.	Regular audit and competition to procedure follow up of 5S Regular training to the all levels of information sharing person.

4.1. Waste from Excess Information (Overproduction)

Excess Information, where valuable is included and overloaded. And much information is non value added. And generating of excess information is considered as a waste to the process information. It may sound surprising, but many companies and supply chain partners are generating more than necessary information contained or communicating the less information because they are not aware about the information required, unavailability of data, storing of excess information data, poor storage method of information data, searching of data. Without any request, preparing of data and information sharing considered as excess information. In this case all the efforts are ending with waste of energy, time and manpower.

Eliminating the excess data and non-value added information by avoiding the excess information and communication of additional and shortage of information. Information management applies the 5S concept to improve the information sharing. And storing of required data and updated data in the correct storing method. The concept of 5S can be practiced to set up guidelines for data storage and describe data in location, size and space. The guidelines have to be widely emphasized to all. So everybody knows the detail of information data of where, what, why, how much and how long. Through this continuous improving the condition, the guidelines updated to the new level of position and follow up the rules.

4.2. Waste from Transfer of information (Transport)
The situation of necessary to transit and movement of information can be initiated by the earlier mentioned wastes. All the movement may not remove, but they have tried to minimize. There is chance in waste of information, when it is sharing from one person to another and multiple person through multiple communication system. Most of the people preferred the more information, but a proper guidelines, discipline and proper transfer system may be enforced to avoid waste and problem.

4.3. Waste from Waiting of Information (Waiting)
Waiting is a concern of lack synchronization among the communication process or lack of preparation of data. Waiting for data, IT application, tools, approval process, rules, itch can be triggered by a lack of guidelines about information, data storage places, when persons have to search everywhere. Are the element you are waiting for really essential? If they're not, if they do not add value to the information process, work or the product, it's wise to try to remove them or at least reduce the searching and waiting time of data. Waiting time occurs due to searching documents or files. The simple way to find the files or documents is sketching a coloured incline line on the back of the files:

- Easily visible the missing file.
- Wrong place of file shows a broken line

This method is better than the file identification number, which needs reading and translating the information. The color management is the part of visual management approach. It is simple and fast to search the files and reduce the searching time.

Table 3. Influences of 5S approach in information sharing

Waste in information Sharing	5S –Influence
Waste from Excess Information (Overproduction)	Easy identification of Data from master file Unwanted data reduced due to duplication avoided
Waste from Transfer of information (Transport)	Time will be reduced due to clearing of unwanted document, application and software
Waste from Waiting of Information (Waiting)	Organize files, folder, software and application the searching time time will be reduced The easy identification of document for information through master file and visual tracking flow
Waste in Lack of Information (Over processing)	Organize files and optimized arrangement Reduce the searching time, waiting time. Easy Identification of required data/information
Waste related to Storing Of Information (Inventory)	Duplication and temp files cleared to avoid the complicated to data retrieval
Useless Searching of Information (Motion)	Due to organized information storage, searching, wrong data, format, multiprocessing will be avoided
Waste from Wrong Information (Defects)	Identification of data and path will be cleared to avoid the wrong data selection

4.4. Waste in Lack of Information (Over processing)

Procedures and rules which are not regularly updated are possible to let non value added processes be executed in the process. Sorting and ordering applies also in the sequential process and the associated documents. This waste is common in the process of office and administrative. The existing procedure still remains follow even the new updates are made due to unaware of new updates. As long as nobody will update the set of procedures, everybody will follow and use the old with application and discipline. Lack of proper knowledge sharing within the team, lack of backup history of the document, complicated communication process and lack of IT tools and knowledge leads to waste. These will influence the over processing

- Using more & multi technology and method to share the simple information.
- Excess information & passing the information to irrelevant people.
- Making more & irrelevant data collected to send the information sharing.
- More iteration and steps to transfer the data.

4.5. Waste related to Storing of Information (Inventory)

As per the 5S concept whatever that is useless is to be removed. In situation of storage of information, the advantage is the value of the information saves and recovers the data and spaces, which should be dedicated preferably to value adding activities. In the form of paper documents and their several copies, files, catalogues and calendars of previous years, and incomplete data, old back up data, unapproved document, etc. all excess inventories.

4.6. Useless Searching of Information (Motion)

The ergonomics of the workplace is certainly the most popular and "visible" application of the 5S. The layout and display of the area will follow the 5S logic, favoring the availability of necessary items, distance of reach, and ease for tending. Among useless motions, do not forget the walks to search for missing items, data, instructions, complementary information. Unnecessary steps and ask switching also the waste motion of information. It will not add the value to the communication process. Through the 5S can possible to improve the information management. Data should be stored in the constant location in the server with standard tracking method.

4.7. Waste from Wrong Information (Defects)

Defect data and insufficient information are considered as the defect of information sharing. It can be happened due to lack of understanding of customer requirements, insufficient product and process knowledge. Some requested communication doesn't have the objective, due to that the sharing of information not satisfied the customer. Poor quality of information sharing due to lack of IT tools and technical skills, complicated approval process and lack of involvement. 5S will improve or avoiding the mistake of data collection and information sharing. Storing of right data in the proper location with identification is reduced waste. When the data has revised, there should have option with updating, revision number and date. Updated data with proper approval procedure and stored in the dedicated location for tracking. Hard copy of data source in the identified location with identification tag should be kept. The same process of communicating through electronic and IT tools has to consider the proper, suitable IT tolls for communication. Select tools have a need to supportable to other interactive partner.

5. Conclusion

This paper studies the concept of lean concept of Toyoto Production System (TPS) comparison with the supply chain information sharing. The study has also discussed the validity of the seven types of waste in the context of information management. Muda of seven wastages are described as in the manufacturing and the identified the equivalent wastages in the information sharing are identified through literature survey. Transfer of information (Transportation) lack a basis in information management because today's organizations will usually transport information by electronic means.

Hence, transportation is an immediate transfer, which adds no time to the process. The study has also discussed the validity of the seven types of waste in the context of information management. And the 5S concept is described and the improvement of reduction of wastes in information sharing through 5S concept. Through this analysis, the possibility to apply the 5S concept to eliminate the wastes in the information sharing and management was addressed. In future research scope possibility for analysis the information sharing losses in the supply chain network through the lean tools. And reduce the waste through the TPS concept and 5S.

6.References

[1] Ballou R H, Gilbert S M and Mukherj A 2000 *Industrial Marketing Management*. 7-18.
[2] Plenert G.2007 Burlington: Butterworth-Heinemann.
[3] Christopher M 2011Harlow: Financial Times Prentice Hall.
[4] Schniederjans M J, Schniederjans D G, and Schniederjans A M 2010 New Jersey: World Scientific.
[5] Womack J P and Jones D T 1996 New York: Free Press. (2003rd ed.).
[6] Hicks B J, Culley S J and McMahon C A 2006 *International Journal of Information Management*. **26(4)**, 267-289.
[7] Hicks B J 2007 *International Journal of Information Management*. **27(4)**. 233.
[8] Ibbitson, A and Smith R.2011 *In Sernenti E.* (Ed.). Ark Group.

Computational Fluid Dynamic Simulation of Flow in Abrasive Water Jet Machining

S Venugopal[1], S Sathish[2],V M Jothi Prakash[3] and T Gopalakrishnan[4]

[1,3]Assistant Professor, Department of Automobile Engineering, Vels University, Chennai, India.
[2]Head of the department, Automobile Engineering, Vels University, Chennai, India.
[4]Assistant Professor, Department of Mechanical Enginering, Vels University, Chennai, India.

E-mail: venugopalmechlec@yahoo.co.in

Abstract. Abrasive water jet cutting is one of the most recently developed non-traditional manufacturing technologies. In this machining, the abrasives are mixed with suspended liquid to form semi liquid mixture. The general nature of flow through the machining, results in fleeting wear of the nozzle which decrease the cutting performance. The inlet pressure of the abrasive water suspension has main effect on the major destruction characteristics of the inner surface of the nozzle. The aim of the project is to analyze the effect of inlet pressure on wall shear and exit kinetic energy. The analysis could be carried out by changing the taper angle of the nozzle, so as to obtain optimized process parameters for minimum nozzle wear. The two phase flow analysis would be carried by using computational fluid dynamics tool CFX. It is also used to analyze the flow characteristics of abrasive water jet machining on the inner surface of the nozzle. The availability of optimized process parameters of abrasive water jet machining (AWJM) is limited to water and experimental test can be cost prohibitive. In this case, Computational fluid dynamics analysis would provide better results.

1. Introduction

AWJM is a well-known non-traditional machining process. It is a machining process where material is removed by smack erosion of high pressure high velocity of water and entrained high velocity of grit abrasives on a work piece [1]. The added abrasives [2] increased the materials range, which can be cut with a Watergate desperately. It is widely used, because of its main advantages
Materials which are cut by AWJM are

Polymers, Honey comb, Metal matrix Composites [3].

Ceramic Matrix Composites, stone, Granites,

Wood, Reinforced Plastics, Metal Polymer Laminates and Glass Fibre metal Laminates.

Page layout (headers, footers, page numbers and margins)

2. Theoretical Formulation

2.1 Numerical Model and Assumptions

The abrasive water suspension mixture[4] is let into the nozzle at inlet and carried through converging cone to focus tube and exits at nozzle, in which the focus tube is used for guiding the flow. The numerical model adopted closely follows the work at Hu et al which liquid-solid two-phase flow is considered.

- Water is a continuous medium and incompressible.
- Flow is considered as two phase flow mixture in which water is liquid phase and abrasives is solid phase.
- Two phase flow is steady and possesses turbulent flow characteristics.

The nozzle head of AWJM is modeled by using pro/E software and it is saved. We modeled the nozzle head of varying taper angles .Then it is imported in ICEM meshing software for meshing the model. Tetrahedron is used for fine meshing. Then it is imported in CFX Pre for giving input parameters and properties of water and abrasive.

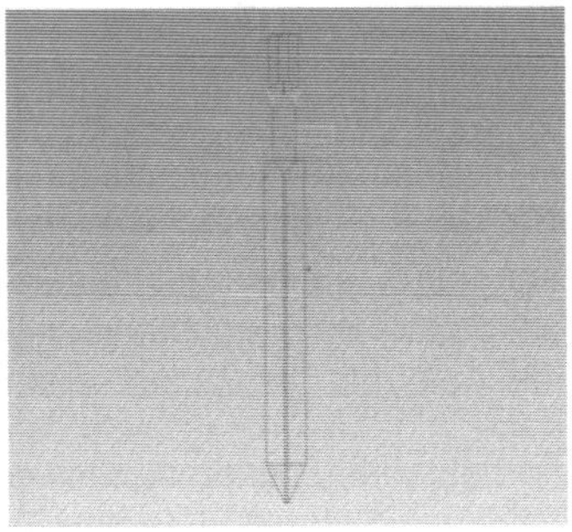

Figure.1 Pro/E model of nozzle head

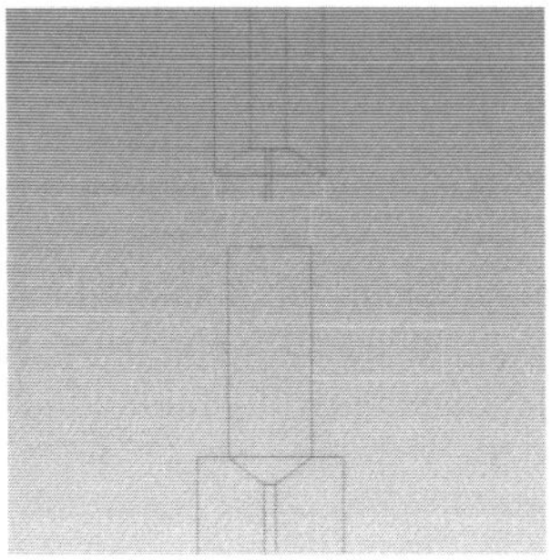

Figure 2.15 deg. Taper Angle

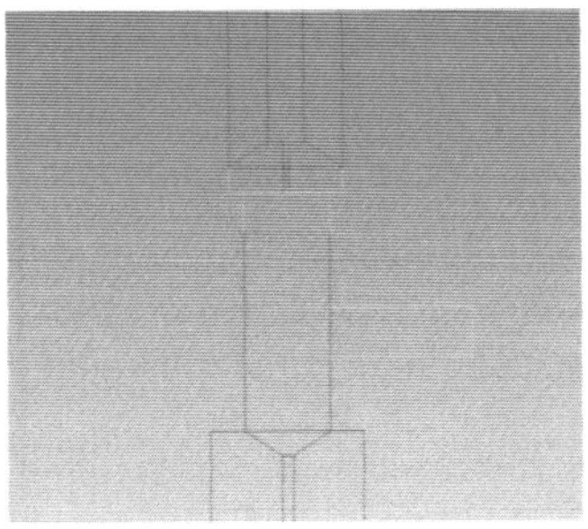

Figure.3 30 deg. Taper Angle

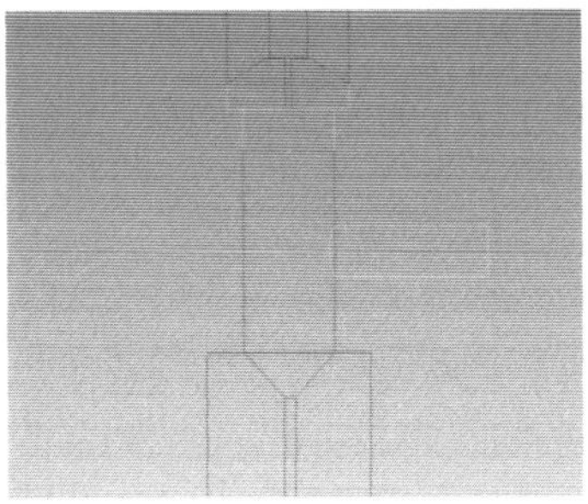

Figure 4. 45 deg. Taper Angle

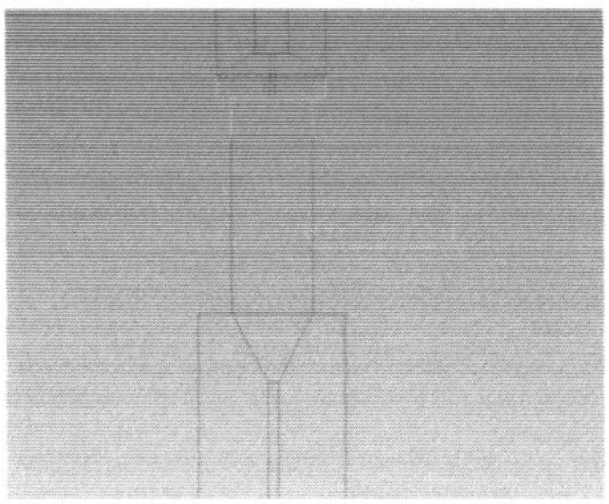

Figure 5. 60 deg. Taper Angle

IOP Conf. Series: Materials Science and Engineering **183** (2017) 012018 doi:10.1088/1757-899X/183/1/012018

Figure 6. 15 deg. Taper Angle

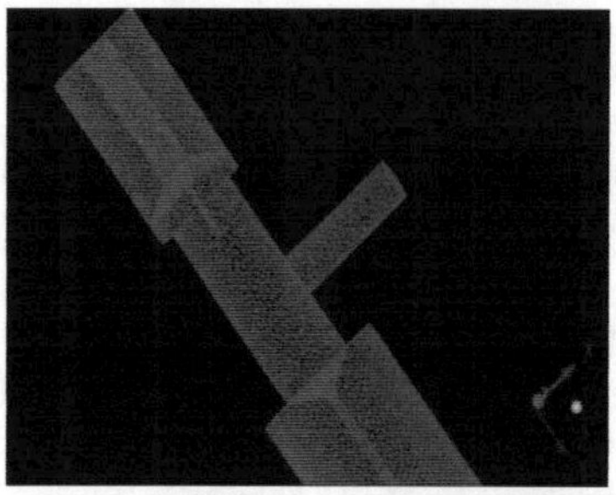

Figure 7. 30 deg. Taper Angle

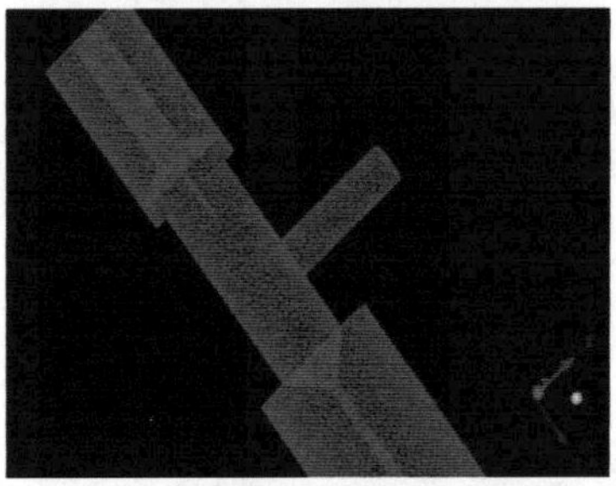

Figure 8. 45 deg. Taper Angle

Figure 9. 60 deg. Taper Angle

2.2 Computational Domain Input

For the meshed model computational domain input should be given in the first. You have to create a water inlet, abrasive inlet and mixture outlet to the meshed model. In CFX preprocessor, you have to give the thermodynamic properties of water and abrasive. The water is taken as continuous liquid, and dispersed solid abrasive size is 1mm. In boundary condition you have to give domain type, name of nozzle head and mixing chamber. You have to name the liquid domain and solid domain of nozzle head and mixing chamber. Then you have to tell at water inlet, water enters and at abrasive inlet abrasive enters and also give the boundary details. Then you have to give domain interface between solid and liquid domain. The outlet will be the velocity. In CFX Solver, the model is imported and no of equations is performed and you have to give no of iterations and run solver and monitor. In CFX post, the model is imported and post processes the results.

2.3 The general specifications of our model

Focus tube (Mixing tube) Diameter : 0.76 mm
Focus tube length : 76 mm
Taper angle of nozzle : 45 deg
Mixing chamber diameter: 6 mm
Mixing chamber length: 12 mm
Orifice diameter : 0.2 mm
Water inlet diameter : 2.5 mm
Abrasive inlet diameter : 3 mm

The pressure of water is taken as 400 bar and Density of water $\rho_W - 1000$ kg/m^3

The velocity and mass flow rate of water and abrasive will be calculated by using the standard dimensions. This will be used for domain input. The properties of water and abrasive will be given.

3. Results and discussions

3.1 Validation of the numerical model (Effect of nozzle angle)

3.1.1 Velocity Variation

Fig 10 shows the velocity variation along the mixing chamber and focus tube length. In the case of Mixing Chamber the velocity reduces gradually and increases when it reaches near the end of tube. In the case of focus tube, the gain in velocity is observed when the flow past the nozzle. Further kinetic energy lost is observed when the flow is along the focus tube for all the cases. This may be due to some of the abrasive particles collide with the wall of the focusing tube. Hence, there is some overall velocity reduction as the jet exits the focusing tube. The kinetic energy lost is high in the cases of taper angle 30° and 60° and it is significantly less in the case of 45° taper angle

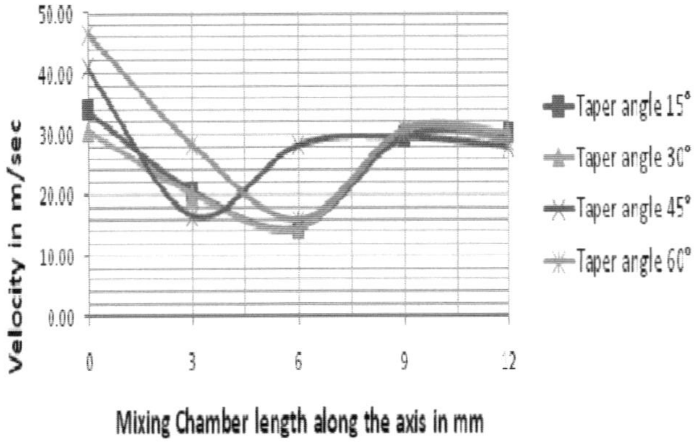

Figure10.(a) Mixing Chamber

3.2 Wall Shear Stress Distribution

Figure 11 shows the wall shear stress distribution along the mixing and focus tube length. Figure 11(a) shows that increased wall shear at the entry of Mixing Chamber for 60° taper angle. The magnitude decreases till the mixing region after that it increases sharply along its flow.

Figure 11(b) shows that the wall shear increases when the tape angle increases. Increased magnitude of wall shear has been observed for 60° taper angle.

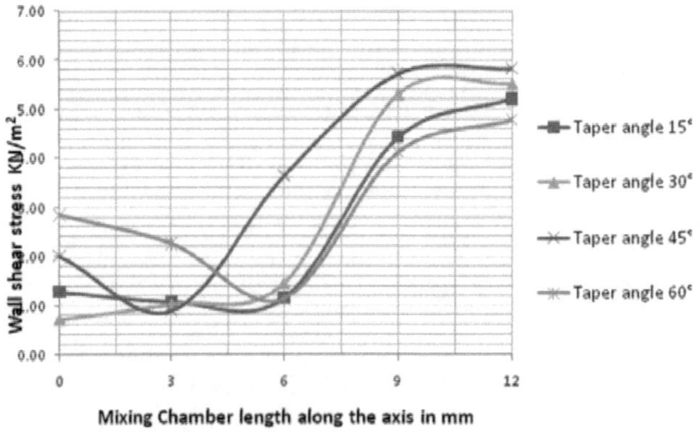

Figure 11(a). Mixing Chamber

3.3 Rate of Energy Dissipation

Figure 12 shows the rate of energy dissipation due to wall shear along the mixing and focus tube length. Figure 12(a) shows rate of energy dissipation that is high at inlet of the Mixing Chamber for the taper angle 30°. Its magnitude decreases sharply along the flow and it increases gradually beyond the mixing region. It is observed from the Figure 12(b) the rate of energy dissipation is high for

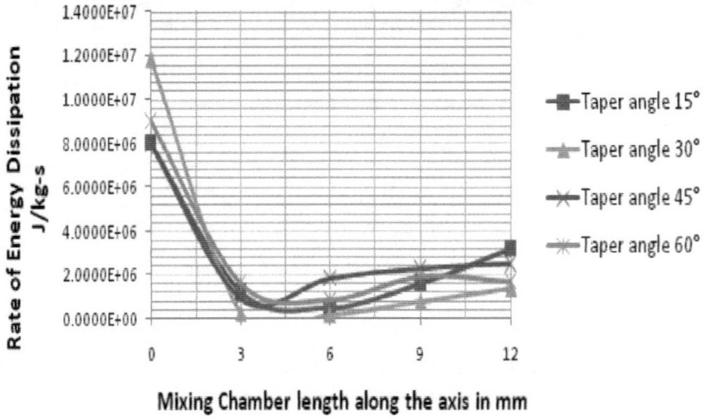

Figure 12(a). Mixing Chamber

4. Conclusions

Therefore, Computational fluid dynamic [CFD] simulation of flow through nozzle head of abrasive water jet Machining has been carried out and the following conclusion has been drawn.

4.1 Effect of nozzle angle

- Loss in kinetic energy has been observed when the flow is along the focus tube .This may be due to some of the abrasive particles do collide with the focusing tube wall. The kinetic energy loss is relatively less for 45° taper angle.
- The magnitude of wall shear stress increases when the taper angle increases. The wall shear in the mixing chamber increases sharply after the mixing region.
- The energy dissipation due to wall shear is relatively low for 30° taper angle.

5. References

[1] MishraP K, Non conventional machining *Narosa publishing house* third reprint – 2005. Non conventional machining version 2 ME, IIT kharagpur

[2] Ray P K,Paul A K -1987, Some studies on abrasive jet Machining, reprinted from the journal of institution of engineers (india) **68**, part PE, 2

[3] Shah R V, Prof Patel R.M – 2012, *Abrasive Water Jet Machining* – 803-806The Review, **2**, Issue 5

[4] Hu Z, zhu W, TaoY and Yuan J - 2008 "Numerical simulation and experimental study of liquid- solid two-phase flow in nozzle of DIA jet", in *proceedings of the 6th IEEE international conference on Industrial informatics (INDIN ` 08)*

[5] David G.Taggart, MadhusarathiNaduri and Thomas J.Kim "Effect of the inlet taper angle on AWJ nozzle wear" – *9th American waterjet conference August 23 -26, 1997.Dearborn;Michigan*

[6] David G.Taggart, MadhusarathiNaduri and Thomas J.Kim" *E*valuation of an accelerated wear test for AWJ nozzles" *water jet laboratory, University of Rhode Island, Kingston,RI USA*

[7] Deepak,D.Anjaiah, VasudevaKaranth K and Yagnesh Sharma N – "Effect of inlet operating pressure and volume fraction on skin friction and exit kinetic Energy" (Manipal University, Manipal 576104, India

[8] CFX – 2009, Software User`s Manual

IOP Conf. Series: Materials Science and Engineering **183** (2017) 012019 doi:10.1088/1757-899X/183/1/012019

Modeling and Prototyping of Automatic Clutch System for Light Vehicles

S Murali [1], V M Jothi Prakash[2] and S Vishal [3]
[1] Associate Professor, Department of Mechanical Engineering, Vels University, Chennai.India.600117.
[2] Associate Professor, Department of Mechanical Engineering, Vels University, Chennai.India.600117.
[3] UG Student, Department of Mechanical Engineering, Vels University, Chennai.India.600117.

E-mail: murali.se@velsuniv.ac.in

Abstract. Nowadays, recycling or regenerating the waste in to something useful is appreciated all around the globe. It reduces greenhouse gas emissions that contribute to global climate change. This study deals with provision of the automatic clutch mechanism in vehicles to facilitate the smooth changing of gears. This study proposed to use the exhaust gases which are normally expelled out as a waste from the turbocharger to actuate the clutch mechanism in vehicles to facilitate the smooth changing of gears. At present, clutches are operated automatically by using an air compressor in the four wheelers. In this study, a conceptual design is proposed in which the clutch is operated by the exhaust gas from the turbocharger and this will remove the usage of air compressor in the existing system. With this system, usage of air compressor is eliminated and the riders need not to operate the clutch manually. This work involved in development, analysation and validation of the conceptual design through simulation software. Then the developed conceptual design of an automatic pneumatic clutch system is tested with proto type.

1.Introduction
Many of automobile manufacturing industries are producing both automatic and manual transmission car. A transmission basically transfers the power from a car's engine to drive shaft and the wheels. Clutch is a device which is used in automobiles, to transmit power from engine to gear box transmission system. In a car, the use of clutch is very essential, because the engine spins all the time, but the car's wheels do not. In order to stop the car without killing the engine, the wheels need to be disconnected from the engine somehow. The clutch is used to engage the spinning engine very smoothly with the non-spinning transmission by controlling the slippage between them [1,2].

2.Research Methodology
At present, clutches are operated automatically by using an air compressor in the four wheelers. It requires more power to operate the clutch. It incurs more running and maintenance cost. Due to the provision of air compressor, the weight of the vehicle also increases and thus the mileage is reduced. Previous studies conducted on automatic clutch system. Aleksandr Blokhin et al. [3] developed a new

technical solution for creation of mechanisms of the friction clutch automatic control for potential heavy-duty trucks and buses. Min Wu and Ze Rong Tang [4] developed automobile intelligent electronic control automatic clutch systems, use the electromagnet controlled by circuit to replace the clutch pedal of manual car. Prasanth et al. developed the semi automatic clutch system with proportional control valve and the actuation by a double acting cylinder[5]. After reviewing the various studies on automatic clutch, it is found that the clutch can also be actuated without the help of air compressors [6] – [9]. This can be done by effective use of the exhaust gas coming out from the turbocharger. By this method the air compressor can be removed from the system and thereby decreasing the weight of the vehicle and increasing the mileage also.

The first stage of the work deals with identification and design of components required for the clutch operating system. In the second stage, the conceptual design is developed and analysed by using ANSYS software. Optimal design is arrived from the analysis which can be generated and drafted by using CATIA software. The final stage of the work involved in making and validating the prototype of the clutch operating system.

In this system, there is one proximity sensor provided with the gear lever to have a smooth operation of clutch. The gear changing lever is connected with relay switch to facilitate the operation of clutch automatically by means of pneumatic cylinder which is powered by existing battery. The clutch shaft is connected with vacuum cylinder. There is an unloader valve fitted in the line before the reservoir and whenever the pressure reaches the maximum, the unloader valve releases and diverts exhaust gas to the catalytic converter. Whenever the gear is changed, the relay switch operates a pneumatic cylinder which causes the air coming out of turbo charger to be sucked, compressed and sent to the vacuum cylinder. This causes the piston moves inside the vacuum cylinder and thus the clutch disengages. The automatic clutch with pneumatic cylinder relates to an automatic transmission more particularly, the invention describes a transmission for producing an automatic movement for light-weight vehicles, such as cars and other type of four wheelers.

3. Construction and working

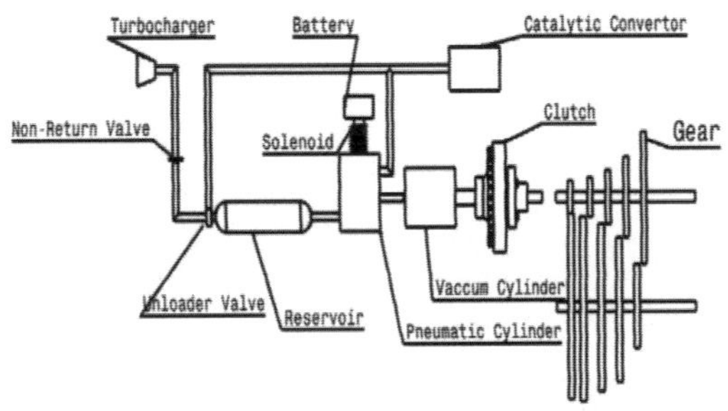

Figure 1. Schematic diagram of automatic clutch system.

Figure.1 shows the schematic diagram of automatic clutch with pneumatic cylinder. This system is working by the exhaust gas coming out from the turbocharger which is sent to the reservoir. From the reservoir, the exhaust gas is sent to the pneumatic cylinder in which the exhaust gas is compressed and the clutch is operated. In the pneumatic cylinder, when the piston moves from top dead centre (TDC) to bottom dead centre (BDC), the exhaust gas from the reservoir is injected and compressed inside the pneumatic cylinder and then sent to the vacuum cylinder. The piston in the vacuum cylinder is attached to the clutch. The exhaust gas moves the piston and thus the gear is engaged smoothly through the clutch. When the piston moves from BDC to TDC, vacuum is created in the cylinder which causes the compressed gas from the vacuum cylinder is sucked into the pneumatic cylinder. From the pneumatic cylinder, the exhaust gas is sent out to the catalytic convertor. There is an unloader valve present in the reservoir. When the reservoir is full the unloader valve opens and the remaining exhaust gas is sent to the catalytic convertor through separate line.

4. Modeling of conceptual design

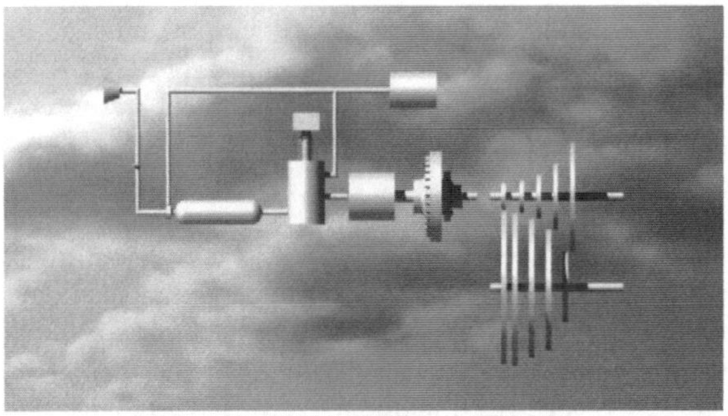

Figure 2. 3D Modelling in CATIA.

Figure 2 shows the 3D model of the proposed clutch system which is developed by using CATIA.

5.Prototyping

The prototype model of the automatic clutch operating system shown in figure 3 consists of non-return valve, unloader valve, solenoid, pneumatic cylinder and vacuum cylinder.

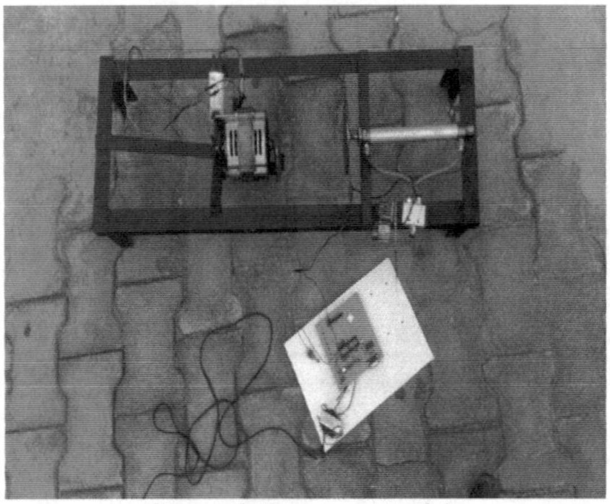

Figure 3. Prototype model.

All the components are assembled as shown in figure 3 and tested for feasibility and effectiveness.

6.Conclusion

Clutch is a mechanism for connecting and disconnecting the engine with the transmission system in a vehicle, or the working parts of any machine. Turbo Charger is a supercharger, driven by a turbine powered by the engine's exhaust gases. The exhaust gas from the turbocharger is used to activate the clutch system which facilitates smooth changing of gears. This work involved in development, analysation and validation of the conceptual design through simulation software. Then the developed conceptual design of an automatic pneumatic clutch system is tested with proto type. With this system, usage of air compressor is eliminated and the riders need not want to operate the clutch manually. As there is no limit for making improvements in any kind of work, this automatic clutch operating system may also be further improved with reducing the wear and tear of parts by filtering the exhaust gas and reducing the weight of the system by using the parts made by composite materials.

7. References

[1] Kirpal Singh 2008 Automobile Engineering **1**, Standard Publishers Distributors. Delhi.

[2] Fischer R, Küçükay F, Jürgens G, Najork R and Pollak B 2015 The Automotive Transmission Book Springer International Publishing.

[3] Aleksandr Blokhin, Lev Barakhtanov, Alla Koshurina, Aleksandr Taratorkin, Pavel Lubichev 2015 Research of the Friction Clutch Automatic Control Performance at the Stand *International Conference on Electrical Electronics and Mechatronics* Atlantis Press.

[4] Wu M and Tang Z R 2014 Study on Automobile Intelligent Automatic Clutch System *Applied Mechanics and Materials* 599-601 1253-1256.

[5] Prasanth N, Arun M, Balagurubaran R, Sabarish D and Kirubashankar R 2014 Semi Automatic Clutch in Heavy Vehicles *Innovative Research in Science Engineering and Technology* **3** pp 210- 214.

[6] Raut S S and Mali D P 2014 Automatic transmission gearbox with centrifugal clutches *International Journal of Research in Engineering and Technology* **3** 307-309.

[7] Khairnar, P G, Kute, A P, Kavade, Y K, Gadekar, P L, Ghodke,A P and Bhane, A B 2015 Automatic Muli-Plate clutch control in two wheeler bike *International Journal of Emerging Technology and Advanced Engineering* **5** 540-541.

[8] Srivatsan S, Naveen R, Siddharth Narayanan C, Shyam Narayan S 2013 Automatic clutch actuation using touch sensor based gear lever and Interconnected Brake *International Journal of Engineering Research & Technology* **2** 2458-2463.

[9] Nilkanth G B and Patil S V 2015 Design and analysis of auto disengagement single plate clutch in automatic transmission and its testing performance *International Engineering Research Journal* **2** 1366-1371.

Improvement of Productivity in TIG Welding Plant by Equipment Design in Orbit

C Gnanavel[1], R Saravanan[2], M Chandrasekaran[3] and J J Jayakanth[4]
[1]Assistant Professor, Department of Mechanical Engineering, Vels University, Chennai, India
[2]Asso. Professor, Department of Mechanical Engineering, Vels University, Chennai, India
[3] Professor, Department of Mechanical Engineering, Vels University, Chennai, India
[4]Assistant Professor, Department of Mechanical Engineering, Vels University, Chennai, India

E-mail: gnanavelmech1986@gmail.com

Abstract. Measurements and improvements are very indispensable task at all levels of management. Here some samples are, at operator level: Measuring operating parameters to ensure OEE (Overall Equipment Effectiveness) and measuring Q components performance to ensure quality, at supervisory level: measuring operator's performance to ensure labour utility at managerial level: production and productivity measurements and at top level capital and capacity utilization. An often accepted statement is "Improvement is impossible without measurement". Measurements often referred as observation. The case study was conducted at Government Boiler factory in India. The scientific approach followed for indentifying non value added activities. Personalised new equipment designed and installed to achieve productivity improvement of 85% for a day. The new equipment can serve 360o around its axis hence it simplified loading and unloading procedures as well as reduce their times and ensured effective space and time.

1. Introduction

The efficiency of production not only depends on sophistication of core machine but also the various supporting equipments, facilities, services and employee's efforts which are located in a plant [1]. The authors focused on plant layout and material handling to achieve up to 85.31% production efficiency and 46% in reduction of transportation lengths. They used simulation software and string diagrams as tool for the same. Shewale et al [2] made attempt to increase the productivity by plant layout optimization through systematic layout planning (SLP). The operation process chart, flow of material and activity relationship chart were employed in their study and achieved shortened material flow length from 663.63 m to 351m. Mayank et al [3] studied about overall productivity improvement in casting and fastening Industry and incorporated changes is plant layout and achieved 5 to 6 minutes per mould. ArashGhodrati and NorzimaZulkifli [4] and Dinesh et al [5] used 5s technique to improve the productivity by reorganized the work place. Mayank et al [6] concentrated on overall productivity improvement in casting industry by applying work and motion study for reducing non-productive time and applying ergonomics for various working position and ensured effective utilization of plant area after incorporated necessary changes in plant and process layouts. Mallikarjun Koripadu and Venkata

Subbiah [7] used day-in-life-DILO (day in a life of...) along with Lean Principles for improving productivity in service organization. DILO is a kind of Time and Motion study technique used for identifying the waste by separating value-add, value-enable and non-value-add activities as per the process analysis. Anand Gurumurthy and Rambabu Kodali [8], Niall Piercy and Nick Rich [9], Alberto and Marco [10] and Manoj Dora et al [11], Krishna Kumari et al [12] and Dayanand Yadav [13] and many research articles published in productivity improvement by use of lean and six sigma principles in industries. This work is unique by achieving productivity by redesigned ancillary equipment.

2. Problem definition

Welding plays a major role in boiler manufacturing. Most of the products such as boiler drum, super heater.pre heater and economizer are manufactured with the help of welding. There are several welding processes available to make the permanent joints, Orbital TIG is one of the welding process used for making permanent joints (refer Figure 1). Orbital TIG is an expensive machine in which the defect rate of the weld is only one percent. The plant belongs to govt sector which produces super heater tubes for steam boilers. As it is government sector and its products' quality and services (even after sales) its p in India there is huge demand for their boilers in and around the nation. Hence the productivity improvement is demanded before lunching expansion plan. This work is part of it. The approach started systematically with Work study, which is a basic scientific procedure to identify the possibilities of improving the production rate or in the words of lean identifying the waste. The work study conducted and the time consumed by for each and every process is recorded by 6 trails and average of them is considered for analysis. The work study helped to identify certain unnecessary operations which lead to poor production rate. Those unnecessary times were eliminated by designing alternate ancillary equipment namely Hoist. The conventional hoist is simply supported type; the implemented new one is cantilever beam type rotary hoist which can rotate 360 degree about its axis.

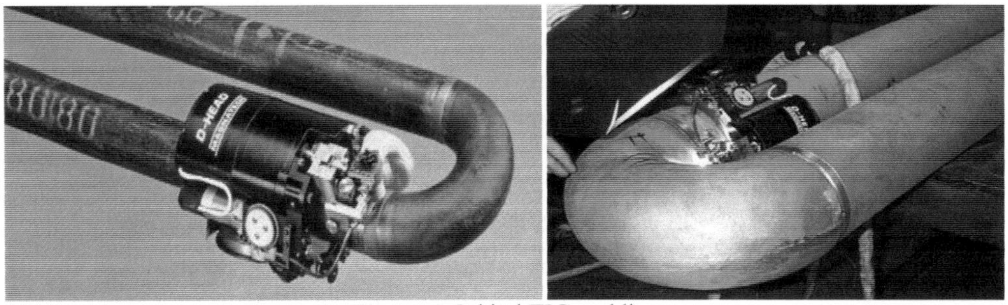

Figure 1. Orbital TIG welding

3. Identifying the waste

The process of identifying the waste was made by means of work-study. Orbital TIG (O-TIG) welding process is focused for this study. The objective is to record all the relevant factors in O-TIG welding process such as loading time, work preparation time, clamping time, welding time, and unloading time are recorded by conducting the stop watch time study which is one of the work measurement techniques.

4. Work measurement

The loading measurement details were tabulated in table 1. The unloading related measurements are given in table 2. Time and method of edge preparation was recorded and tabulated in table 3. Work piece setting time was recorded and noted in table 4. Self explanatory flow chart (refer figure 1) prepared as per standard format which consolidate above measurement. The details of welding process were studied and recorded systematically and shown such measurement in table 5.

Table 1. Conventional Loading Time

Description	Time Taken (Minute) *						
	Trial 1	Trial 2	Trial 3	Trial 4	Trial 5	Trial 6	Average
Move from hanger to extreme end	4.46	4.4	4.32	4.43	4.45	3.89	4.33
Waiting for hoist	18.5	20.5	17.5	20	15.5	19	18.5
Transport empty	1.07	1.09	1.05	1.11	1.08	1.14	1.09
Grasping	0.55	0.58	0.57	0.55	1	0.56	0.57
Transport load	1.53	10.55	1.5	1.56	1.65	1.56	1.56
Release load	0.26	0.27	0.28	0.3	0.26	0.28	0.28
Allowances							1.32
Total loading time							27.65

Table 2. Conventional Unloading Time

Description	Time Taken (Minute) *						
	Trial 1	Trial 2	Trial 3	Trial 4	Trial 5	Trial 6	Average
Transport empty	1.07	1.09	1.05	1.11	1.08	1.14	1.09
Grasping	0.58	0.55	0.57	0.55	0.6	0.56	0.57
Transport load	1.25	1.31	1.12	1.21	1.1	1.2	1.2
Release load	0.26	0.27	0.29	0.3	0.26	0.28	0.28
Total unloading time							3.14

Table 3. Edge Preparation Time

Description	Time Taken (Minute) *						
	Trial 1	Trial 2	Trial 3	Trial 4	Trial 5	Trial 6	Average
Fixing the machine	1.8	1.1	1.4	1.2	1.7	1.3	1.4
Operation	2.15	1.55	2.09	2.13	1.49	1.56	2.25
Removing the machine	1.4	0.8	1.1	1	0.8	0.9	1
Total time taken							4.65

Table 4. Work Piece Setting Time

Description	Time Taken (Minute) *						
	Trial 1	Trial 2	Trial 3	Trial 4	Trial 5	Trial 6	Average
Work piece setting time	7.85	7.95	8	7.9	7.9	7.92	7.92

Table 5. Time for O-TIG Welding Process

Description	Time Taken (Minute) *						
	Trial 1	Trial 2	Trial 3	Trial 4	Trial 5	Trial 6	Average
Machine setting time	6.3	6.4	7	11.7	8.5	7.2	7.8
Arc time	11.8	12.5	15.4	12.4	12.07	13.4	12.93
Machine removal time	4	5	4.2	6.6	4.5	5.5	5
Allowances							1.44
Total welding time							30.17

5. Cantilever beam type rotary hanger

The existing hanger is simply supported beam type, which can handle only single row of pipes in the layout. For increasing productivity the new type hanger systematically designed, fabricated and tested well. This new hanger is cantilever type can rotate around 306o and simple in construction which is illustrated in figure 2. In the existing lay out (ref. figure 3) of pipes are single row and a stand provided to place them but in the newly suggested layout (shown in figure 4), the pipes can arrange two rows (both side of the proposed type hanger) by using its advantage of rotary type (It can works for 360o). So that time taken for (loading) moving the work piece to the work centre is reduced (76.67%) from 15 minutes to 3.5 minutes as well as the time taken for (unloading) moving the work piece for Temporary storage was reduced (80.77%) from 13 minutes to 2.5 minutes.

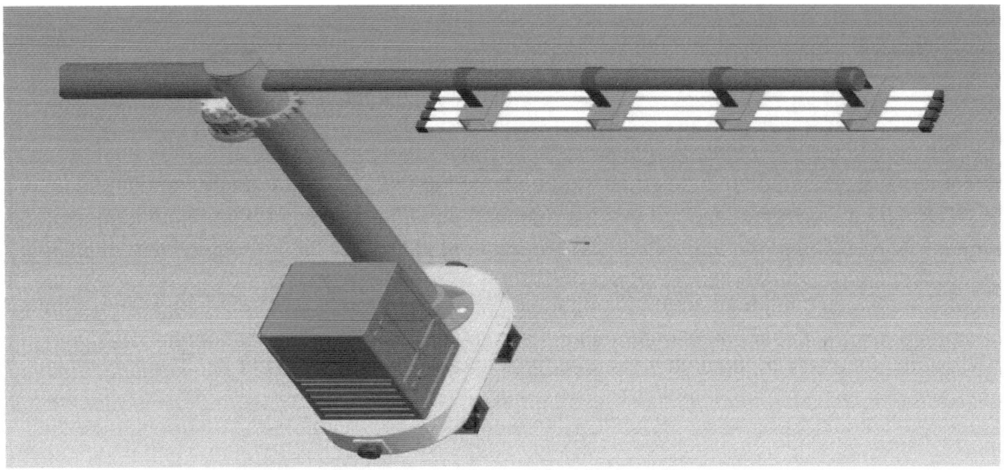

Figure 2. Cantilever Beam Type Rotary Hanger

Figure 3. Coventional layout of pipes to be welded Pipe

Figure 4. New layouts of pipes to be welded

Table 6. New status of Loading

Description	Time Taken (Minute)*						
	Trial 1	Trial 2	Trial 3	Trial 4	Trial 5	Trial 6	Average
Transport empty	1.07	1.09	1.05	1.11	1.08	1.14	1.09
Grasping	5.55	5.89	5.78	5.56	5.78	5.65	5.7
Transport load	0.77	0.78	0.75	0.77	0.83	0.78	0.78
Release load	0.26	0.27	0.28	0.3	0.26	0.28	0.28
Allowances							0.39
Total loading time							8.24

Table 7. New status of Unloading

Description	Time Taken (Minute)*						
	Trial 1	Trial 2	Trial 3	Trial 4	Trial 5	Trial 6	Average
Transport empty	1.07	1.09	1.05	1.11	1.08	1.14	1.09
Grasping	0.58	0.55	0.57	0.55	0.6	0.56	0.57
Transport load	0.45	0.5	0.43	0.41	0.42	0.5	0.45
Release load	0.29	0.27	0.28	0.31	0.26	0.28	0.28
Total Unloading time							2.51

Hence the new equipment reduced the time for loading from 27.65 minutes to 8.24 minutes per joint 70.199% (shown in table 6) similarly in case of unloading the time the new equipment and method reduced 20.064% of unloading time per joint (shown in table 7). A boiler tube requires 4 joints. Performance of new hoist during loading is reduced to 8.24 minutes (ref. Figure 5) and also the performance of new hoist during unloading is reduced to 2.51 minutes (ref Figure 6)

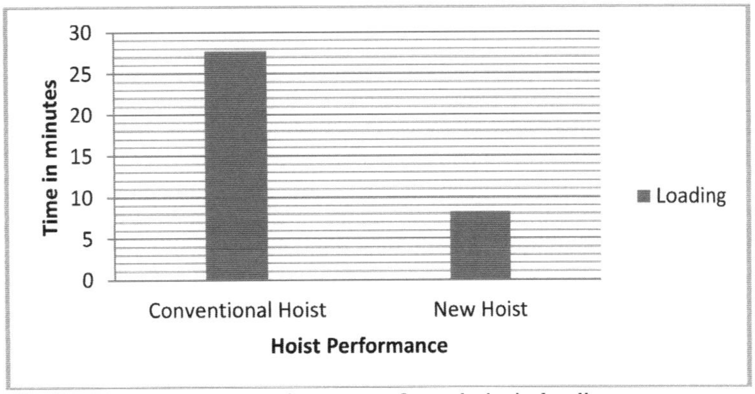

Figure 5. Performance of new hoist in loading

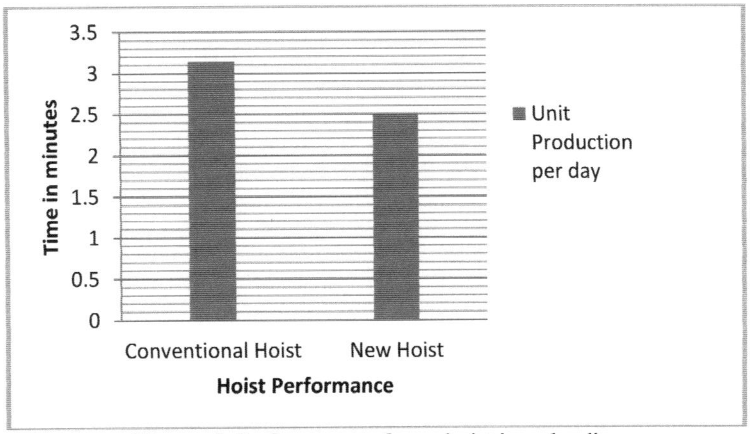

Figure 6. Performance of new hoist in unloading

6. Productivity improvement

6.1 Conventional Status

Working hour per shift is 480 minutes (8 hour shift), Time spent for organization matter such as Enter daily work consumed 7minutes, Official discussion 15minutes, Time spent for personal needs like Tea time (twice) 20minutes, Lunch hour 30minutes, Personal allowances 24minutes (5% of shift time), Fatigue allowances 24minutes (5% of shift time), hence total time consumed for non productive work is 120minutes. (shown in Table 8)

So Net Available working time per shift (T) = 480-120 = 360 minutes

Total processing time per weld joint (tp)= 27.64+3.14+4.65+7.92+30.17 = 73.51minutes

Total no. of Boiler Tubes can be welded = (T / tp) * no. of joints per boiler tube piece

= (360 / 73.53) * 4 = 19.59 $\cong$ 19boiler tube piece per shift

6.2 New status

In case of new equipment the loading and unloading time was reduced significantly(shown in Figure 7). Hence the Total processing time per weld joint (tp) = 8.24+2.51+4.65+7.92+30.17 = 53.49minutes

Total no. of Boiler Tubes can be welded = (T / tp) * no. of joints per boiler tube piece

= (360 / 53.49) * 4 = 26.92 $\cong$ 26 boiler tube piece per shift

Table 8. Improvement Status

S. No.	Description of Productivity Criteria	Conventional Status	New status	Difference	Productivity Improvement	Remarks
1	Productivity in loading & unloading Tasks (minutes)	30.79	10.75	20.04	65.09%	Loading & Unloading Time Reduced
2	Productivity per joint (minutes)	73.51	53.49	20.04	27.25%	Total time per joint reduced
3	Productivity per Boiler tube (minutes)	294.12	213.96	80.16	27.25%	Total time per boiler tube manufacturing reduced
4	Net Productivity per shift (Nos)	19	26	7	36.84%	No of boiler tube manufactured increased
5	Productivity per day (2 shift) (Nos)	38	52	14	85.71%	Production per day increased

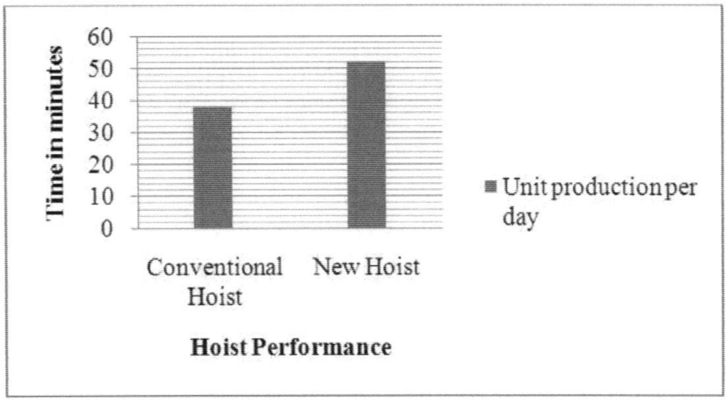

Figure 7. Improvement in day production

7. Conclusion

The O-TIG welding on pipes to produce super heater tubes for the boiler is discussed in this article. Based on the lean principle the conventional method was studies with work study principles and identified the waste. For eliminating the identified waste new hanger was suggested and implemented. The new method decreased the loading time 70% and unloading time by 20% thereby increases (85.71%) of number of super heater tubes production for a day (two shift per day) from 38 to 52 tubes per day. Hence the significant productivity was achieved in O-TIG welding plant.

8. References

[1] Vivekanand S Gogi, Rohith D, ShashiKiran K and Suhail M Shaikh, 2014 Efficiency improvement of a plant layout *International Journal of Innovative Research in Science Engineering and Technology* **3(4)** 11203-11209.

[2] Pramod P Shewale, Manmath S Shete and Prof. DR. S. M. Sane, 2012 Improvement in plant layout using systematic layout planning (SLP) for increased productivity *International Journal of Advanced Engineering Research and Studies (IJAERS)* **1(3)** 259-261.

[3] MayankDev Singh, RavalApurv J, Patel Dhaval G and ParthVarmora J, 2013 Overall productivity improvement in casting and fastening industry *International Journal of Mechanical and Production Engineering (IJMPE)* **1(1)** 43-48.

[4] ArashGhodrati and Norzima Zulkifli, 2013The impact of 5s implementation on industrial organizations performance *International Journal of Business and Management Invention* **2(3)** 43-49.

[5] Dinesh B Shinde, and Prashant N Shinde, 2014 Improvement of plant layout by using 5s technique-an industrial case study *International Journal of Modern Engineering Research (IJMER)* **4(2)** 141-146.

[6] MayankDev Singh, Swati Singh, DerasariKeyur, SoniSaumil, Patel Niki and PanchalHarshal, 2015 Overall Productivity Improvement in Casting Industry by Using Various Industrial Engineering Techniques *International Journal of Innovative Research in Science Engineering and Technology* **1(4)** 18713-18721.

[7] Mallikarjun Koripadu K and Venkata Subbiah, 2014 Productivity Improvement by Applying DILO (Time and Motion) and Lean Principles *Int. Journal of Engineering Research and Applications* **10(4)** 12-14.

[8] Anand Gurumurthy and RambabuKodali, 2009 Application of benchmarking for assessing the lean manufacturing implementation *Benchmarking:An International Journal* **2(16)** 274 – 308.

[9] Niall Piercy and Nick Rich, 2009 Lean transformation in the pure service environment: the

case of the call service centre *International Journal of Operations & Production Management* **1(29)** 54 – 76.

[10] Alberto Portioli-Staudacher and Marco Tantardini, 2012 Lean implementation in non-repetitive companies: a survey and analysis *Management Journal* **11** 1744-2370.

[11] Manoj Dora , Dirk Van Goubergen ,Maneesh Kumar , AdriennMolnar and Xavier Gellynck , 2014 Application of lean practices in small and medium-sized food enterprisesBritish Food Journall **116** 125 – 141.

[12] S. Krishna Kumari and A.N.Balaji, R.Sundar, 2014 Productivity Improvement of an Industry by Implementing Lean Manufacturing Principles *International Journal of Innovative Research in Science, Engineering and Technology* **3(3)** 1442-1446.

[13] DayanandYadav, 2014 Study of Productivity Improvement Using Lean Six Sigma Methodology *International Review of Applied Engineering Research* **1(4)** 33-38.

IOP Conf. Series: Materials Science and Engineering **183** (2017) 012021 doi:10.1088/1757-899X/183/1/012021

Computational Fluid Dynamics Analysis of Nozzle in AbrasiveWater Jet Machining

S Venugopal [1]M Chandresekaran[2] V Muthuraman[3] and S Sathish[4]
[1]Assistant Professor, Department of Automobile Engineering, Vels University, India.
[2]Professor and Director, Department of Mechanical Engineering, Vels University, India.
[3]Associate Professor and Head, Department of Mechanical Engineering, Vels University, India.
[4]Assistant Professor and Head, Department of Automobile Engineering, Vels University.

Email id:venugopalmechlec@yahoo.co.in

Abstract: Abrasive water jet cutting is one of the most recently developed non-traditional manufacturing technologies. The general nature of flow through the machining, results in rapid wear of the nozzle which decrease the cutting performance. It is well known that the inlet pressure of the abrasive water suspension has main effect on the erosion characteristics of the inner surface of the nozzle. The objective of the project is to analyze the effect of inlet pressure on wall shear and exit kinetic energy. The analysis would be carried out by varying the inlet pressure of the nozzle, so as to obtain optimized process parameters for minimum nozzle wear. The two phase flow analysis would be carried by using computational fluid dynamics tool CFX. The availability of minimized process parameters such as of abrasive water jet machining (AWJM) is limited to water and experimental test can be cost prohibitive.

1.Introduction

AWJM is a well-establishednontraditional machining process. It is a machining process where material is removed in 1980 and 1983 the first commercial system with abrasive involvement in the jet becomes available.The added abrasives increased the range of materials, which can be cut with a Watergate drastically. It is widely used, because of its main advantages Materials which are cut by AWJM are Polymers, Honey comb, Metal matrix Composites.Ceramic Matrix Composites, Concrete, stone, Granites.

2.Theoretical Formulation

2.1 Numerical Model and Assumptions

The abrasive water suspension mixture [1]is let into the nozzle at inlet and carried through converging cone to focus tube and exits at nozzle, in which the focus tube is used for guiding the flow. The numerical model adopted closely follows the work at Hu et al [2] which liquid-solid two-phase flow is considered. Water is a continuous medium and incompressible.Flow is considered as two phase flow mixture in which water is liquid phase and abrasives is solid phase, but mixed with the liquid phase. Two phase flow is steady and possesses turbulent flow characteristics[3]. The nozzle head of AWJM [4] is modeled by using pro/E software and it is saved. We modeled the nozzle head of varying inlet pressure of nozzle, then it is imported in ICEM meshing software for meshing the model. Tetrahedron is used for fine meshing. Then it is imported in CFX Pre for giving input parameters and properties of water and abrasive by impact erosion of high pressure high velocity of water and entrained high velocity of grit abrasives on a work piece. The invention of the abrasive water jet by impact erosion of high presssure high velocity of water and entrained high velocity of grit abrasives on a workpiece

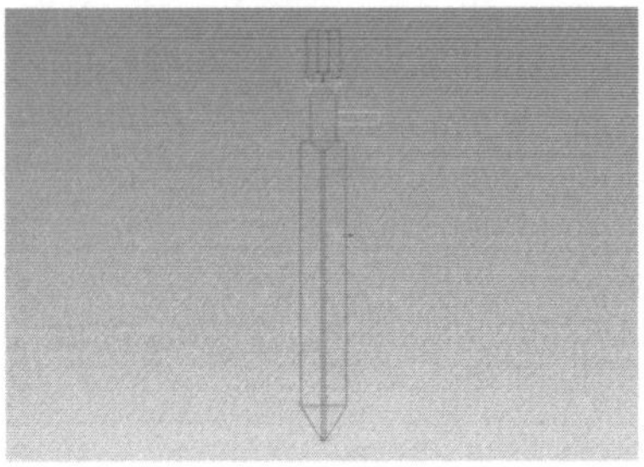

Figure 1 Pro/E model of nozzle head

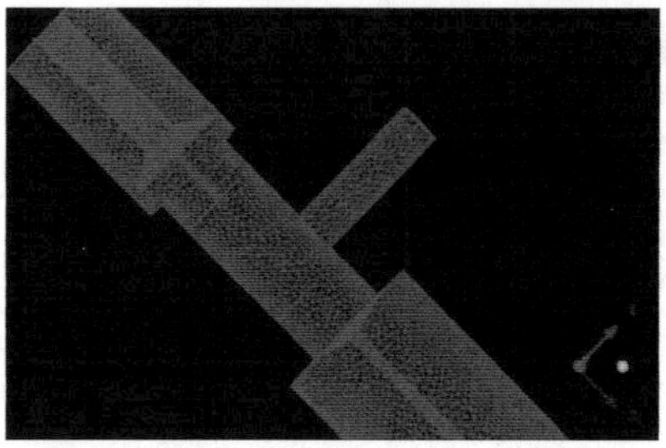

Figure 2 Meshed Model of Nozzle head

2.2 Material Domain Input

For the meshed model computational domain input should be given in the first . You have to create a water inlet, abrasive inlet and mixture outlet to the meshed model. In CFX preprocessor, you have to give the thermodynamic properties of water and abrasive. The water is taken as continuous liquid, and dispersed solid abrasive size is 1mm. In boundary condition you have to give domain type, name of nozzle head and mixing chamber. You have to name the liquid domain and solid domain of nozzle head and mixing chamber.Then you have to tell at water inlet, water enters and at abrasive inlet abrasive enters and also give the boundary details. Then you have to give domain interface between solid and liquid domain. The outlet will be the velocity. In CFX Solver, the model is imported and no of equations is performed and you have to give no of iterations and run solver and monitor. In CFX post, the models is imported and post process the results.

3 Results and Discussions

3.1 Validation of the numerical model (Effect of nozzle inlet pressure)

Velocity Variation

In the case of mixing tube the velocity reduces gradually and it reaches maximum in the mixing region and then gradually decreases when it reaches at the end of tube. The magnitude of velocity is high at higher water inlet pressure. In the case of focus tube length, the gain in velocity is observed when the flow past the nozzle. The velocity gain is more for 300 MPa pressure. Further kinetic energy lost is observed when the flow is along the focus tube for all the cases. This may be due to some of the abrasive particles do collide with the focusing tube wall.

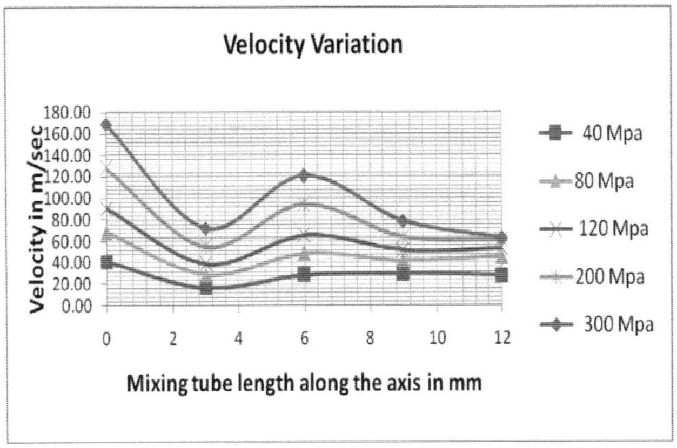

Figure 3 shows the velocity variation along the mixing chamber.

3.2 Wall shear stress distribution

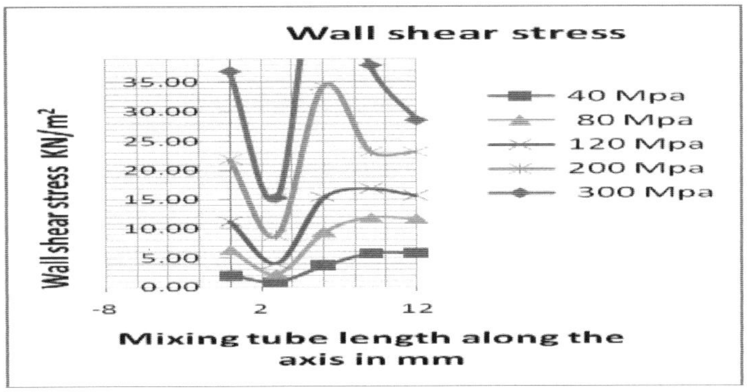

Figure 4 shows the wall shear stress distribution along the mixing chamber

The above diagram shows that increased wall shear for at higher water inlet pressure. Wall shear increases maximum at the mixing region and it decreases along its flow. It has been observed from the graph that when the water pressure increases the magnitude of wall shear increases sharply in the mixing region. The wall shear is relatively low in the case of 40 MPa water pressure.

3.3 Shear Strain Rate Variation
The above diagram shows that strain rate is high at entry of mixing tube for both cases and it decreases gradually along its flow and it increases in the mixing region.

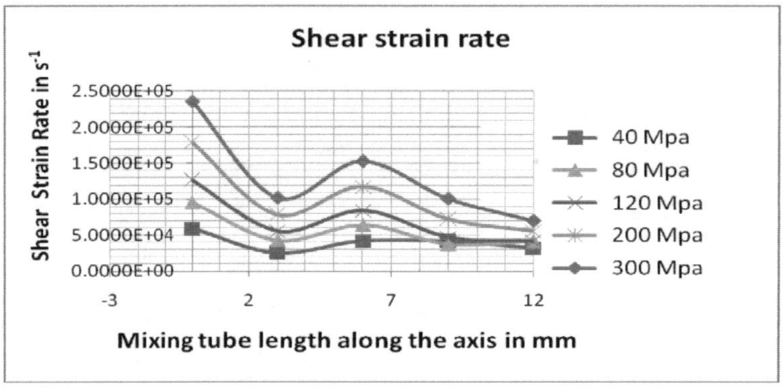

Figure 5 shows the shear strain rate distribution along the mixing chamber

4. Conclusion

The effect of pressure of the abrasive fluid suspension has significant effect on the erosion characteristics of the inner surface in the nozzle. The general nature of flow through the abrasive fluid jet machining, results in rapid wear of the nozzle which affects the cutting performance. High pressure of water and abrasive particles may cause severe wall shear stress due to wear. Various flow parameters are considered wall shear stress, velocity, shear strain rate. While taking parameters, the outlet velocity should be high and wall shear stressdue to wall shear, and shear strain rate should be low. These phases of flow through nozzle head of abrasive fluid jet Machining has been carried out and analysed by using Computational fluid dynamic [CFD]software CFX. The water inlet pressure increases, its velocity increases, wall shear stress also increases and all parameters increases. Without affecting wall shear stress, we take inlet pressure. So, it is optimum to take velocity and with out affecting wall shear stress, energy dissipation of flow, rate of energy dissipation of flow, shear strain rate.Here we obtain the optimal inlet pressure as 120 Mpa and 200 Mpaof nozzle head of abrasive fluid jet machining.

5. Reference

[1] Azmir M A, Ahsan A K, Rahmah A, Noor M M and Aziz AA 2007-"Optimization of abrasive water jet machining process parameters using orthogonal array"with grey relationalanalysis" *Regional Conference on Engineering Mathematics,* Mechanics, Manufacturing & Architecture (EM3ARC)

[2] Deepak D 2011 ,"Numerical Analysis of Flow through Abrasive Water Suspension Jet: The effect of Inlet Pressure on Wall Shear and Jet Exit Kinetic Energy",*Proceedings of World Congress of Engg III WCE*

[3] Deepak,D.Anjaiah.D, VasudevaKaranth.K and YagneshSharma.N – 2012 "Effect of inlet operating pressure and volume fraction on skin friction and exit kinetic Energy" *(ManipalUniversity,Manipal,India)*

[4] DewanHasan Ahmed, Rowan Deam (IRIS), Jamal Naser and Frank chen (IRIS) - 2001 Manipal 576104, India." Numerical simulation of abrasive wter jet for different inlet taper angles "

[7] DewanHasan Ahmed, Rowan Deam (IRIS), Jamal Naser and Frank chen (IRIS) 2003 Optimising the focus tube length for abrasive water jet using Numerical simulations

[5] Hu G, zhu W,Tao Y and Yuan.J 2008 "Numerical simulation and experimental study of liquid-solid two-phase flow in nozzle of DIA jet", in *proceedings of the 6th IEEE international conference on Industrial informatics (INDIN ` 08)*

[6] Jigneshpatel and Ankitpatel , "Abrasive water jet machining and water jet machining" Module 9, Lesson 37, Non conventional machining, version 2 ME, IIT kharagpur John David Anderson "*Basic concepts of CFD and its applications*"

[7] Liu H, "Computational Fluid Dynamic (CFD) Simulation of Ultrahigh Velocity Abrasive Water jet" 2003 ,*Key Engineering Materials*, **233-236**

[8] MadhusarathiNaduri, David Taggart G and ThomasKim J, 1997 "Effect of the inlet taper angle on AWJ nozzle wear" – 9th *American waterjet conference*

Optimization of CO_2 laser cutting parameters on Austenitic type Stainless steel sheet

A Parthiban [1], S Sathish [2], M Chandrasekaran[3] and R Ravikumar[4]
[1]Assistant professor, Department of Mechanical Engineering, School of Engineering, Vels university, Chennai-117.
[2]Assistant professor, Department of Automobile Engineering, School of Engineering, Vels university, Chennai-117.
[3]Professor, Department of Mechanical Engineering, School of Engineering, Vels university, Chennai-117.
[4]Professor, Department of Mechanical Engineering, Ariyalur Engineering College, Melakaruppur-621707.

E-mail: [1]parthibana@yahoo.co.in

Abstract. Thin AISI 316L stainless steel sheet widely used in sheet metal processing industries for specific applications. CO_2 laser cutting is one of the most popular sheet metal cutting processes for cutting of sheets in different profile. In present work various cutting parameters such as laser power (2000 watts-4000 watts), cutting speed (3500mm/min – 5500 mm/min) and assist gas pressure (0.7 Mpa- 0.9Mpa) for cutting of AISI 316L 2mm thickness stainless sheet. This experimentation was conducted based on Box- Behenken design. The aim of this work is to develop a mathematical model kerf width for straight and curved profile through response surface methodology. The developed mathematical models for straight and curved profile have been compared. The Quadratic models have the best agreement with experimental data, and also the shape of the profile a substantial role in achieving to minimize the kerf width. Finally the numerical optimization technique has been used to find out best optimum laser cutting parameter for both straight and curved profile cut.

1. Introduction

CO_2 Laser cutting is one of the advanced processes to achieve high precision and accurate assembly with minimum consumption of sheet metal. Austenitic stainless steel is playing vital role in modern industries. Now a day's AISI 316L stainless steel sheet has significant role in automobile industries. However AISI 316L is one of the important materials due to its inherent properties such as high strength, good corrosion resistance. Still machining of AISI 316L with laser cutting is a difficult task due to different profile cutting. Consequently kerf dimensions play the most significant role in determining the productivity and the quality of a product produced with AISI 316L. So this work takes AISI 316L stainless steel as a work piece material for CO_2 laser cutting. With regard, to the laser power, cutting speed and gas pressure are considered as a predominant principal parameter in laser cutting. Lot of researcher and investigations has been done in analyzing the cost and quality of laser cutting.

.

However with regard to the reduction of wastage that is caused in kerf dimension has not been taken up for serious study. Hence this study considers the principal parameters on kerf width for straight and curved profile.

The Nd: YAG laser is a solid state laser, working at a wavelength of 1.06 μm. CO_2 laser is a gas type laser that radiates at wavelength of 10.6 μm [1]. Nd: YAG lasers have low beam power, but when operating in pulsed mode, and to obtain maximum peak power is able to cut still thicker sheet metals [2]. CO_2 laser is proper for well cutting of sheet metals at high speed because it delivers the maximum average beam power, superior efficiency and good quality of beam. Conversely, Due to maximum peak power at high speed of

CO_2 laser, it is reflected to a lesser extent by metallic surfaces and this high absorptive of the CO_2 laser enable to process even highly reflective materials with relatively high cutting speed [3]. Therefore, CO_2 laser is suitable for cutting of metals in general and reflective materials in particular.
Laser cutting of sheet metals has a most important research area for marking out the best quality of cut. [4]. the quality of cuts mainly depends upon the variety of process parameters such as laser power, gas pressure, cutting speed and sheet metal thickness. [5,6] have been investigating about the outcome of laser cutting parameters on kerf dimension and surface quality of cut. Most of the researchers to varying the only one parameter at a time approach to study the effect of process parameters on responses. However, this approach consumes more time and money in favor of a large number of experimental runs since only one parameter is varied in each experimental run, and then keeps other parameters as constant. As well as in this method the interaction effects process parameters are not considered. To beat such that problems, Design of experiments (DOE) commonly used in experimental approach [7,8]. From the litrature survey no more researcher not consentrate about laser cutting of different profiles. So that the present work consentrate about optimization of laser cutting of different profiles like straight and curved profile.

2. Methodology

The CO_2 Laser cutting process considered in this work AISI 316L Stainless steel sheet for straight and curved profile is discussed following section.

2.1. Experimental procedure

The experiments were carried out on AMADA make CO_2 Laser cutting machine as shown in figure.1 and the specification of machine as shown in table 1. The work piece considered for this work is stainless steel AISI 316L the sheet size was 500 X 500 X 2 mm each specimen to cut straight profile 20mm and curved profile radius of 20mm. Cutting operation carried out on work piece with straight and curved profile as shown figure.2 the work piece was carefully clamped on work table[9]. The input parameters considered are power, cutting speed and gas pressure. Box-Behnken Design was selected for the three levels and 17 run experiments were carried out both straight and curved profile cut. The considered parameters ranges are the laser power in between (2000watts, 3000watts and 4000watts), Cutting Speed levels are within (3500mm/min, 4500 mm/min and 5500mm/min) and Gas pressure was (0.7Mpa, 0.8 Mpa and 0.9Mpa) [10]. The 17 experimental runs were conducted based on Box-behnken approach. The collected experimental data were given in table 2.

2.2. Measurement of Responses

To measure the kerf width for straight and curved profile by indirect measurement using Tool makers microscope with 10X magnification factor, the work piece hold on the work table.

3. Result and Discussion

3.1. Response surface methodology

Response surface methodology is a collection of mathematical and statistical techniques, useful for the mathematical modelling and analysis of problems in which a response of interest is influenced by several parameters and the objective is to optimize this response [11-12]. RSM models were developed and evaluated with Design expert software. Second order polynomial equations were obtained for predict the data [13]. The generalized form of response surface model is shown in equation 1.
Table 3 shows the model summary of kerf width for stright profile. The table shows that the quadratic model is best for predict the kerf width value. The significant parameters are identified by using ANOVA table.The Model F-value of 14.17 implies the model is significant. There is only a 0.10% chance that a Model F-Value is large could occur due to noise.Values of "Prob > F" less than 0.0500 indicate model terms are significant.In this case power, cutting speed, power^2 and Gas pressure^2 are significant model terms.and the model summary of kerf width for curved profile. The table shows that the modified model is best for predict the kerf width value. The significant parameters are identified by using ANOVA table 3.The Model F-value of 11.32 implies the model is significant.There is only a 0.21% chance that a Model F-Value is large could occur due to noise.Values of "Prob > F" less than 0.05 indicate model terms are significant.In this case power, cutting speed, power^2 and Gas pressure^2 are significant model terms.

The figure 3(a) shows the effects of power and cutting speed on kerf width for straight profile. The Power and cutting speed are carried out on X and Y axis. The response for kerf width straight profile carried on Z axis. This graph is mainly plotted for make inference on kerf width responses against cutting speed and gas pressure. It is increasing Laser power increase the kerf width. The figure 3(b) is mainly plotted for kerf width for straight profile responses against power and gas pressure. The graph contains some interaction effects, because the graph does not lie in linear plane. But it is a curved plane due to the interaction effects of gas pressure from graph laser power increase kerf width for straight profile also increase and the figure 3(c) shows the effects of power and cutting speed on kerf width for curved profile. The Power and cutting speed are carried out on X and Y axis. The response of kerf width for curved profile considered on Z axis. These graphs are also same like and power response surface graphs it also contains some interaction effects, because the graph is not a linear one. Then it is an interaction plots. The power and cutting speed are not a linear one kerf width for curved profile. From graph power and cutting speed increases kerf width for curved profile also increase. The figure 3(d) is mainly plotted kerf width for curved profile responses against power and gas pressure. The graph contains some interaction effects, because the graph does not lie in linear plane. But it is a curved plane due to the interaction effects of power a from graph laser power and gas pressure increase kerf width also increases that the developed mathematical model should be a second order polynomial equation. The developed mathematical models are as shown in Equation 2 and 3.

3.2 Performance Evaluation of Developed Models

The Figure 4 shows that the kerf width for straight and curved profile response surface model values and Experimental values were compared on the basis of their prediction. The models were validated with 17 data sets of Box-behenken design used for the experimental data collection [14]. The plots represent the experimental value nearer to the Response Surface models.

$$Ra = b_0 + \sum_{i=1}^{4} b_i X_i + \sum_{i=1}^{4} b_{ii} X_{ii}^2 + \sum_{i=1}^{4}\sum_{j=1}^{4} b_{ij} X_i X_j + \varepsilon \quad (1)$$

$$\begin{aligned} \textit{Kerf width for straight Profile} &= +0.2087 - 9.34e^{-5} \times P + 1.66e^{-5} \times v + 0.606 \times p \\ &+ 1.325e^{-9} \times P \times v + 7.5e^{-7} \times P \times p + 2.5e^{-6} \times v \times p + 1.528e^{-8} \times P^2 \\ &- 2.99e^{-9} \times v^2 - 0.3945 \times p^2 \end{aligned} \quad (2)$$

$$\begin{aligned} \textit{Kerf width for curved Profile} &= +0.1362 - 8.22e^{-5} \times P + 2.02e^{-5} \times v + 0.6787 \times p \\ &+ 1.075e^{-9} \times P \times v - 7e^{-6} \times P \times p + 3e^{-6} \times v \times p + 1.4662e^{-8} \times P^2 \\ &- 3.287e^{-9} \times v^2 - 0.4262 \times p^2 \end{aligned} \quad (3)$$

Figure 1. AMADA CO_2 laser cutting Machine

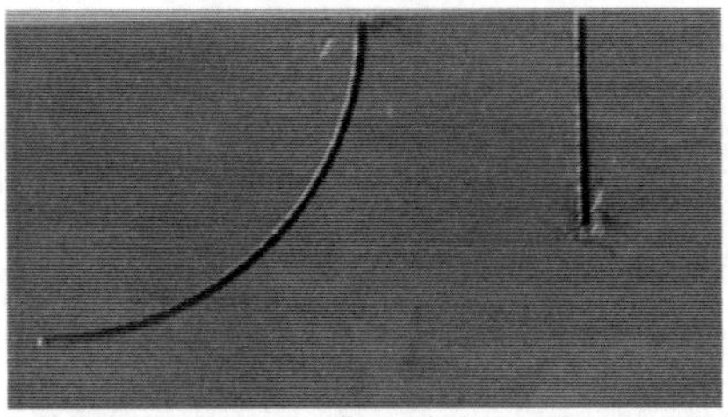

Figure 2. Photo image of straight and curved profile

Table 1. Specification of machine

Model	Amada FO-3015NT
Laser type	carbon di oxide (CO_2)
Mode	Continious Wave(CW)
Power	4000 Watts
Focal length	127.5mm
Assist gas	Nitrogen
Focal position	-0.5

Table 2. Experimental data

S.I No	Power (Watts)	Cutting Speed (mm/min)	Gas pressure (Mpa)	Straight Profile Kerf Width (mm)	Curved Profile Kerf Width (mm)
1	3000	5500	0.7	0.3282	0.3117
2	3000	3500	0.9	0.3393	0.3218
3	3000	4500	0.8	0.3448	0.3291
4	4000	4500	0.9	0.3552	0.3374
5	2000	4500	0.7	0.3495	0.3312
6	3000	4500	0.8	0.341	0.3255
7	3000	4500	0.8	0.339	0.3225
8	4000	3500	0.8	0.3594	0.3419
9	2000	5500	0.8	0.3444	0.3279
10	2000	4500	0.9	0.3472	0.3307
11	3000	5500	0.9	0.3262	0.3099
12	3000	4500	0.8	0.3387	0.3222
13	3000	3500	0.7	0.3423	0.3248
14	4000	5500	0.8	0.3582	0.3417
15	2000	3500	0.8	0.3509	0.3324
16	3000	4500	0.8	0.3412	0.3237
17	4000	4500	0.7	0.3572	0.3407

Table 3. ANOVA table for kerf width for straight profile

Source	Sum of Squares	DF	Mean Square	F Value	p-value Prob>F
Model	0.0013993	9	0.0001555	14.172779	0.0010
A	0.0001805	1	0.0001805	16.454072	0.0048
B	0.0001523	1	0.0001523	13.878965	0.0074
C	1.081E-05	1	1.081E-05	0.9855351	0.3539
AB	7.023E-06	1	7.023E-06	0.6401591	0.4499
AC	2.25E-08	1	2.25E-08	0.0020511	0.9651
BC	2.5E-07	1	2.5E-07	0.0227896	0.8843
A^2	0.0009831	1	0.0009831	89.614708	< 0.0001
B^2	3.777E-05	1	3.777E-05	3.4429145	0.1059
C^2	6.553E-05	1	6.553E-05	5.9734688	0.0445
Residual	7.679E-05	7	1.097E-05		
Lack of Fit	5.304E-05	3	1.768E-05	2.9772931	0.1597
Pure Error	2.375E-05	4	5.938E-06		
Cor Total	0.0014761	16			

Table 4. ANOVA table for kerf width for curved profile

Source	Sum of Squares	DF	Mean Square	F Value	p-value Prob>F
Model	0.0013092	9	0.0001455	11.317291	0.0021
A	0.000195	1	0.000195	15.17289	0.0059
B	0.0001103	1	0.0001103	8.5780195	0.0221
C	9.245E-06	1	9.245E-06	0.7192354	0.4244
AB	4.622E-06	1	4.622E-06	0.3596177	0.5676
AC	1.96E-06	1	1.96E-06	0.1524826	0.7078
BC	3.6E-07	1	3.6E-07	0.028007	0.8718
A^2	0.0009052	1	0.0009052	70.42333	< 0.0001
B^2	4.551E-05	1	4.551E-05	3.5402345	0.1019
C^2	7.65E-05	1	7.65E-05	5.9515391	0.0448
Residual	8.998E-05	7	1.285E-05		
Lack of Fit	5.794E-05	3	1.931E-05	2.4110487	0.2073
Pure Error	3.204E-05	4	8.01E-06		
Cor Total	0.0013992	16			

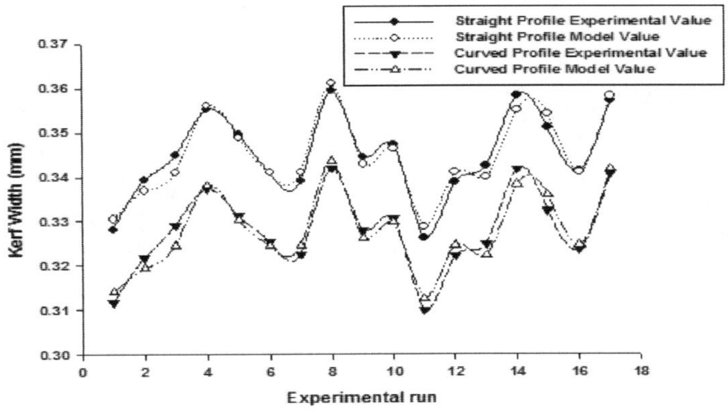

Figure 3. kerf width for straight and curved profile (Experimental data Vs Model data)

Table 5. Criteria for Numerical optimization

Name	Goal	Lower Limit	Upper Limit	Importance
Laser power (watts)	is in range	2000	4000	3
Cutting speed (mm/min)	is in range	3500	5500	3
Gas pressure (Mpa)	is in range	0.7	0.9	3
Kerf Width in straight profile(mm)	minimize	0.3262	0.3594	5
Kerf Width in curved profile (mm)	minimize	0.3099	0.3419	5

Table 6. Optimization result for cutting conditions

Si. No	Laser Power (watts)	Cutting speed (mm/min)	Gas pressure (Mpa)	Kef width in straight Profile (mm)	Kef width in curved Profile (mm)
1	2595.59	5500	0.9	0.3287	0.3128
2	2587.66	5500	0.9	0.3287	0.3128
3	2603.12	5499.89	0.9	0.3286	0.3127
4	2577.47	5500	0.9	0.3288	0.3129
5	2684.49	5500	0.9	0.3283	0.3123
6	2503.08	5500	0.9	0.3294	0.3135
7	2589.96	5500	0.9	0.3290	0.3131
8	2598.72	5500	0.9	0.3291	0.3132
9	2590.4	5500	0.89	0.3292	0.3133
10	2586.65	5500	0.7	0.3306	0.3138

IOP Conf. Series: Materials Science and Engineering **183** (2017) 012022 doi:10.1088/1757-899X/183/1/012022

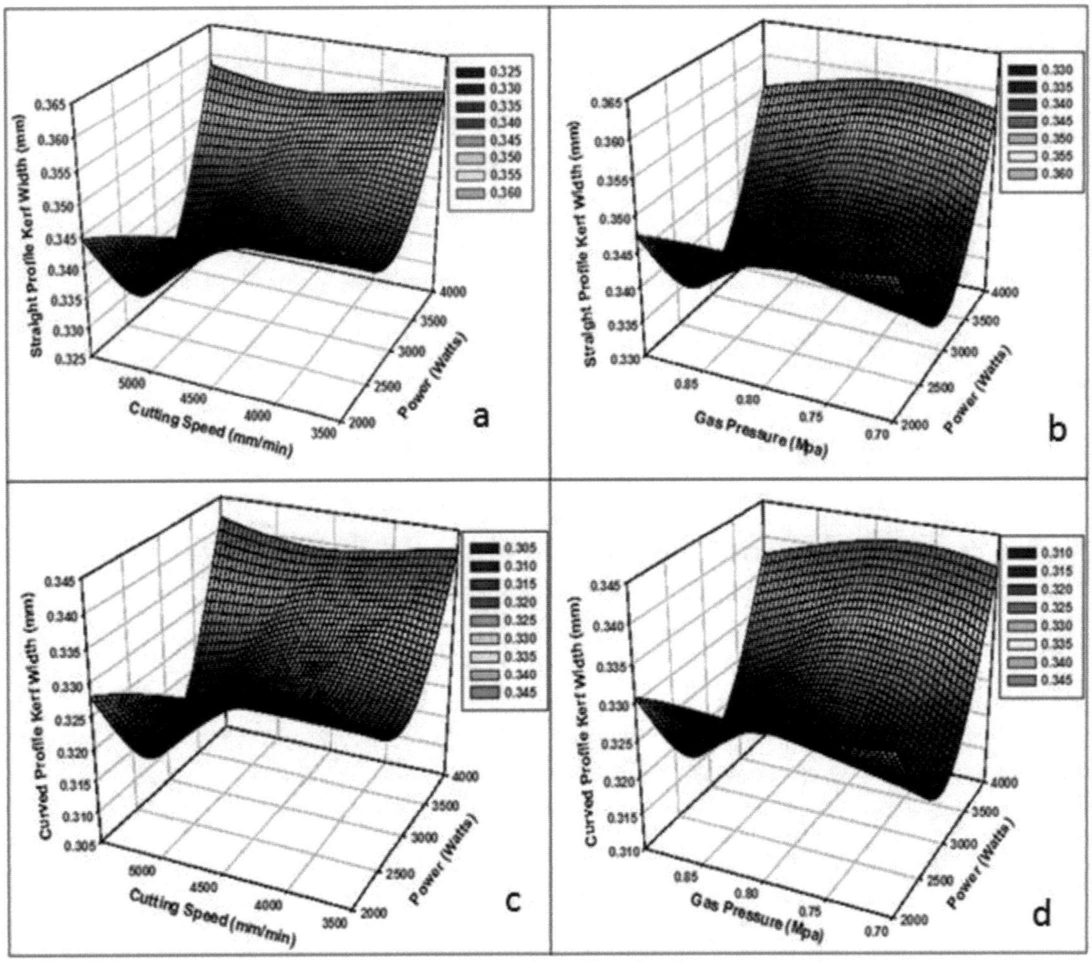

Figure 4. Surface model of interaction effect for CO_2 Laser cutting parameter

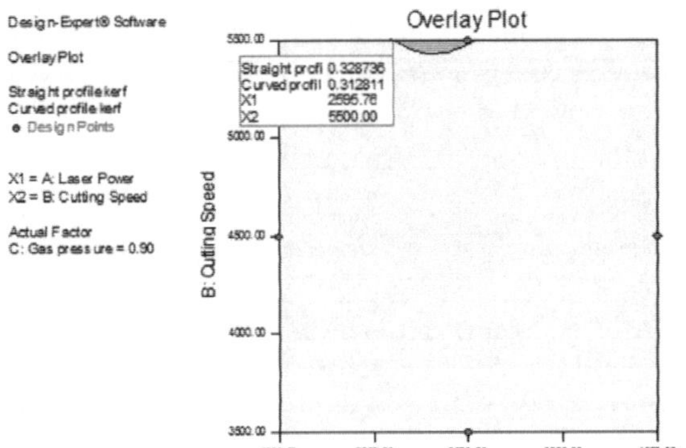

Figure 5. Over lay plot shows the region of optimal cutting condition

4. Optimization

Optimization can be defined as the process of finding the situations that give the maximum or minimum value of a function. Optimization, in its broadest sense, can be applied to solve any engineering problem [15].

In this work optimization of straight profile and curved profile cut. The kerf width is a quality, the cutting parameter as Laser power, Cutting speed and gas pressure are required different cutting geometry. But this work is tried to achieve same cutting parameter at cut all type of geometry. The numerical optimization was used; the multi response optimization is solved desirability approach by using design expert software. The numerical optimization is set the goal as shown in Table 5 to minimize the response to achieve high quality. Table 6 shows the optimized value for AISI 316 L 2mm thickness sheet of input parameters to the corresponding response value. the finally to predicted the response value kerf width for straight profile 0.3287 mm and kerf width for curved profile 0.3128 mm has been achieved. And the figure 9 shows overlay plot drawn between the laser power and cutting speed the Assist gas pressure at 0.9 Mpa. These ranges of graphical optimization result shows on the figure 9.the shaded area on the overlay plot regions are to be achieved for proposed criteria and select the optimal laser cutting parameters.

5. Conclusion

The stainless steel 316 L as work piece material for straight and curved profile and CO_2 laser processes were carried out in this work. Based on experimental work the following conclusions were made

- The developed second-order response surface models for Kerf width for straight profile and Kerf width for curved profile using the Box-behenken design have been found adequate.
- The developed empirical models can be used to predict the response values within the range of experimental values. From the response plot it has been observed laser power, gas pressure has been less effect of kerf dimensions as compared to cutting speed.
- For achieving smaller value of kerf width a moderate value of cutting speed is required for straight profile cut. Similar to curved profile cut.
- These models may be used satisfactorily for the prediction of Kerf width for straight and curved profile cut rate in laser cutting of thin steel sheet.
- This second-order response surface models model can reduce the experimental cost and time in manufacturing industries. This proposed methodology can be used to predict the other response values in CO_2laser cutting operation and other non tradition machining operations
- The minimum of kerf width for straight profile and kerf width curved profile is obtained from the study were 0.3281 mm and 0.3120 mm, when the process parameters such as Laser power, cutting speed and gas pressure were maintained at 2807.45 watts, 5500 mm/min, and 0.9 Mpa. It is possible to obtain the minimum kerf width using the above values of process parameters to achieve both straight and curved cut profile geometry.

Acknowledgments

The authors wish to thank southern engineering Coimbatore for the technical support of this work.

6. References

[1] Johan Meijer 2004 Laser beam machining (LBM) state of the art and new opportunities, *Journal of Materials Processing Technology* **149** 2–17
[2] Avanish Kumar Dubey and Vinod Yadava 2008 Laser beam machining—A review *International Journal of Machine Tools & Manufacture* **48** 609–628

[3] Shang-Liang Chen 1999 The effects of high-pressure assistant-gas flow on high - power CO_2 laser cutting Journal of Materials Processing Technology **88** 57–66

[4] Yilbas B S 2008 Laser cutting of thick sheet metals Effects of cutting parameters on kerf size variations *Journal of materials processing technology* **201** 285–290

[5] Paulo Davim J Carlos Oliveira Nuno Barricas and Marta Conceicao 2008 Evaluation of cutting quality of PMMA using CO_2 lasers *International Journal of Advanced Manufacturing Technology* **35** 875–879

[6] Yilbas B S and Arif A F M 2011 Laser cutting of steel and thermal stress development *Optics & Laser Technology* **43** 830–837

[7] Eltawahni H A Olabi A G and Benyounis K Y 2010 Effect of process parameters and optimization of CO_2 laser cutting of ultra high-performance polyethylene *Materials and Design* **31** 4029–4038

[8] Chong Zhian Syn Mohzani Mokhtar Chin Jeng Feng Yupiter H P and Manurung 2011 Approach to prediction of laser cutting quality by employing fuzzy expert system *Expert Systems with Applications* **38** 7558–7568

[9] Wang J and Wong W C K 1999 CO_2 laser cutting of metallic coated sheet steels *Journal of Materials Processing Technology* **95** 164-168

[10] Shang-Liang Chen 1998 The effects of gas composition on the CO_2 laser cutting of mild steel *Journal of Materials Processing Technology* **73** 147–159

[11] Azmir M A and Ahsan A K 2009 A study of abrasive water jet machining process on Glass epoxy composite laminate *Journal of Materials Processing Technology* **209** 6168–6173

[12] Hocheng K R and Chang 1994 Material removal analysis in abrasive water jet cutting of ceramic plates *Journal of Materials Processing Technology* **40** 287-304

[13] John Rozario Jegaraj J and Ramesh Babu N 2005 A strategy for efficient and quality cutting of materials with abrasive water jets considering the variation in orifice and focusing nozzle diameter *International Journal of Machine Tools & Manufacture* **45** 1443–1450

[14] Wang J and Wong W C K 1999, A study of abrasive water jet cutting of metallic coated sheet steels *International Journal of Machine Tools & Manufacture* **39** 855–870

[15] Mahabalesh Palleda 2007 A study of taper angles and material removal rates of drilled holes in the abrasive water jet machining process *Journal of Materials Processing Technology* **18** 292–295

Case Study of Cycle Time Reduction by Mechanization in Manufacturing Environment

C Gnanavel[1], R Saravanan[2], M Chandrasekaran[3] and R Pugazhenthi[4]
[1]Assistant Professor, Department of Mechanical Engineering, Vels University, Chennai, India
[2]Professor, Department of Mechanical Engineering, Ellenki College of Engineering and Technology, Hyderabad, Telangana, India
[3]Professor, Department of Mechanical Engineering, Vels University, Chennai, India
[4]Associate Professor, Department of Mechanical Engineering, Vels University, Chennai, India.

E-mail: gnanavelmech1986@gmail.com

Abstract. Research and Innovation offer facilitation, fulfil the day to day requirements, and bring sophisticated things to the society. In industries such facilitation reduces the fatigue, employee morale and improves productivity. The research is focused on such requirement for gear box manufacturing industries. The Six sigma approach of Eliminate, Combine, Rearrange, Simplify (ECRS) method and cycle time analysis were carried out to understanding the problem and solution requirements. The newly mechanized equipment was installed and tested. The result shows that cycle time reduced significantly and reduced employee fatigue and raised morale to a great extent. The designed equipment is a general purpose to improve the productivity of similar industries.

1. Introduction

Nowadays, industrial gearboxes are vital in many industries, notably in Automobile industries, Cement industries, sugar industries, Paper industries, Steel plants, mining industries, Food processing industries, Pharmacy industries, and Pulp industries etc. This research focused on gear box manufacturing industry in particular testing section. The various kinds of gear boxes manufacture of different assembly line and arrive randomly to the Testing section. Hence the delivery schedule was probably affected by the bottleneck area (Testing Section). The facilitation in such bottleneck area will considerably increase the productivity. From the literature evident that six sigma approach can be used for improving quality in injection moulding process, in metal casting foundry industries in semiconductor foundry in a manufacturing industry [1 and 2]. A specific solution for engine overheating problem was solved by using six sigma approaches [3]. Apart from manufacturing industries, Shanmugaraja et al [11 and 12] used for various applications and proved quality with productivity achieved by using six sigma in project industries.

The lean Six Sigma is a powerful business improvement methodology for improving the efficiency and effectiveness made attempts to implement it in higher education institutions. Antony et al [4] discussed the feasibility of six sigma implications for service industries and Jiju Antony, and Darshak Desai A [5] for software industries and provided assessing the status of Six Sigma implementation in the Indian industry. Darshak A Desai [9, 10, 13 and 16] furnished the impact of Six Sigma in a developing economy analysis on benefits drawn by Indian industries. Gijo E V et.al

[14] discussed important concepts, surveys and case studies in continuous improvement methodology and highlighted future implications.

The authors noticed that the literature on this classification is limited. Kunal Ganguly [17] highlighted the role and responsibilities of academic institutions in developing six sigma by research. Jiju Antony [5] highlighted the pros and cons of six sigma concept with the academic perspectives. Neha Gupta [15] presented overview on six sigma quality improvement program. Jiju Antony [5] distinguished that even though Six Sigma and lean methodologies are focused on process and quality improvement, Lean has special features of formalization and codification of experience and judgment where as in Six Sigma has not. However Six Sigma emphasizes variation, defects and process evaluation, but Lean emphasizes speed and waste. But integration both concepts have some added advantages.

Rishi Pareek and Jaiprakash Bhamniya [18] provided a conceptual model by integrating lean principles with Six Sigma methodology as a consistent approach to providing continuous improvement. Jose Mehdiuz Zalan et.al [19] deployed compatibility of the green, lean and Six Sigma concepts and insisted that all these should be an integrated approach. In this work The Six sigma approach of Eliminate, Combine, Rearrange, Simplify (ECRS) method and cycle time analysis were carried out to understanding the problem and solution requirements.

2. Problem definition

The factory produces the gearboxes are tested under three categories like horizontally mounted gearboxes, vertically mounted gearboxes and angularly mounted gear boxes. The activities involved in the existing method of setting up for testing are as follows.

- Lifting the gear box
- Mounting the gearbox in bed
- Motor tilting
- Clamping the gear box in bed
- Clamping the motor in bed/ frame
- Oil filling
- Gearbox Testing
- Removing gear box clamping
- Oil drains.

In which the motor tilting and clamping the motor with proper alignment are not only consuming more time, but also causes for Frequent injuries to the labors and gives fatigue to workers by its frequency. Hence, these activities increase the overall cycle time of testing and lower the productivity. The research aims to study and solve this problem systematically to increase the time productivity, employee morale and their safety. It was planned to design supporting equipment instead of using standard supporting blocks, crane, and eliminate the practice of manually guessed tilt and checking of the angle of tilt, motor clamping time etc.

3. Methodology

- Recording of the customer demand, according to the production cycle time.
- Observing the corresponding customer requirement trend.
- Observing the production cycle time of the testing process and motion.
- Finding the root cause of reducing the process cycle time.
- Making the action plan to implement the root cause.
- Revising the Testing process sequence as the result of analysis in accordance with ECRS method.
- Designing the new fixture for Tilting the Motor as a result of analysis.
- Analyze the new design

- Create a part drawing & assembly drawing of the new Tilting machine as the design considered.
- Fabricate the Tilting machine as per the drawing.
- Inspecting all the parameters of Machine in manufacturers end.
- Commissioning the Tilting machine in the plant
- Take a trail of Testing process cycle time.
- Recording the cycle time continuously with minimum number of cycles.
- Monitoring the productivity result with the condition of before/after.
- Ensue the sustaining of the proposed procedure.

4. The motor tilting machine

Initially the proposed equipment was designed by using the software of solid house, version 2014. The figure 1 shows the principal part of a newly designed machine for tilting the testing motor with common clamping for all types of gearboxes. The Table 1 is Part list of MTM. The MTM consists of horizontal & vertical frame. Lead screw provided to achieve for horizontal and vertical motions to set the motor as desired location.

The lead screws are driven by electric motors. The vertical frame contains L bracket in which the face plate is attached to provide the desired angle of rotation of the motor. The motor is mounted on the shaft of face plate in the L bracket. As the faceplate is a critical part in the machine, the face plate is designed well and analyzed by FEM by using ANSYS software before it fabricated. The figure 2 shows such details.

Table 1. Part List of MTM.

Part No	Description	Qty
1	Horizontal Frame	1
2	Vertical Frame	1
3	Horizontal Lead Screw	1
4	Vertical Lead Screw	1
5	Face Plate	1
6	Face Plate Flange	1
7	Up Down Plate	1
8	LM Rail 30	2
9	LM Black 30	4
10	Gear Box	1
11	Motor 0.75 KW	1
12	Gear Box	1
13	Motor 1.1 KW	1
14	Gear Box	1
15	Plummer Block UCFL	2
16	Plummer Block UCP	4
17	Motor 22KW	1

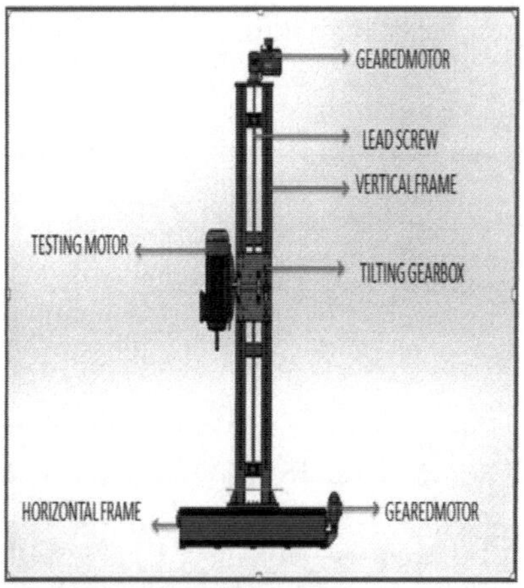

Figure 1. Tilting testing motor

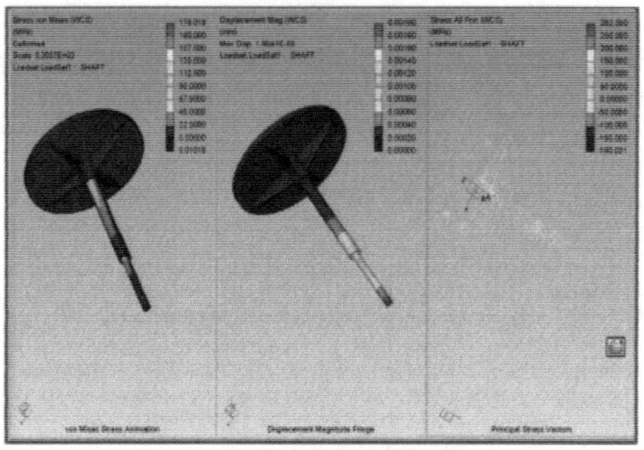

Figure 2. Design of the base

5. Performance evaluation

The Motor tilting machine fabricated commissioned and its performances were evaluated.

Figure 3. Before and after Vertical type gear box testing setup

Figure 4. Before and after Horizontal type gear box testing setup

6. Result And Discussion

The problem identified by using ECRS method the same was used here to compare performance of MTM. The time productivity can be observed from Table 2. The sustainability of the results was monitored and shown in figure 5 -8.

Table 2. ECRS observation and Before and after cycle Time Comparison

Sl. No	Activity	Before 'T' min	E	C	R	S	After 'T' min
1	Lifting the gear box	5	X	X	X	X	5
2	Mounting the gearbox in bed	5	X	X	X	X	5
3	Motor tilting	30	X	0	X	X	3
4	Clamping the gear box in bed	10	X	X	X	X	10
5	Clamping the motor in bed/ frame	30	X	0	X	X	0
6	Oil filling	10	X	X	X	X	10
7	Gearbox running time	60	X	X	X	X	60
8	Removing gear box clamping	10	X	X	X	X	10
9	Oil draining	10	X	X	X	X	10
	Total Time	170	X	0	X	X	113

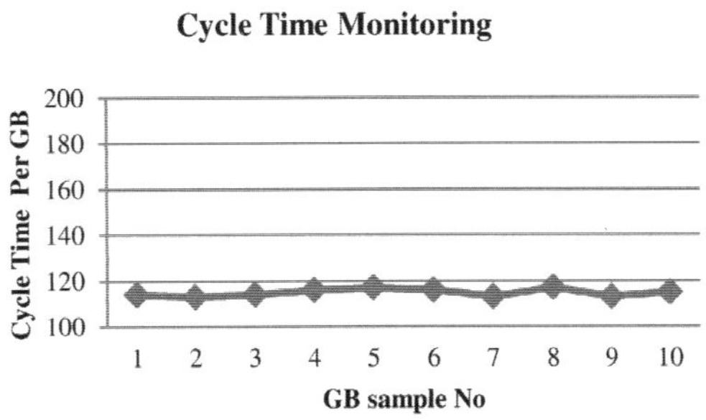

Figure 5. Cycle Time performance per week

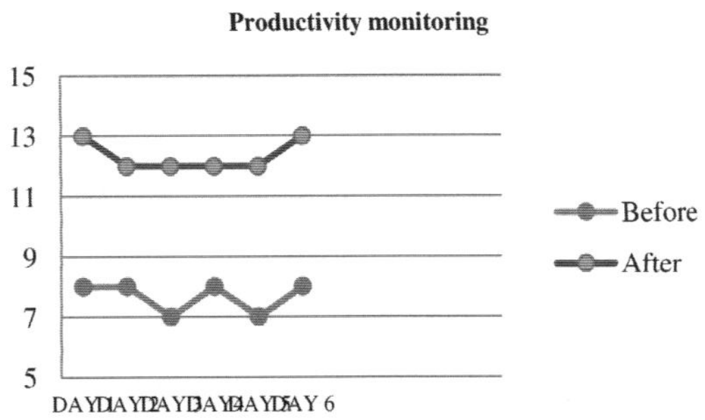

Figure 6. Productivity performance per week

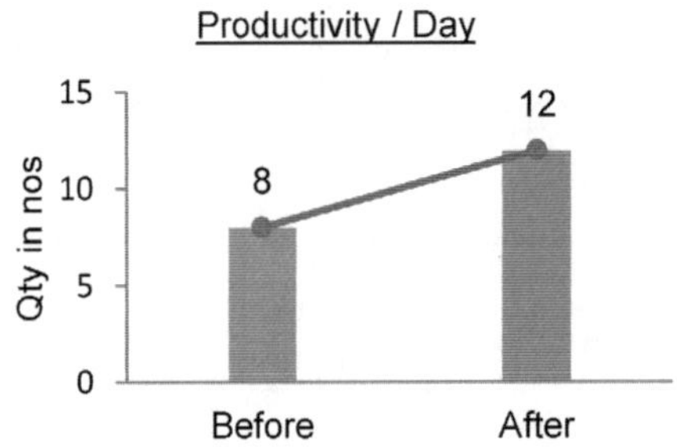

Figure 7. Average Productivity statistics for a week

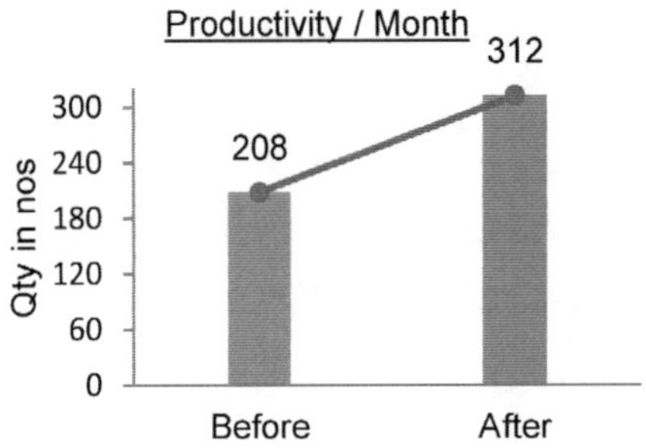

Figure 8. Average Productivity statistics for a month

7. Conclusion

The identified problem of high time and human effort consuming in gear box setting for various configurations were rectified with designing equipment. After Designing major parts like lead screws and face plate they were analyzed by ANSYIS software. The new equipment helped to reach the average testing rate from 208 gear boxes per month to 302 gear boxes per month by reducing cycle time by 57 minutes per gear box in average. The designed Machine will be useful for similar industries and increasing testing speed.

8. References

[1] VikasDwivedi Mohd and Anas Mohd Siraj 2014 Six Sigma: As applied in Quality Improvement for injection Molding Process *International Review of Applied Engineering Research* **4(4)** 317-324.

[2] Bharati P K, Khan M I and Harvinder Singh 2011 Six Sigma Approach for Quality Management in Plastic Injection Molding Process: A Case Study and Review *International Journal of Applied Engineering Research* **3(6)** 303-314.

[3] Vijayaram T R, Sulaiman S, Hamouda A M S and Ahmad M H M 2006 Foundry quality control aspects and prospects to reduce scrap rework and rejection in metal casting manufacturing industries *Journal of materials processing technology* **178** 39-43.

[4] Antony J, Kumar M, and Tiwari M K, 2005 An application of Six Sigma methodology to reduce the engine-overheating problem in an alternative company *Proceeding of the Institution of Mechanical engineers Part B Journal of Engineering Manufacture **219(8)** 633-646.*

[5] Jiju Antony, and Darshak Desai A, 2009, Assessing the status of Six Sigma implementation in the Indian industry *Management Research News* **5(32)** 413-423.

[6] Vikas Tayal and Jitendra Kumar 2012 Improvement in production rate by reducing the defects of injection moulding *International Journal of computer science and comm.* Engg. IJCSCE Special Issues on Emerging Trends in Engg ICETIE.

[7] Saravanan Rathinasamy and Raju R, 2010 Sequencing and scheduling of non-uniform flow pattern in parallel hybrid flow shop *International Journal of Advanced Manufacturing Technology* **49** 213–225.

[8] Darshak A Desai and Mulchand B Patel 2009 Impact of Six Sigma in a developing economy analysis on benefits drawn by Indian industries *Journal of Industrial Engg. And Mana* **2(3).**

[9] Darshak A Desai, 2012 Increasing Bottom-line through Six Sigma quality improvement drive: case of small scale foundry Industry Udyogpragati **2** 36.

[10] Darshak A Desai 2008 Improving productivity and profitability through Six Sigma: Experience of a small scale jobbing industry *International Journal of Productivity and Quality Management* **3(3).**

[11] Shanmugaraja M, Nataraja M and Gunasekaran N 2011 Quality and Productivity improvement using Six Sigma and Taguchi method *International Journal of Business Excellence* **5(4).**

[12] Shanmugaraja M, Nataraja M and Gunasekaran N 2012 Literature snapshot on Six Sigma project selection for future research *International Journal Services and Operations Management* **3(11)**.

[13] Darshak A Desai, 2012 Quality and Productivity improvement through Six Sigma in foundary industry *International Journal of Productivity and Quality Management* **2(9)**.

[14] Gijo E V, ShreerangaBhat and Jnanesh N A 2014 Application of Six Sigma methodology in a small-scale foundary industry *International Journal of Lean Six Sigma* **2(5)**193-211.

[15] Neha Gupta, 2013 An overview on Six Sigma: Quality Improvement Program *International Journal of Technical Research and Application* **1(1)** 29-39.

[16] Darshak A Desai 2006 Improving customer delivery commitments the Six Sigma way: case study of an Indian small scale industry *International Journal of Six Sigma and Competitive Advantage* **1(2)**.

[17] Kunal Ganguly 2012 Improvement process for rolling mill through the DMAIC six sigma approach *International Journal for quality research* **3(6)** 221-231.

[18] Rishi Pareek and Jaiprakash Bhamniya, 2013 Optimization of Injection Moulding Process using TAGUCHI and ANOVA *International Journal of Scientific & Engineering Research* **1(4).**

[19] Mehdiuz Zalan, Sujitkumar Pattanayak and Arun Chandra Paul, 2013 Study of feasibility of Six Sigma implementation in a manufacturing industry: A case study *International Journal of Mechanical and Industrial Engineering (IJMIE)* **1(3)**.

Optimization of the Machining parameter of LM6 Alminium alloy in CNC Turning using Taguchi method

S Arunkumar [1], V Muthuraman [2], and V P M Baskaralal [3]

[1, 3,] Assistant Professor Department of Mechanical Engineering, Vels University, Chennai, India -600117

[2] Associate Professor and Head, Department of Mechanical Engineering, Vels University, Chennai, India -600117.

E-mail:gct.arunkmar@gmail.com

Abstract. Due to widespread use of highly automated machine tools in the industry, manufacturing requires reliable models and methods for the prediction of output performance of machining process. In machining of parts, surface quality is one of the most specified customer requirements. In order for manufactures to maximize their gains from utilizing CNC turning, accurate predictive models for surface roughness must be constructed. The prediction of optimum machining conditions for good surface finish plays an important role in process planning. This work deals with the study and development of a surface roughness prediction model for machining LM6 aluminum alloy. Two important tools used in parameter design are Taguchi orthogonal arrays and signal to noise ratio (S/N).Speed, feed, depth of cut and coolant are taken as process parameter at three levels. Taguchi's parameters design is employed here to perform the experiments based on the various level of the chosen parameter. The statistical analysis results in optimum parameter combination of speed, feed, depth of cut and coolant as the best for obtaining good roughness for the cylindrical components. The result obtained through Taguchi is confirmed with real time experimental work.

1. Introduction

Two important issues in manufacturing process of modeling and optimization. A greater attention is given to accuracy and surface roughness of product by the industry these days. Surface finish has been one of the most important considerations in determining the machinability of materials. The predictive modeling of machining operations requires detailed prediction of the boundary conditions for stable machining. Most surface roughness prediction models are empirical and are generally based on experiments in the laboratory. In addition, it is very difficult in practice; to keep all factors under controls as required obtaining reproducible results. Generally these models have a complex relationship between surface roughness and operational parameters not only increase the utility for machining economics, but also the product quality to a great extent.Column number represents the influencing factor and row number represents the trial number. The other numbers represent the level of the factor represented by their column number.

Optimization process involving seven parameters at two levels. Experiments have been conducted based on the orthogonal array and these data obtained have been analyzed using ANOVA and response curves [1].Statistical model for surface roughness estimation in a high-speed flat end milling process under wet cutting conditions. The variables taken are spindle speed, feed rate, depth of cut, and step over. This work deals with obtaining parameter settings for both minimum surface roughness and long tool life [2].Optimized the turning operation with multiple performance characteristics. Taguchi method of optimization was applied to optimize three control factors such as cutting speed, feed rate, and depth of cut considering the performance characteristics such as Tool life, cutting force, surface Roughness which was designed in L9 orthogonal array [3].Optimal value of cutting force by an optimal setting of turning process parameters while machining En24 alloy steel with TiC coated carbide inserts. The effects of the selected parameters have been accomplished using Taguchi's parameter design approach. The results indicate that the selected process parameters –speed, feed, and depth of cut, as well as interaction between cutting speed and depth of cut significantly affect the mean and variance of cutting force [4]. Predict the surface roughness of machined parts in the end milling process was developed to assure product quality and increase reproduction rate by predicting the surface finish parameter in real time [6].

2. Orthogonal array

Column number represents the influencing factor and row number represents the trial number. The other numbers represent the level of the factor represented by their column number which is shown in Table 1.

Table 1. Standard L9 (3^4) Array

	COLUMN NUMBER			
Trial number	1	2	3	4
1	1	1	1	1
2	1	2	2	2
3	1	3	3	3
4	2	1	2	3
5	2	2	3	1
6	2	3	1	2
7	3	1	3	2
8	3	2	1	3
9	3	3	2	1

2.1Details of experiments conducted

Taguchi experiments to identify the optimum surface roughness and to find the percentage contribution of parameters namely speed, feed rate, depth of cut and coolant flow rate shown in Table 2. Taguchi L 9(34) orthogonal array was used to design the experiments as shown in Table 3.

Table 2. Variables factor levels

Controllable factors	Level 1	Level 2	Level 3
Depth of cut (mm)	0.4	0.6	0.8
Spindle speed (rpm)	1400	2000	2600
Feed (mm)	0.05	0.15	0.25
Coolant flow rate (lit/sec)	0.02	0.05	0.1

Table 3. Orthogonal array L_9 (3^4)

Exp. no	Control factor assignment and column number				Noise factor (measurement location)		
	Depth of cut	Speed	Feed	Coolant flow rate	1	2	3
1	0.4	1400	0.05	0.02	0.69	0.72	0.67
2	0.4	2000	0.15	0.05	2.22	2.40	2.20
3	0.4	2600	0.25	0.1	2.74	2.56	2.87
4	0.6	1400	0.15	0.1	2.03	2.06	2.02
5	0.6	2000	0.25	0.02	1.90	1.80	2.14
6	0.6	2600	0.05	0.05	0.71	0.68	0.76
7	0.8	1400	0.25	0.05	1.75	1.76	1.75
8	0.8	2000	0.05	0.1	0.68	0.69	0.74
9	0.8	2600	0.15	0.02	0.84	0.73	0.88

3. Results and discussion

The results of the experiments for nine trials conditions with three measurement locations the average response vales and the S/N ratio were recorded in Table 4.

Table 4. Result of the L_9 (3^4) experiment

Experiment number	(R_{aij})j =1,2,3 Response value 1	2	3	(R_{ai}) Average response vale	(dBi) S/N value
1	0.69	0.72	0.67	0.69	3.17
2	2.22	2.40	2.20	2.27	-7.14
3	2.74	2.56	2.87	2.72	-8.71
4	2.03	2.06	2.02	2.03	-6.17
5	1.90	1.80	2.14	1.94	-5.80
6	0.71	0.68	0.76	0.71	2.88
7	1.75	1.76	1.75	1.75	-4.87
8	0.68	0.69	0.74	0.70	3.09
9	0.84	0.73	0.88	0.82	1.70
				$\bar{y}$=1.4,8	$\bar{\eta}$=-21.85

The average effects and S/N ratios for each level of cutting parameter are summarized and referred to in the average effects response table and S/N ratios response table roughness (Ra), as shown in Table 5 and 6.

Table 5. Average effect response the raw data

Levels	Depth of cut (A)	Spindle speed (B)	Feed rate (C)	Coolant flow rate (D)
1	1.89	1.49	0.70	1.45
2	1.56	1.62	1.70	1.57
3	1.09	1.41	2.13	1.82
Max- Min	0.80	0.22	1.42	0.66
Rank	2	4	1	3

Table 6. Average effect response for S/N ratio

Levels	Depth of cut (A)	Spindle Speed (B)	Feed rate (C)	Coolant flow rate (D)
1	-5.46	-2.62	3.05	-0.31
2	-3.04	-3.28	-3.87	-3.04
3	-0.02	-1.37	-7.86	-3.93
Max-Min	5.44	1.91	10.91	3.62
Rank	2	4	1	3

4. Pareto anova

It is a quick and easy method for analyzing result. This method enables the significance of factors and allows the optimal levels of factors to be obtained, as shown in Table 7.

Table 7. Pareto anova

Factor		A	B	C	D	Total
	1	5.67	4.47	2.10	3.45	
Sum at factor levels	2	4.68	4.91	5.12	4.73	13.62
	3	3.27	4.25	6.41	5.45	
Sum of squares of differences		8.72	0.67	29.46	6.14	44.88
Contribution ratio (%)		19.42	1.49	65.37	13.68	100
Overall optimum conditions for all factors		A_3=0.8 mm B_3=2600 rpm C_1=0.05 feed rate mm D_1= 0.02 lit/sec				

Table 8. Pareto anova for raw data factor

Pareto ANOVA for raw data Factor		A	B	C	D	Total
Sum at factor levels	1	-12.62	-7.86	9.07	-0.79	
	2	-9.03	-9.82	-15.06	-9.09	-21.69
	3	-0.04	-4.01	-19.32	-11.80	
Sum of squares of differences		251.82	51.66	1290.14	197.43	1791.05
Contribution ratio (%)		14.05	2.88	72.03	11.023	100
Overall optimum conditions for all factors		A_3 =0.8 mm B_3 =2600 rpm C_1=0.05 feed rate mm D_1= 0.02 lit/sec				

4.1. Response graphs are drawn with parameters and levels

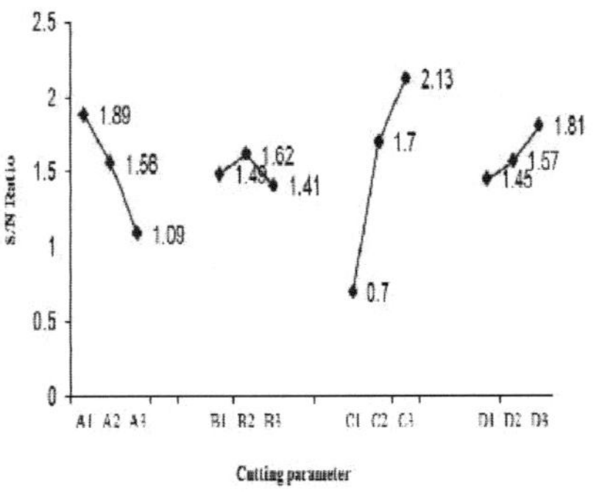

Figure1. Average effect response chart for the raw data

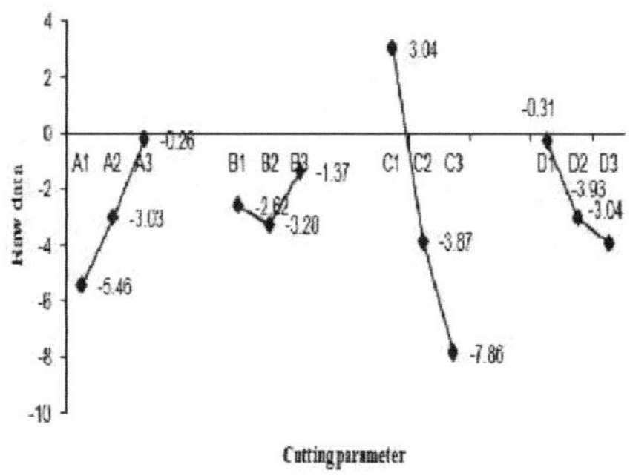

Figure 2. Average effect response chart for S/N ratio

IOP Conf. Series: Materials Science and Engineering **183** (2017) 012024 doi:10.1088/1757-899X/183/1/012024

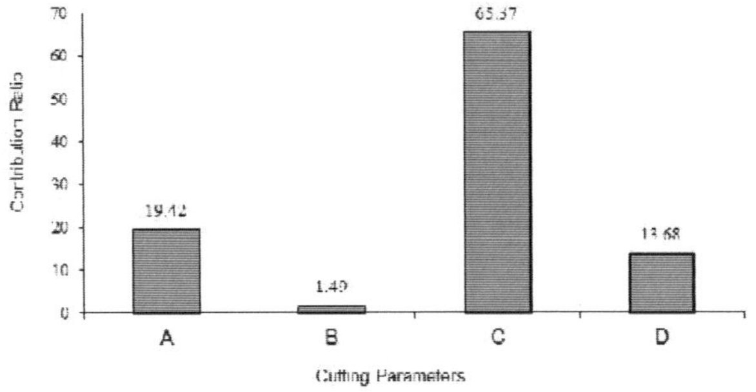

Figure 3. Pareto ANOVA for Raw Data

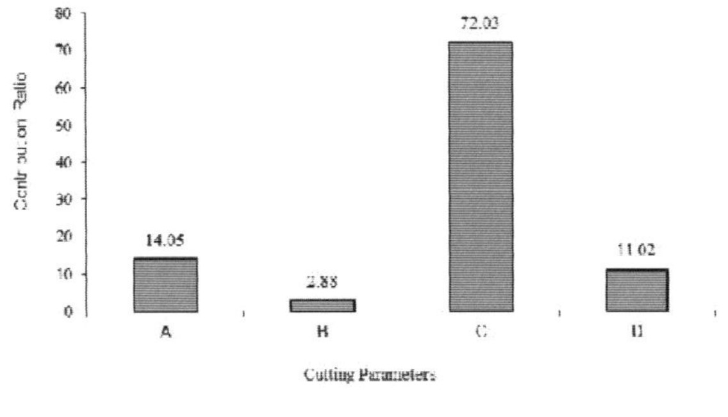

Figure 4. Pareto ANOVA for S/N Ratio

Table 9. ANOVA for raw data

Square of variation	Sum of square	Degree of freedom	Mean of square	F_{cal}	F_{tal}
Depth of cut	2.95	2	1.47	32.22	6.01
Spindle speed	0.22	2	0.11	2.51	6.01
Feed rate	9.74	2	4.87	108.20	6.01
Coolant flow	1.56	2	0.78	17.30	6.01
rate	0.81	18	0.04		
Error	15.28	26			
Total					

In order to determine which cutting parameter significantly affect the quality characteristic for S/N Ratio. Table 9 shows the results of ANOVA analysis of S/N Ratio for surface roughness. From Table 10 it is apparent that the F values of factor A (depth of cut), and factor C (feed rate) were all greater than $F_{(0.01,2,18)}$ =6.01. Figure 1 and Figure 2 represents average effect response chart for raw data and S/N ratio respectively. Figure 3 and Figure 4 represents Pareto ANOVA for the raw data and S/N ratio respectively.

Table 10. ANOVA for raw data

Square of variation	Sum of square	Degree of freedom	Mean of square	F_{cal}	F_{tal}
Depth of cut	83.63	2	41.81	13.02	6.01
Spindle speed	17.40	2	8.7	2.71	6.01
Feed rate	431.0	2	215.5	67.13	6.01
Coolant flow rate	8.09	2	4.04	1.25	6.01
Error	58.33	18	3.27		
Total	598.45	26			

4.2. Predicting optimum performance

The optimal surface roughness and S/N ratio are predicted at the selected optimal setting of process parameter is shown in Table 11. The significant parameters with optimal levels are selected as:

Predicted optimal surface roughness $= \bar{A}_3 + \bar{C}_1 + \bar{D}_1 - 2(\bar{y})$ (1)

$=1.09 + 1.70 + 1.45 - 2(1.48)$

$=0.30\,\mu m$

Predicted optimal S/N ratio $= \bar{A}_1 + \bar{C}_3 - (\bar{y})$ (2)

$= (-5.46) + (-7.86) - (-21.85)$

$= 8.23 dB$

4.3. Confirmation Experiment.

Conformation experiment was contacted at the optimum setting Ra values at optimum are

Table 11. The values obtained in the experiments are in the Predicted limits

Trial 1	Trial 2	Trial 3	mean	S/N Ratio
0.59	0.49	0.55	0.54	5.72

5. Conclusion

Table 12. The Optimal setting process parameter for optimal surface roughness

S.NO	Parameter	Optimal levels
1	Depth of cut	0.8
2	Spindle speed	2600
3	Feed rate	0.05
4	Coolant flow rate	0.02

Table 13. The percent contribution of raw data

1	Parameter	Percent contribution on surface roughness
1	Depth of cut	19.42
2	Spindle speed	1.49
3	Feed rate	65.37
4	Coolant flow rate	13.68

The predicted optimal surface roughness = $0.01 \leq Ra \leq 0.59$ (µm)
The calculation C.I for confirmation runs =±0.29.
The predicted optimal range (99% C.I) of the surface roughness for confirmation runs: $Ra<0.59$(µm).
The average value of surface roughness for confirmation runs at the optimal setting of turning process parameters was found to be 0.54µm. this result was within the predicted optimal surface roughness therefore, the optimum surface roughness was verified in turning operations is shown in Table 12 and 13.

Table 14. The percent contributions of S/N ratio parameters in affecting variation in surface roughness while machining aluminum turning operation.

S.NO	Parameter	Percent contribution on surface roughness
1	Depth of cut	14.05
2	Spindle speed	2.88
3	Feed rate	72.03
4	Coolant flow rate	11.02

The predicted optimal surface roughness = 5.39 $\leq$Ra $\leq$11.37 dB

The calculation C.I for confirmation runs = ±2.84

The predicted optimal range (99% C.I) of the surface roughness for confirmation runs: Ra<11.37dB

The S/N Ratio value of surface roughness for confirmation runs at the optimal setting of turning process parameters was found to be 5.72dB. This result was within the predicted optimal S/N Ratio therefore, the optimum surface roughness was verified in turning operations are shown in Table 14.

6. References

[1] Avani Gandhi 2003, Problem solving using Taguchi DOE techniques *Industrial engineering Journal* No.**12** p 16-25.

[2] Babur Ozcelik Mahmut Bayramoglu 2005 The Statistical modeling of surface roughness in high speed flat end milling *International Journal of machine tool and manufacture.* No. **14** p65-72.

[3] Nian.C Y ,Yang W H ,Tarng Y.S, 1999, Optimization of turning operations with multiple performance characteristics *Journal of material processing and technology* No.**24** p90 – 96.

[4] Hari Singh & Pradeep Kumar, 2005, Optimizing cutting force for turned parts by Taguchi's parameter design approach *Indian journal of engineering & Material sciences*, No.**12**, p97-103.

[5] Juan.H,Yu.S.F,Lee B.Y, 2003,The Optimal cutting parameter selection of production in HSM for SKD61 tool steels, *International Journal of Machine Tools and manufacture*,No.**43** p679-686.

[6] Yu-Hsuan Tsai Joseph C Chen Shi-Jer Lou 1999 An in process surface recognition system based on neural networks in end milling cutting operations *International journal of machine tool and manufacture* No.**39** p583-605.

[7] John L Yang & Joseph C Chen, 2004 A systematic approach for identifying surface roughness performance in end milling operations *Journal of industrial technology* No **17** p125-134.

[8] El Baradie 1993 Surface roughness model for turning grey cast iron (154BHN) Proceeding of Institution of Mechanical Engg Part B, *Journal of Engineering Manufacture 207*No.**23**, p43-54.

Optimal Solution for an Engineering Applications Using Modified Artificial Immune System

S Padmanabhan[1], M Chandrasekaran[2], S Ganesan[3], Mahamed Naveed Khan patan[4] and Polina Navakanth[5]

[1,3] Associate Professor, School of Mechanical Engineering, Sathyabama University, Chennai, Tamil Nadu, India.

[2]Professor, Department of Mechanical Engineering, Vels University, Chennai, Tamil Nadu, India.

[4,5]Students, Department of Mechanical Engineering, Sathyabama University, Chennai, Tamil Nadu, India.

E- mail: padmanabhan.ks@gmail.com

Abstract. An Engineering optimization leads a essential role in several engineering application areas like process design, product design, re-engineering and new product development, etc. In engineering, an awfully best answer is achieved by comparison to some completely different solutions by utilization previous downside information. An optimization algorithms provide systematic associate degreed economical ways that within which of constructing and comparison new design solutions so on understand at best vogue, thus on best solution efficiency and acquire the foremost wonderful design impact. In this paper, a new evolutionary based Modified Artificial Immune System (MAIS) algorithm used to optimize an engineering application of gear drive design. The results are compared with existing design.

1. Introduction

Most engineering optimization problems normally have continuous and discrete design variables with usually conflicting with each other. These problems with several objectives, are called "multi objective" optimization problems, and were initially studied as nonlinear objective functions. Constraints are extremely vital in engineering design problems with nonlinear constraints, usually obligatory on the statement of the issues and generally are terribly exhausting to satisfy, that makes the search tough and inefficient.

A computational paradigm system called Artificial Immune System (AIS) which is inspired by the biological immune system. It is based on the principal of the natural system. This system is based on the general Immune system as in our body. This system finds application in the future computerized industries, Robotics, Aerospace, automotive, etc., because of its immune system which develops its own antibodies for the virus or foreign bodies which will harm the system and also because of its identification system which identifies the previous attack or problem and produce solution for that .These are the main reason for the attraction towards the system AIS.

Yap, David et al [1] proposed the clonal selection algorithm is modified using the best solutions for each exposure (iteration) namely Single Best Remainder (SBR) and the proposed algorithm is able to improve the conventional CSA in terms of accuracy and stability for single and multi objective functions. Deb at al [2] discussed immune system for his first-generation algorithms use simplistic models of immunology as the initial inspiration, for example negative and clonal selection.

Coello and Cortes [3] developed multiobjective immune system algorithm (MISA) for its employment of an AIS in a multiobjective evolutionary algorithm context. Woldemariam [4], his research exploits the fact embedded in the immune system to communicate between innate immune response and adaptive immune response levels. Luh et al. [5] Proposed a Multi-objective Immune Algorithm (MOIA), MO-AIS research grew into maturity. Following that tens of algorithms were proposed.

Garrett [6] developed Adaptive Clonal Selection (ACS) by improving a number of processes and parameter initialization. Research based on aiNet extended aiNet to opt-aiNet. Wu, J.Y. [7] Proposed algorithm which combines the metaphor of clonal selection and idiotypic network theories to design an AIS method.

The somatic hypermuation and receptor editing operators explore a search space of solutions to an optimization problem. Juan Carlos Herrera–Lozada et al [8] proposed a new cloning system and for the maturation stage of the clones, two simple and fast mutation operators are used in a nominal convergence that works together with a re-initialization process to preserve the diversity. In this paper, a new immune clonal system based Modified Artificial Immune System (MAIS) algorithm used to optimize the helical gear drive design..

2. Artificial Immune System Algorithm

Artificial Immune Systems square measure new intelligent drawback determination techniques that square measure getting used in varied engineering optimization issues. AIS are outlined as procedure systems impressed by theoretical medicine, ascertained immune functions, principles and mechanisms so as to resolve issues. Because the biological processing algorithms, artificial system has been with success applied to a range of optimization issues.

AIS square measure a procedure intelligence paradigm impressed by the biological system that has found application in pattern recognition and machine learning. Alternative ways of AIS for optimization because the immune network theory and therefore the organism choice principle are projected and enforced by totally different researchers. The steps of the Modified Artificial Immune system as followed:

Generate randomly a population of "N" size with design variables. In the initial generation, these variables are copied directly to the working population and nominal convergence is controlled by the number of generations.

1. Finding the Objective Function for those variables and continued for sorting and ranking.
2. Use selection based on ranking. The variables with the highest affinity will be the best individual.
3. Perform the cloning of the antibodies using, $N_C = \sum(n-(i-1))$ Where NC is the number of clones to be generated for each variable, n is the total number of variable in the population and i is the current variables starting from the variables with the highest affinity [8].
4. All the clones in the set of clones which are the copies of variables with good affinity degrees, undergo a mutation process [8]. This is a two phased mutation process

$$x' = x + \frac{(\alpha \times \text{range} \times \text{generation})}{N_c}$$

and

$$x' = x + \frac{(\alpha \times \text{range})}{\text{generation} \times N_c}$$

5. After the first mutation operator, In the case of having a better solution in mutated string, then the clone is replaced with the new one, else the second operator is used.
6. In the second mutation method, if there is no improvement in the mutated string, then the original solution remains with no change.
7. A model of receptor editing mechanism of the immune system was used, i.e., a proportion of the worst solution eliminated and new ones are generated in placed after predefined iterations.
8. Again find the objective function for the clones and sort it, rank it, mutate it then extract first best 10 solutions from the list for the further design process and visualization.

3. Engineering Design Application

In this paper, an engineering design problem has been adopted from Jayakumar [9]. The design problem states that, to design a pair of helical gears to transmit 12.5 kW at 1200 rpm. The transmission ratio is 3.5 and material adapted as Alloy steel material 40Ni 2Cr1Mo28. The helix angle is 15° and pressure angle is 20°. The material properties of gear drive are tabulated [11] in Table 1.

Table 1. Material Properties of Gear drive.

Material	Density(ρ) kg/mm3	Bending Stress(σb) kg/mm2	Compressive Stress (σc) kg/mm2	Young's Modulus (E) N/mm2
Alloy Steel	8.839x10-6	173.133	852.6	2.15x105

3.1. Optimal Helical Gear Drive Design

The design variable such as module, gear thickness and number of teeth are determined in order to optimize the helical gear drive design and number of objective functions include: Maximization of Power transmitted (f1),Minimization of Weight (f2), Maximization of Efficiency (f3) and Minimization of Centre distance (f4). The above gear drive objectives should satisfies with the design constraints of allowable bending stress, allowable compressive stress, minimum module and minimum centre distance.

The complete optimized problem of helical gear drive in terms of design variables P, mn, b and Z1 for the above problem with Alloy steel material, after simplification the helical design problem is in terms of non linear objectives and non linear constraints as follows:

Maximize $f_1 = P$ (1)

Minimize $f_2 = 98.58 \times 10^{-6} \times b \times (m_n Z_1)^2$ (2)

Maximize $f_3 = 100 - P_L$ (3)

Where, $P_L = 3.97 \times \dfrac{(H_s^2 + H_t^2)}{(H_s + H_t)}$ (4)

$$H_s = 4.5 \times \left\{ \left[\left(1 + \frac{0..552}{Z_l} \right)^2 - 0.883 \right]^{0.5} - 0.342 \right\} \quad (5)$$

$$H_t = 1.286 \times \left\{ \left[\left(1 + \frac{1.931}{Z_l} \right)^2 - 0.883 \right]^{0.5} - 0.342 \right\} \quad (6)$$

Minimize $f_4 = 2.33\ m_n Z_1$ (7)

Subject to,

$m_n Z_1\ b^{0.5}\ P^{-0.5} \geq 65.741$ (8)

$$m_n^2\,(3.5Z_1 + 20)\, b\, P^{-1} \geq 250.51 \qquad (9)$$

$$m_n\, Z_1\, P^{-0.333} \geq 18.35 \qquad (10)$$

$$m_n^3\,(3.5Z_1+20)^{0.333}\, P^{-0.333} \geq 2.94 \qquad (11)$$

The efficiency equation (3) has been adopted from Dudley [10] and all other equations followed from Design Data book [11].

Since all these objective functions are on different scales, these factors are to be normalized to the same scale. The normalized objective function is obtained as follows:

COF=

$$\left[\left(\frac{power}{\max.power} xNW_1\right) + \left(\frac{\min.weight}{weight} xNW_2\right) + \left(\frac{efficiency}{\max.efficiency} xNW_3\right) + \left(\frac{\min.cent.dist}{cent.dist} xNW_4\right)\right]$$

5. Results and Discussion

The above algorithm was developed using Microsoft Visual C#, with design parameters (P, mn, b and Z1) boundary values as inputs. After iteration of AIS for the specified helical gear reducer, the optimized results were tabulated in table 2 and compared with existing design [9]. The AIS result shows, significantly enhanced values with respect to weight reduction and center distance for the helical gear drive.

Table 2. Comparison of gear drive optimized results with MAIS.

Parameters / Material	Existing System	MAIS
Power (P) kW	12.5	12.52
Normal Module (mn) mm	3.00	2.76
Gear Thickness (b) mm	42.00	35.62
No. of teeth on pinion (Z1)	20	18
Centre Distance (a) mm	139.76	115.72
Gear Weight (kg)	14.9	8.66
Efficiency (%)	98.67	98.54

The above results shows, a slight increment in power and 41.88% of weight reduction for the existing alloy steel material with respect to MAIS result, and also significant reduction of 17.2% center distance in compared with existing system. From the results, MAIS produces a vital enhancement in its optimal design values to its objectives.

6. Conclusion

The engineering design problem results shows that, by using the AIS algorithm we can get the modified better solution for the practical problem and also for the theoretical problem.

In future, AIS will play the major role in the industries like automotive, aerospace, robotics, manufacturing etc., and it's also easy to understand and explain to the followers as it contains the natural human immune system as it base. And also by using AIS we can also solve the multi objective problems which contains different nonlinear, non-differentiable and multi variables objective functions. By using AIS we can assure that the better solution are extracted using better values for the betterment of the design.

Notations

H_s :Specific sliding velocity at start of approach action
Ht : Specific sliding velocity at end of recess action.
P : Power transmitted in kW
Z_1, Z_2: Number of teeth in pinion, gear
m_n : Normal Module in mm
i : Gear (or) transmission ratio
b : Thickness of gear and pinion in mm
a : Centre distance between shafts in mm
ρ : Density of the material in kg/mm^3
E : Young's modulus in N/mm^2
σ_b : Induced bending stress in N/mm^2
$[\sigma_b]$: Allowable bending stress in N/mm^2
σ_c :Induced compressive stress in N/mm^2
$[\sigma_c]$: Allowable compressive stress in N/mm^2
η : Percentage Efficiency
P_L: Percentage of power loss

7. References

[1] Yap, David F W, Koh S P and Tiong S K 2011 Mathematical function optimization using AIS antibody remainder method *International Journal of Machine Learning and Computing* **1 (1)** 13-19.

[2] K Deb, J Sundar, N Udaya Bhaskara Roa and S Chaudhuri 2006 Reference point based multi-objective optimization using evolutionary algorithms*Int. Journal of computational Intelligence Research* **2(3)** 273-286.

[3] Coello C and Cortés N 2002 Solving Multi objective Optimization Problems Using an Artificial Immune System *Genetic Programming and Evolvable Machines* **6** 163-190.

[4] Woldemariam K M and Yen GG 2010 Constrained optimization using artificial immune system *IEEE Congress on Evolutionary Computation (CEC)* 1 – 8.

[5] Luh G C, Chueh C H and Liu W W MOIA 2003 Multi-objective Immune Algorithm*Engineering Optimization***35 (2)** 143-164.

[6] GarrettSM 2004Parameter-free, adaptive clonal selection *In Proc of Congress on Evolutionary Computing* Portland Oregon USA **1** 1052-1058.

[7] Wu JY 2007 Artificial Immune System for Solving Constrained Global Optimization Problems *IEEE Symposium on Artificial Life* 92-99.

[8] Juan Carlos Herrera–Lozada, Hiram Calvo and Hind Taud 2011 A Micro Artificial Immune system *Polibits* **43** México ene.

[9] V Jayakumar2012 Design of transmission Systems Fourth Edition *Lakshmi Publications*Chennai.

[10] Dudley DW 1994 Handbook of Practical Gear Design*McGraw –Hill* Book Co: New York.

[11] Design Data Book 2008 Faculty of Mechanical Engineering, *PSG College of Technology* Coimbatore- 641004 Tamil nadu, India.

IOP Conf. Series: Materials Science and Engineering **183** (2017) 012026 doi:10.1088/1757-899X/183/1/012026

Optimization of heterogeneous Bin packing using adaptive genetic algorithm

R Sridhar[1], M Chandrasekaran[2], P Sriramya[3]and Tom Page[4]

[1]Assistant Professor and Research Scholar, Department of Mechanical Engineering,Vels University, Chennai, TN, India.
[2]Professor, Department of Mechanical Engineering,Vels University, Chennai, TN, India.
[3]Associate Professor, Department of Computer Science and Engineering,Saveetha Engineering College, Chennai, TN, India.
[4]Lough borough University, U.K.

E-mail: srisampangy@gmail.com

Abstract. This research is concentrates on a very interesting work, the bin packing using hybrid genetic approach. The optimal and feasible packing of goods for transportation and distribution to various locations by satisfying the practical constraints are the key points in this project work. As the number of boxes for packing can not be predicted in advance and the boxes may not be of same category always. It also involves many practical constraints that are why the optimal packing makes much importance to the industries. This work presents a combinational of heuristic Genetic Algorithm (HGA) for solving Three Dimensional (3D) Single container arbitrary sized rectangular prismatic bin packing optimization problem by considering most of the practical constraints facing in logistic industries. This goal was achieved in this research by optimizing the empty volume inside the container using genetic approach. Feasible packing pattern was achieved by satisfying various practical constraints like box orientation, stack priority, container stability, weight constraint, overlapping constraint, shipment placement constraint. 3D bin packing problem consists of 'n' number of boxes being to be packed in to a container of standard dimension in such a way to maximize the volume utilization and in-turn profit. Furthermore, Boxes to be packed may be of arbitrary sizes. The user input data are the number of bins, its size, shape, weight, and constraints if any along with standard container dimension. This user input were stored in the database and encoded to string (chromosomes) format which were normally acceptable by GA. GA operators were allowed to act over these encoded strings for finding the best solution.

1. Introduction

Due to industrial revolution, today most of the industries are focusing towards globalization. Globalization paved way for their customers to purchase the goods as per their requirements all around the world under a single roof. In turn, industries have substantially increases the scope and magnitude of their global production and distribution networks all around the world. So a necessity arises for those firms to deliver and distribute their goods to their customers, warehouses, and to the distribution centers all around the world in time. Thus the globalized market results in rapid development of international trade and creates rigorous competitions among the industries. Recently, Economical research proved that the survival of a firm in this competitive globalized environment will be possible only if those firms reduce their product cost without compromising the quality. So a necessity arises for those multinational firms to reduce their product cost by optimizing various factors which influencing the cost. The major factors which influences the product cost are generally categorized as Material cost, Manufacturing cost, Inspection cost, packing cost and Distribution cost.

In the logistic industries, determining which box should be packed on which container by which sequence can mean the difference between significant profit and debilitating loss. So this research work concentrates on perfect packing of goods into the container by considering major practical constraints using evolutionary algorithm which is essential for a firm to survive in today's the competitive globalized environment.

2. Literatures on Bin Packing

When packing problems were first proposed, simple models were used with exact methods for solving the problem. Given a problem, the exact methods find the best solution every time they are run. However, as constraints were added, the difficulty of solving the problem increased, and simply finding a good feasible solution. In addition, many methods take too long when applied to problems of significant size. Many methods focused exclusively on packing and assumed that the plan was static with specific assumptions. Later methods attempted to integrate more than one algorithm in an effort to find optimization possibilities that take advantage of the inherent best properties from those algorithms. Some methods focus on precedence or temporal constraints. Stochastic algorithms may find a different solution each time they are run. Review of the relevant literature confirms that the evolutionary approach works very well when compared to the traditional optimization procedures and computational time also reduced significantly [1]. Genetic algorithm is the best among the evolutionary algorithm. Interestingly majority of the Genetic approaches are hybrid / heuristic, i.e. combining GA with other algorithm or by modifying GA [2].

3. Existing solution

Artificial Intelligence (AI) is a very great innovation in the field of Engineering now-a-days. This algorithm uses human intelligence like techniques in most cases to identify the solution for a new set of input pattern. Recent research proved that the AI techniques is a most reliable and important part in this transportation domain. Most commonly used Artificial Intelligence techniques are [3]

- Artificial Neural Network
- Case Based Reasoning
- Fuzzy Logic
- Expert Systems

3.1 Artificial Neural Network

An Artificial Neural Network (ANN) is an information processing system that is inspired by the way of biological nervous systems, such as the brain and process information. The key element of this structure is the information processing system which is formed with a large number of highly interconnected processing elements called neurons. Even though it resembles to human brain, it requires some training set of data to train the neural network. But in the packing problems it may not be possible to define the training set. As the type, dimensions of the boxes, constraints and conditions vary from time to time. So it may not be feasible to use these techniques in the bin packing applications.

3.2 Case Based Reasoning

Case based reasoning is a technique of Artificial Intelligence that attempts to solve new problems by using past experience. The main task involves is matching the given problem details against records of previous cases. This technique is effective and less time consuming, but the box varieties were limited and the input can be predicated. So it may not be suitable for wide varieties of box types with varying nature.

3.3 *Fuzzy Logic*

Fuzzy Logic is a mathematical technique for imprecise data and problems that have many solutions rather than one. Although it is implemented in digital computers which ultimately make only yes-no decisions, fuzzy logic works with ranges of values, solving problems in a way that more resembles to human logic. Fuzzy logic is used for solving problems with expert systems and real-time systems that must react to an imperfect environment of highly variable, volatile or unpredictable conditions. Some of the researchers used this technique for the bin packing applications.

3.4 Expert Systems

Expert Systems are Artificial intelligence based system that converts the knowledge of an expert in a specific subject into a software code. These techniques may not be suitable for the bin packing applications, as it lacking to solve the new constraint facing in the real time packing.

4. Evolutionary algorithms

An evolutionary algorithm uses some mechanisms inspired by biological evolution, reproduction, mutation, recombination, and selection. Candidate solutions to the optimization problem play the role of individuals in a population, and the fitness function determines the environment within which the solutions "live". Evolution of the population then takes place after the repeated application of the above operators. Evolutionary algorithms often perform well approximating solutions to all types of problems because they ideally do not make any assumption about the underlying fitness landscape [4]. These families of approaches include, but are not limited to,

- Genetic Algorithm
- Ants Colony Algorithm
- Tabu Search
- Simulated Annealing
- Quantum Annealing
- Stochastic Tunneling
- Reactive Search Optimization
- Stochastic Gradient Descent Algorithm
- Cross Entropy Method
- Particle Swarm Optimization
- Intelligent Water Drops
- Bee's Algorithm

4.1 Genetic algorithm

A genetic algorithm is a search technique used in engineering science to find approximate solutions to optimization and search problems. Genetic algorithms are a particular class of evolutionary algorithms that use techniques inspired by evolutionary biology such as inheritance, mutation, natural selection, and recombination. This was found that the genetic approach will provide best suitable solution for the multi objective and multi constraint problems.

4.2 Ants Colony Algorithm

In operations research, the ant colony optimization algorithm(ACO) is aiming to search for an optimal path based on the behavior of ants seeking a path between their colony and a source of food. If other ants find such a path, they are likely not to keep travelling at random, but to instead follow the trail, returning and reinforcing it if they eventually find food. They have an advantage over simulated annealing and genetic algorithm approaches of similar problems when the graph may change dynamically. It is clear that the ACO is used to identify the shortest path, more over it may not produce the better solution for the bin packing problems and it can be best suited for container transportation and distribution problems.

4.3 Tabu Search

Tabu search is a mathematical optimization method, belonging to the class of trajectory based techniques. Tabu search enhances the performance of a local search method by using memory structures that describe the visited solutions, once a potential solution has been detected, it should be marked as "taboo". Thus the algorithm will not be allowed to visit those possibilities repeatedly, thereby reducing the search space and converging towards the optimal solution. As the algorithm climbs over the better solution to reach the peak, it may not be suitable for bin packing optimization problems.

4.4 Simulated Annealing

The name and inspiration come from annealing in metallurgy, a technique involving heating and controlled cooling of a material to increase the size of its crystals and reduce their defects. After making many trades, lowering the temperature several times to a low value, one may then "quench" the process by accepting only "good" trades in order to find the local minimum of the cost function.
Simulated annealing may be more efficient to find an acceptably better solution in the predefined time instead of producing best solution with large set of parameters. So it may not be used for bin packing applications, as the application required only the best solution.

4.5 Quantum Annealing

It uses "quantum fluctuations" instead of thermal fluctuations to get through high, but thin barriers in the target function. So this methodology also may not be suitable for the bin packing problem.

4.6 Stochastic Tunneling

This method attempts to overcome the increasing difficulty of the simulated annealing runs have in escaping from local minima as the temperature decreases, by 'tunneling' through barriers. As this methodology works similar to simulated annealing, it may not be used for bin packing.

4.7 Reactive Search Optimization

This algorithm focuses on combining machine learning with optimization, by adding an internal feedback loop to self-tune the free parameters of an algorithm to suit with the characteristics of the problem. As the bin packing problem taken in this research is not a dynamic problem, so it doesn't have any feedback parameters to optimize.

4.8 Stochastic Gradient Descent Algorithm
It runs many greedy searches from random initial locations and achieves the good solution. But in bin packing the input are well known.

4.9 (i) Cross Entropy Method
This method generates solutions via a parameterized probability distribution. The parameters are updated via cross-entropy minimization function, so as to generate better samples in the next iteration. This methods consumes more times as the parameter increases and not suitable for large sized problems.

4.10 Particle Swarm Optimization
Particle swarm optimization (PSO) is an exciting new methodology in evolutionary computation that is somewhat similar to a genetic algorithm in that the system is initialized with a population of random solutions. Unlike other algorithms, however, each potential solution (called a particle) is also assigned a randomized velocity and then flown through the problem hyperspace. Particle swarm optimization has been found to be extremely effective in solving a wide range of engineering problems.

4.11 Intelligent Water Drops (IWD)
IWD which mimics the behavior of natural water drops to solve optimization problems.

4.12 Bee's Algorithm
Bee's algorithm is based on the foraging behavior of honey bees. It has been applied in many applications such as routing and scheduling. These algorithms may provide better solution for particular category of the problems and can not be used universally to all types of problems.

5. Heuristic and Hybrid Methods
Heuristics make few or no assumptions about the problem being optimized and can search very large spaces of candidate solutions. However, Heuristics do not guarantee an optimal solution is ever found. As the bin packing problems having more number of parameters and the constraints, the algorithm used should be modified accordingly to produce optimal solution [5].
Generally, the parameters of the bin packing problems and its constraints can not be feed directly in to the intelligent evolutionary algorithms. These data should be encoded to a format acceptable by the evolutionary algorithms. Similarly, the output from the algorithm needs to be decoded to user understandable format. For encoding and decoding the data, separate algorithms required and also the procedure vary from application to application. Some time a need arises to synchronize the application with the algorithm. So a separate algorithm can be developed and integrated with the intelligent evolutionary algorithms to create a hybrid algorithm to solve an application.

6. Bin Packing
Bin-packing has been defined in several different forms such as stock cutting, vehicle loading, air cargo loading, container loading, scheduling, knapsack problem, etc depending on the application. In

this research work, bin packing problem was taken for consideration. The various terminologies and the standards used in this research work will be explained in this section

Bin packing is a finite collection of items with varying specifications to be packed into one or more containers or baskets [6]. Each container can hold any subset of the collection of objects without exceeding its capacity. Bin packing is also called as container loading, box packing, cargo loading, knapsack, etc. It not only denotes the packing of boxes in to the container but also arranging of good and items into the bags, baskets, pallets. The figure shows the sample set of boxes packed into the container.

7. Bin packing problem

The common problems faced by the bin packers are to pack the boxes in to the container in such a way that to utilize maximum volume of the container and the constraints satisfaction. This problem may be thought of as either a decision making problem or an optimization problem. In the decision making problems, a determination must be made, whether all of the objects fit into the container? In other words, the decision problem states that enough free space exists to hold the objects or not and having a Boolean answer. Alternatively, the optimization problem attempts to minimize the number of containers or minimize the amount of wasted space inside the container. This way of framing the problem attempts to optimize the containers utilization and having 'n' number of numeric answers, so mathematicians and computer scientists consider the bin packing problem as an optimization problem. General problems faced in bin packing are [7]

- Load the container with least empty space.
- Packed boxes should not exceed the container boundary.
- Total weight of the packed boxes should be less the pay load capacity of the container.
- Boxes to be unloaded first should be placed in the front portions of the container and vice-versa.
- Boxes of heavier weight should be kept at the bottom layers of the containers and vice-versa.
- Fragile items and items with constraints should be kept at the top layers of the container.
- Changing in the orientation of the boxes may leads to variation in the packed volume.
- For some of the items like bottles, liquid containers should allow to change the orientation.
- Boxes to be packed should not overlap with each other and with the container.
- Output from the software should be user understandable format.
- Base of a box to be packed should be completely lying on the face of bottom box.
- Surface profile of upper box to be packed should be the same as the lower box surface profile.

8. Types of bin packing

Form the above section it become very clear that the optimization of the bin packing problems need to satisfy many constraints are shown in Figure 1.

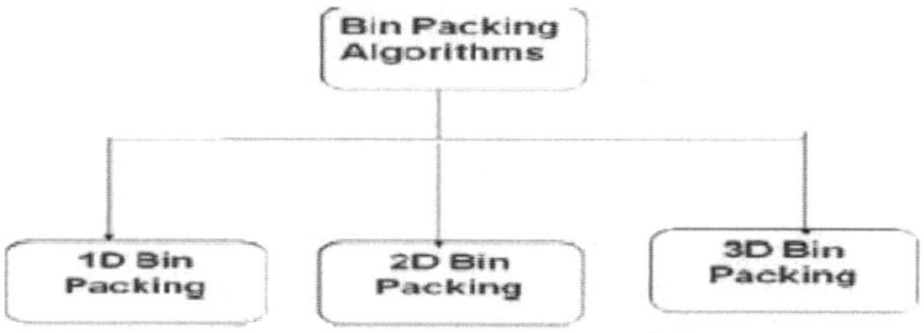

Figure 1. Types of Pin Baking Problem

So the researcher decomposes the problem into simple and builds to the complex by adding the parameter step by step. Generally, Bin packing problem are classified based on the dimension as

9. 1D Bin Packing

One dimensional bin packing problem optimizes only one parameter at a time. For example, writing maximum number of songs on the disc of standard memory size. In this example, by selecting the suitable song length, memory can be utilized properly. For bin packing problems, each object had fixed width and variable height or variable width and fixed height [8]. 1D problem will optimize either the length or the width of the box in a direction. Commonly used 1D Bin Packing Algorithms are:

- First Fit Algorithm (FFA).
- First Fit Decreasing Algorithm (FFDA).
- Refined First Fit Algorithm (RFFA).
- Best Fit Algorithm (BFA).
- Best Fit Decreasing Algorithm (BFDA) and many more.

10.2D Bin-Packing

Two dimensional bin packing problem is expanding the one dimensional bin packing problem by adding a variable to optimize. In the two dimensional version, both parameters will be vary to produce different solutions and need to optimize both the parameters simultaneously. The two dimensional bin-packing problem may be thought of as placing rectangles on a flat surface or inside a rectangle. Rectangle packing equates to efficiently fitting varying sized flat rectangles into a larger flat rectangular space most efficiently [9]. The classic video game 'Tetris' relates to the problem of rectangle bin-packing. However, unlike Tetris, classic two dimensional packing requires each object to orthogonal, so objects may or may not be rotated. Commonly used 2D Bin Packing Algorithms are as follows.

- Slide Pack Algorithm
- Skyline Pack Algorithm
- Nesting Algorithm
- Branch and Bound Algorithm.
- Knapsack Algorithm.

11.3D Bin Packing

The three-dimensional bin-packing problem retains the difficulty of lesser dimensional bin-packing problems, but holds unique and important applications. As one would expect, each object and bin exists in three dimensions. These objects and box represents triplets containing three values: width, length and height. Each box should fit into the container most efficiently [10]. If two dimensional bin-packing equates to rectangle-to-floor plan packing, three dimensional bin-packing equates to box-to-room packing. The 3D bin packing problems are classified as follows

A. Based on number of Container used
    - Single Container Loading problem
    - Multiple Container loading problems.
B. Based on type of box to be packed
    - Homogenous Problems.
    - Heterogeneous problems.

3D bin-packing may involve a single container or multi container problems which is shown in Figure 2. The singular container packing problem involves only one container with either definite or infinite volume. Containers with infinite volume are defined with finite length and width, but with height extending to infinity. This allows packing solutions to pack until the set of boxes are exhausted. Solutions dealing with infinitely sized containers generally care most to compressing objects effectively. But most of the real time applications use the container of definite volumes. In the definite volume container, the boundaries of the container were fixed and the boxes should be placed within that boundary. In this some of the boxes may left unpacked. Another way to approach this problem is by considering multiple containers. Each container has definite volume. In this way, if the volume of the objects exceeds the volume of a container, an algorithm must make choices of which boxes to include in the packing and which to throw away to another container. In this work, heterogeneous of varying sizes were taken to pack into a single container of finite fixed volume.

3D-Bin packing Algorithms

There are various algorithms and methodologies available for solving the 3D bin packing problems. Some of them may consider the constraints or few. Commonly used 3D Bin Packing Algorithms of varying methodologies are

(A)Smile Algorithm.
(B) Wall building Algorithm
(C) Wall building Algorithm along Width
(D) Wall building Algorithm in L
(E) Tree Generating Heuristic
(F) Geometric Method
(G) Strip Packing
(H) Stability method
(I) Hybrid Algorithms
(J) Heuristic Algorithms
(K) Intelligent Algorithms
(L)Evolutionary Algorithm

12.Genetic algorithm

Recent revolutions in molecular genetics made clear that the modular organization of genes is highly important for evolution of complexity. Evolutionary concept of Genetic Algorithm (GA) was introduced by John Holland in 1975 at the University of Michigan. First stage in using Genetic Algorithm is initial population generation randomly based on probability logics and each individual in the randomly generated population is represented by chromosomes in biology. The method of random generation may vary from researcher to researcher. Chromosomes have many segments and 'n' number of strings depends on the complexity of the given problem [2]. These chromosomes were allowed for crossover. Thus the generated children have the properties of both the parents. Mutation operator may applied to the generated off spring to prevent from the stagnation condition. Finally the

IOP Conf. Series: Materials Science and Engineering **183** (2017) 012026 doi:10.1088/1757-899X/183/1/012026

fitness function will be calculated and the worst parents need to be replaced by the best parents. This process need to be continued till the optimal solution reached. The various stages of the genetic algorithm are explained in the following sections.

12.1 Population generation

First stage in using the Genetic Algorithm is an initial population generation with 'n' number of strings generated randomly, based on probability logics [1, 2]. In a genetic algorithm, a population of strings called chromosomes or the genotype of the genome, which encodes candidate solutions called individuals, creatures, or phenotypes to an optimization problem, evolves toward better solutions are shown in Figure 3.

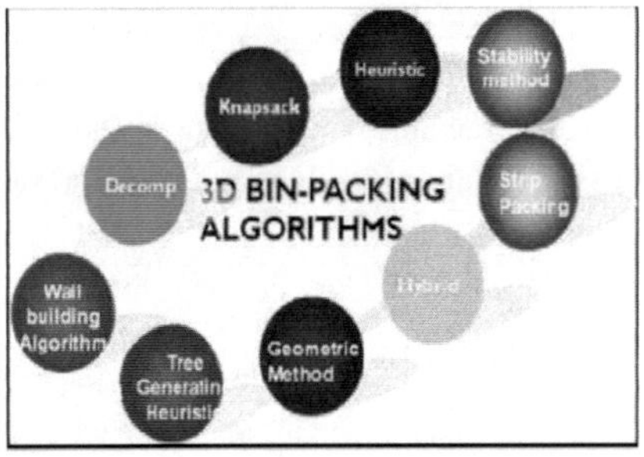

Figure 2. 3D Bin Packing Algorithm

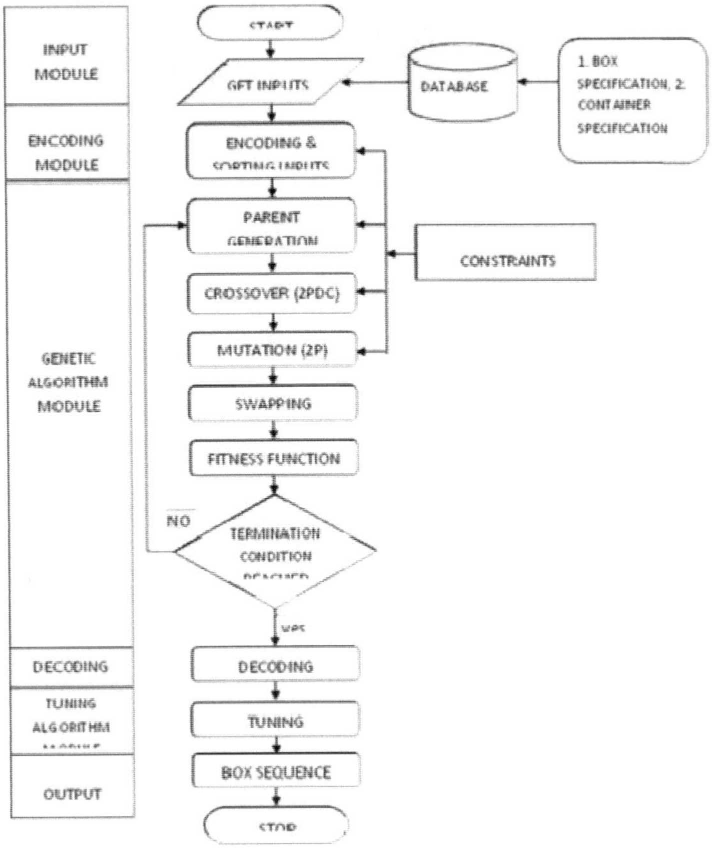

Figure 3. Over view of the solution method

Traditionally, solutions are represented in binary as strings of 0's and 1's, but other encodings are also possible. Decimal numbers, alphabets, alphanumeric values, ASCII values can also be used. The population size depends on the nature of the problem, but typically contains several hundreds or thousands of possible solutions [11]. Traditionally, the population is generated randomly, covering the entire range of possible solutions (the search space)..

12.2 Reproduction

Reproduction or cross over is the second stage of GA, where the properties of the parents mixed to form best solution. The number of cross over site will depend up on the size of each parent. As the size increases, the number of cross over site should be increased to avoid the stagnation at a point. I.e. the set of strings may repeat in all the generation. Percentage of cross over also very from application to application. Many crossover techniques exist for organisms which use different data structures to store themselves.

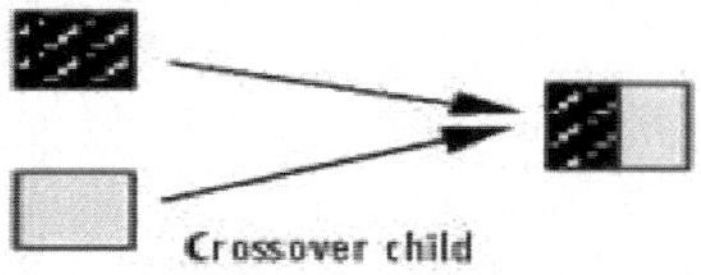

Figure 4. Crossover child

The various crossover techniques commonly used by the researchers are shown in Figure 4

- One Point Crossover
- Two Point Crossover
- Multi Point Crossover
- Uniform Crossover
- Half uniform Crossover
- Arithmetic Crossover

12.3 Mutation

Sometimes, the Fitness function value of the obtained offspring's may seem to stagnate around the optimal point [1]. Because the same strings may repeats again and again in the offspring leads to stagnation. This problem can be solved by applying the mutation operator, which helps to attain the best optimal point. One of the best known mechanisms for producing variations is mutation, where new trial solutions are created by making small, random changes in representation of prior trial solutions. If binary representation is used, the

Mutation is achieved by randomly flipping one bit (figure).

Generated Offspring	:	000|00000	before mutation
Mutant Offspring	:	000|10000	after mutation

Figure: Mutation

Fitness Function

The fitness function measures the quality of the generated solution. The fitness function is always problem dependent. Once the fitness function was defined, GA proceeds to generations repetitively by applying crossover, mutation, inversion and selection operators to produce optimal solution. Fitness function is a minimization or a maximization function used to calculate the relative value of the obtained solution with the expected.

12.4 Selection

Selection operation plays a major role in reducing the computational time. Proper selection of the parent will leads to faster convergence. Selection process may be by random process or by sequential process. Commonly used selection method is Roulette–wheel selection.

12.5 Swapping

Sometimes even after many iterative results, the strings in the starting position of the parents in the population remains same, even after many generations and may lead to stagnation. This can be avoided by using swapping operator. Swapping is a special type of Genetic operator used in this research, to widen the search space instead of being focused towards particular local optima. In this research, a single point random swapping was implemented. Strings beyond the swapping point should swap to the front, so that the strings get interchanged with in the same parent.

12.6 Termination

This is the stage in which the optimal solution was identified. Commonly, the algorithm terminates when either a maximum number of generations has been produced, or a satisfactory fitness level has been reached for the population. The termination conditions may vary from application to application and some of the commonly used termination conditions are.

1. If the solution reached the minimum criteria, fitness function value reaches the desired minimum value.
2. If the fixed number of generation reached.
3. If the successive iterations no longer produce the better results, fitness function value remains the same for continuous iterations.
4. If allotted time/money/cost reached. By manual inspection.

13. Conclusion

In this Work, heuristic GA used for effective packing was described to solve multi constrained packing problem. It seems like, that these methods can be effectively used for these kinds of logistic problems. In the previous chapters, it was described, why to use and why not to use these algorithms. Solving the multi constrained bin packing problems are really hard. That is the reason in this work; GA is used to create a effective solution to solve optimization problems in a rather effective way.

14. References

[1] Goldberg D E 1988 Genetic Algorithm in Search*Optimization and Machine Learning* Addison Wesley.

[2] Mitchell M 1996 An Introduction to Genetic AlgorithmMIT Press.

[3] Hopper E and Turton B 1997Application of Genetic Algorithm to Packing Problems – A Review *Springer Verlag*, London, 279 -288.

[4] Korf R 2002 A New Algorithm of Optimal Bin PackingIn *proc. AAAI* 731-736.

[5] Korf R 2003 A Improved Algorithm for Optimal Bin Packing In proc. IJCAI 1252 – 1258.

[6] Pisinger D2002 Heuristic for Container Loading Problem*European Journal of Operation Research***141**292 – 382.

[7] Gehring H and Bortfeldt A 1997 A Genetic Algorithm for Solving the Container Loading Problem*International transactions in Operation Research***44** 401 – 418.

[8] Dowsland K A and Herbet E A 2004 Using Tree Search Bounds To Enhance A Genetic Algorithm Approach Two Rectangle Packing Problems *European Journal of Operation Research***168**390-402.

[9] Martello S and Pisinger D 2000 The Three Dimensional Bin Packing Problem *Operation research***48**256-267.

[10] Bischoff EE 2004 Three Dimensional Packing of Items with Limited Load Bearing Strength *European Journal of Operation Research***168** 952-966.

[11] Bortfeldt A and Gehring H2001 A Hybrid Genetic Algorithm for Container Loading Problem *European Journal of Operation Research***131**143 -16.

CRITICAL MACHINE BASED SCHEDULING -A REVIEW

P Vivek[1], R Saravanan[2], M Chandrasekaran[3] and R Pugazhenthi[4]
[1]Assistant Professor, Department of Mechanical Engineering, Vels University, Chennai, Tamilnadu, India.
[2]Professor, Department of Mechanical Engineering, Ellenki College of Engineering and Technology, Hyderabad, Telangana, India.
[3]Professor, Department of Mechanical Engineering, Vels University, Chennai, Tamilnadu, India.
[4]Associat Professor, Department of Mechanical Engineering, Vels University, Chennai, Tamilnadu, India.

E-mail: vivek.acpalanisamy@gmail.com

Abstract. This article aims to identify the natural occurrence of the critical machines in scheduling. The exciting scheduling in the real time manufacturing environment is focused on considering equal weight-age of all the machines, but very few researchers were considered the real time constraint(s) like processor/ machine/ workstation availability, etc.,. This article explores the gap between the theory and practices by identifying the critical machine in scheduling and helps the researcher to find the suitable problem in their case study environment. Through the literature survey, it is evident that, in scheduling the occurrence of the critical machine is in nature. The critical machine is found in various names and gives a various range of weight-age based on the particular manufacturing environment and it plays a vital role in scheduling which includes one or more circumstances of occurrence in the production environment. Very few researchers were reported that in manufacturing environment, the critical machine occurrence is in nature, but most of the researchers were focused to optimize the manufacturing environment by only reducing the cycle time. In real-time manufacturing environment, the scheduling of critical machine(s) was keenly monitored and some weight-age was considered.

1. **Introduction**

Nowadays the customer delight is an important criterion in the business environments. The customization products arise diversity of products. Such a global vigorous industrialized scenario intends the manufactures to produce a diversity of products within shortest makespan. On the other hand, utilization of resources, availability of space, fund, and time bound requirements, etc. demands the real time executable manufacturing plan. The Sequencing and Scheduling play imperative role in manufacturing planning in the industries. To avoid the loss of revenues or heavy penalty, etc., for not meeting the deadline.

Today the customer's priority for providing business is based on the earliest delivery. Manufacture a product is not a simple thing it involves many tasks like systematize, prefer, and timing resource to bring out all necessary activities to it in volume (desired output) at the desired time (on or before the due date), while satisfying a large number of times related constraints among the involved activities and the resources [1,2]. Scheduling problems criteria can be broadly classified as throughput related and cost related [2,3]. The Scheduling problem generally defined by the job which require a number of operations, Number of Jobs to be processed, the nature of machines and its configuration in the environment, single or multiprocessors, if the parallel machines available means which are identical/similar/Unrelated, number of stages, etc. job assignment priorities/ disciplines and the objective or criteria of scheduling [4,5]. Graham et al., and Allahverdi et al., furnished some convenient notations to refer the scheduling models with different conditions [6,7]. In general flowshop problems are using the following notation for denoting the standard fields $\alpha / \beta / \gamma$. The first field α describes the scheduling type; the second field β is reserved for the information and conditions of scheduling, while the third field γ contains the performance criteria.

2. Scheduling environments

The scheduling environment natures are day-by-day incarnation as a form of a new version. The chronological order according to the literature is: Single machine scheduling, Parallel machine scheduling, flow shop and job shop scheduling, hybrid flowshop scheduling, Parallel flowshop schedules and parallel hybrid flowshop scheduling. The insertion of new facilities based on the job's operation requirements in the order is the reason behind the evolution of flowshops. The inclusion of standby machine or rental machine or additional machine to line balancing, etc., are the reason of flexible flow shops (Hybrid of Jobshop and Flowshop). The various scheduling environments and their description furnished in Table 1.

3. Scheduling challenges

The Challenges in scheduling for the manufacturing environment still exist. Even though the single machine scheduling is live in the research. Yoav Ben-Yehoshua and Gur Mosheiov [8] dealt single machine scheduling problem and they proved that the problem is NP – Hard. They introduce a pseudo-polynomial dynamic programming algorithm, for verifying NP-hardness of the problem in the ordinary sense. In the parallel machine environment, the parallel jobs scheduling is a new kind of environment applicable to software industries. Recently Yongsheng Hao et al., [9] proposed an adaptive algorithm used in scheduling the modular non-linear parallel jobs in meteorological Cloud. The adaptive algorithm has a distinctive parallelism that only can be configured at the very beginning of the execution.

Table1. Scheduling Environments

Scheduling Environments	Description	Reference
Single machine	Assigning a set of tasks to a single resource or machine	8
Parallel jobs on parallel machines.	Some jobs can only be processed on several processors in parallel.	9
Identical Parallel Machines	The processing time of a task is independent of the machine where it is processed	10
Uniform Parallel Machines	Each machine has a different processing speed.	11
Unrelated Parallel Machines	There is no particular relationship between the processing times in the different machines, i.e., there is no proportionality between the processing time of a task on a given machine and the time of the same task on another machine.	12
Permutation Flow shop	Jobs are processed by a series of 'm' machines in exactly the same order	13
Non-Permutation Flow shop	Jobs are processed by a series of 'm' machines not in the same order	14

Job shop	Multiple entries and exits.	15
Hybrid Flow shop with uniform parallel machines	The processing time of a job on a particular machine is the ratio of the processing time of the job on a machine with a standard speed to the speed of the particular machine	16
Hybrid Flow shop with unrelated parallel machines	Different parallel machines at every stage, and speeds of the machines are dependent on the jobs	17
Hybrid Flow shop with identical parallel machines	Parallel machines at a stage are identical in their processing speeds	18
Parallel Flowshop	Parallel-machine Flowshop	19
Parallel Hybrid Flow shop	Hybrid flowshops are connected parallel at the critical machine for the common operation/ process.	20
Open shop	There are 'm' machines and there is no restriction in the routing of each job through the machines. In other words, there is no specified flow pattern for any job.	21
Closed shop	It is a job shop; however, all production orders are generated as a result of inventory replenishment decisions. In other words, the production is not affected by the customer order.	22
Static Type Shops	Jobs available at shop at zero time of execution.	23
Dynamic Type shops	The decision of scheduling a job can only be made online or job arrives over time to the shop.	24

The identical, uniform and unrelated (Non-uniform) parallel machines scheduling also recently solved with the new kinds and combination of constraints [10-12]. The flowshops with permutation [13] and non-permutation [14] were solved with the objective of Minimization of total flow time. The flexibility of job shop leads to multiple entries at a time can be used for High variety low volume production strategies [15]. The hybridization of flow shop and job shop/ Parallel Machines natures to enhance the flexibility of mass production environment incarnates flexible flow shop or Hybrid flow shop. The parallel machines nature like identical [16], uniform [17] and unrelated [18] parallel machines natures are the additional constraints in the flexible flow shop problem.

The evolution of flexible flow shop is the machine insertion with critical machine. The day by day the manufacturing environment scheduling is getting more complex by bulk additional constraints or tremendous changes in existing configurations. The parallel flow shops scheduling (flowshops with parallel machines) and parallel Hybrid flowshops are example for them [19, 20]. The open shop and closed shop nature, dynamic job/ machine [21-23] (stochastic) availability and are static (deterministic) also the notable scheduling environments [24]. The most of the problems in the literature solved with the ultimate criteria of throughput and cost measures and assumed that machines/ facilities are available continuously, set up times neglected, etc. critical machine based scheduling is a new aspect to provide an executable schedule to the environment.

4. Research contribution

There is a huge contribution of research in scheduling. The large volumes of literature were referred and estimated the promotional contribution in environment wise. The specific contribution of the environment is shown in the Figure 1, the criteria wise contribution in Figure 2 and the deterministic and stochastic scheduling contribution in Figure 3. However, in each case of scheduling, there was one or more critical machine(s) and they are not explored transparently in most of the literature.

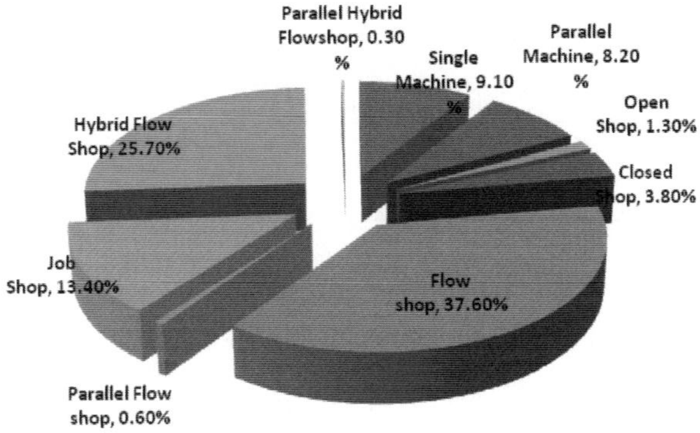

Figure 1. Scheduling Environment wise Research Contribution

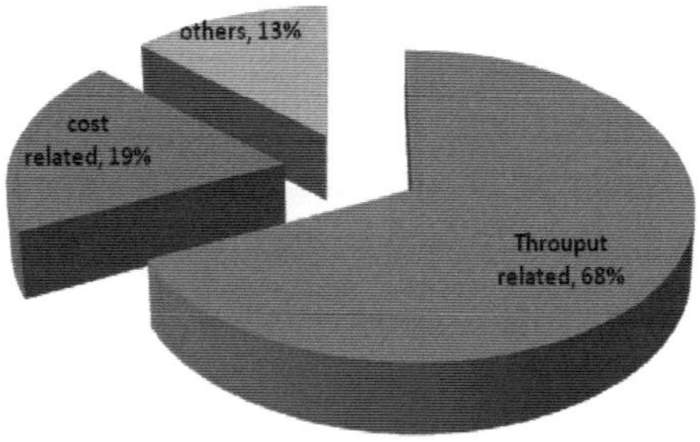

Figure 2. Scheduling Criteria wise Research Contribution

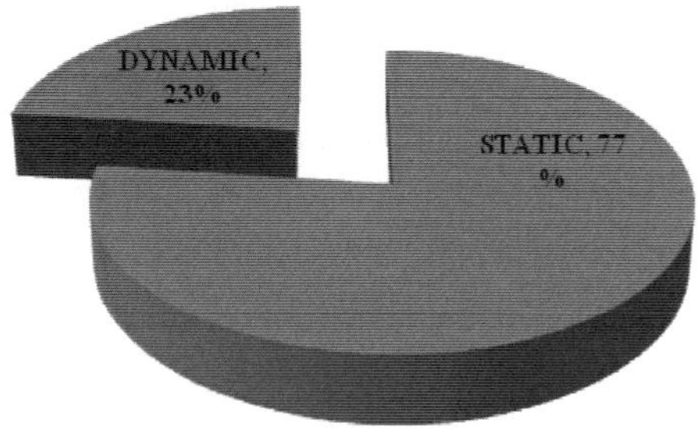

Figure 3. Job/Machine Nature wise Research Contribution in Scheduling

5. Critical machines based scheduling

The critical machine is one, which is most crucial or vital machine in a shop floor, which agrees with the objective and effectiveness of the scheduling environment. It arises in the source for many scheduling environments based on the manufacturing method employed, raw material availability, to avoid penalty for tardiness, require to meet the due date, optimize the cost, and so on. Some example cases are provided in the Table 2. The critical machines based scheduling is has arises a special interest, in recent era. The automatic machine scheduling in one of the examples of critical machine based scheduling. Jiabin Luo et al., [25] investigated the integration of the two problems focusing on the unloading process in an automated container terminal, where all or part of the equipment are built in automation and proposed a genetic algorithm to solve the small and large sized problems. Blender and feeder are very important devices in a polymer processing plant. The accuracy in delivering the polymer to the plastification unit is a fundamental property for the quality of the final product. Nonlinear gain scheduling controller was proposed to maximize accuracy and precision gravimetric feeder are the best available solution, which measure the weight of the hopper and the feeder [26].

The Blender and feeder are very important devices in a polymer processing plant. The accuracy in delivering the polymer to the plasticisation unit is a fundamental property for the quality of the final product. Hence Belloli et al., [27] used a nonlinear gain scheduling controller to maximize accuracy and precision gravimetric feeders are the best available solution, which measure the weight of the hopper and the feeder. Jing-hua Hao et al., [28] dealt a hybrid approach for solving the bottleneck stage scheduling problem (BSP) using the surrogate modelling technique. Sascha Herrmann and Christoph Schwindt [29] considered the continuous process case of the short-term production planning of multi-product continuous plants in the process industries. Due to constraints on material availability and storage capacity for intermediate products, classical schedule-generation schemes cannot be applied to the scheduling problem and they proposed a new two-phase approach dealing with the two types of constraints separately.

In many continuous process industries, machine breakdowns may affect the planned activities, especially when maintaining high shop performance is desired. Liao and Chen [30] considered a practical problem encountered in a textile company where the machine breakdown occurs frequently and provided a longer setup time (or equivalently, a longer idle time) to reduce the breakdown rate. They used branch-and-bound algorithm for the scheduling. The neglect of buffering requirements in a classical job shop scheduling system often results in inapplicability in many complex real-world applications. So the Shi Qiang Liu and Erhan Kozan [31] made an attempt to overcome this inapplicability, a new and more generalised scheduling problem is proposed under different stage-dependent buffering requirements and parallel use of identical-function machine units at each processing stage in job shop environments. The authors proposed DBPMJSS methodology has a potential to analyse, model and solve many industrial systems with the requirements of buffering conditions, particularly in manufacturing, railway, healthcare and mining industries.

A heuristic approach was used by Enrique Gerstl and Gur Mosheiov [32] to solve at two-stage flowshop, where each job processes in the first (critical) machine, and then continues to one of two second-stage (dedicated) machines. The measure of the performance is to minimize the makespan, total load, and weighted flow-time. Omar Souissi et al., [33] investigated on the single machine scheduling problem taking into account preventive maintenance failures and proposed a hybrid approach by combining mixed integer program and longest processing time first) heuristic in order to solve that problem. The authors used simulation approaches to deal with the stochastic failures.

Table-2. Some samples of occurrence of Critical Machine

S. No.	Critical Machine	Sample Environment	Reference
1	Automated machine	Decentralized scheduling with heterogeneous machines and competing job sets.	25
2	Blender Machine	Gain Scheduling Control of a Gravimetric Blender with Vibrant Duct	26
3	Bottleneck machine	Scheduling bottleneck stages	27
4	Continuous processing machine	Machine breakdown in a continuous process industry	28
5	Breakdown nature machine	Scheduling under machine breakdown	29
6	Buffer or delay required machine	Stage-dependent buffering requirements	30
7	Dominant machine	Deteriorating jobs on no-idle dominant machines	31
8	First stage machine	Scheduling with a critical machine	32
9	High maintenance required machine	Scheduling with preventive maintenance	33
10	Critical machine	Scheduling critical / rental machine with optimal makespan	34

Vivek et al., [34] analyzing the critical machine based permutation flow environment through the heuristic optimal makespan obtained. Pugazhenthi R and Xavior M A [35 and 36] deals the rental/critical machine in a flowshop; in this a new heuristic was proposed to minimise the idle time of the machine. The effect of grouping technique was studied in a real time multi stage hybrid flowshop environment; the critical machine was considered and an optimum makespan obtained. Pugazhenthi and Anthony Xavior were highlighted the critical machines in various names i.e., High bound machines, Maximum lateness machine, Minimum weighted machine, Slowest processing machines, Parallel machine, Rental machine, Machines in critical path, Identical machine, Maximum work load machine, Semi-automated machine, High maintenance required machine, Minimum tardiness machine, Pivoted machine, High cost machine, Maximum processing machine, Identical parallel machine, Repeatedly shifting machine, Short run machine, Special machine, High operating cost machine, etc., [21].

6. Finding for further research

The present industrial scenario is to produce High Variety at Low Volume with Highest Quality at lowest prices to sustain in the world class market. Hence scheduling plays key role in those industries and includes one or more circumstances of occurrence of the critical machines in the manufacturing environment. For example, it is not economical to procure high cost machines for low volume of production, so the machine may hire on rental basis.

- The cost based criteria paid less attention in the literature and the cost based criteria may fulfil by changing the orientation of scheduling in a critical machine based scheduling.
- The dynamic scheduling, paid less attention by the researcher, the computerization made many things on online and availability of machine may not be dedicated for a single purpose. The stochastic scheduling environment may obvious in forthcoming days. So need more attention in stochastic scheduling (dynamic scheduling).
- Many cases, the occurrence of critical machine(s) are evidenced even though in the scheduling environment, they were not utilized well for meeting the objectives. The new aspect of scheduling may save cost and time effectively.

7. References

[1] Baker K R 1974 Introduction to Sequencing and Scheduling *Wiley New York.*

[2] Morton T E and Pentico D W 1993 Heuristic scheduling systems *John Wiley and sons* New York.

[3] Rinnooy Kan A H G 1976 Machine Scheduling Problems: Classification, Complexity, and Computations *Martinus Nijhoff* The Hague.

[4] French S 1982 Sequencing and Scheduling: An Introduction to the Mathematics of the Job-Shop *New York John Wiley and Sons.*

[5] Conway R W, Johnson B M and Maxwell W L 1960 An Experimental Investigation of Priority Dispatching *Journal of Industrial Engineering* **11(3)** 221-229.

[6] Graham R L, Lawler E L, Lenstra J K, Rinnooy Kan A H G 1979 Optimization and approximation in deterministic sequencing and scheduling: a survey *Ann. Discrete Math.* **5** 287–326.

[7] Allahverdi J N D, Gupta T and Aldowaisan 1999 A review of scheduling research involving setup considerations *Manage. Sci.* **27** 219–239.

[8] Yoav Ben-Yehoshua and Gur Mosheiov 2016 A single machine scheduling problem to minimize total early work *Computers & Operations Research* **73** 115-118.

[9] Yongsheng Hao, Lina Wang and Mai Zheng 2016 An adaptive algorithm for scheduling parallel jobs in meteorological *Cloud Knowledge-Based Systems* **98** 226-240.

[10] Stefan Schwerdfeger and Rico Walter 2016 A fast and effective subset sum based improvement procedure for workload balancing on identical parallel machines *Computers & Operations Research* **73** 84-91.

[11] Qian Cao and Zhaohui Liu 2016 Semi-online scheduling with bounded job sizes on two uniform machines *Theoretical Computer Science.*

[12] Omid Shahvari and Rasaratnam Logendran 2017 An Enhanced Tabu search algorithm to minimize a bi-criteria objective in batching and scheduling problems on unrelated-parallel machines with desired lower bounds on batch sizes *Computers & Operations Research*, **77** 154-176.

[13] Hamid Abedinnia, Christoph H Glock and Andreas Bril 2016 New simple constructive heuristic algorithms for minimizing total flow-time in the permutation flowshop scheduling problem, *Computers & Operations Research* **74** 165-174.

[14] Behdin Vahedi-Nouri, Parviz Fattahi and Reza Ramezanian 2013 Minimizing total flow time for the non-permutation flow shop scheduling problem with learning effects and availability constraints *Journal of Manufacturing Systems* Vol.32 (1) Pp. 167–173.

[15] Kuhpfahl and Jens 2015 Job Shop Scheduling with Consideration of Due Dates: Potentials of Local Search Based Solution Techniques. *Springer*.

[16] Sushil Verma and Maged Dessouky 1999 Multistage Hybrid Flowshop Scheduling with Identical Jobs and Uniform Parallel Machines *Journal of Scheduling* 2 1-16.

[17] Chun-Lung Chen and Chuen-Lung Chen 2009 A bottleneck-based heuristic for minimizing makespan in a flexible flow line with unrelated parallel machines *Computers & Operations Research* **36(11)** 3073-3081.

[18] He Long-min and Sun Shi-jie 2007 A hybrid two-stage flexible flowshop scheduling problem with m identical parallel machines and a burn-in processor separately *Journal of Shanghai University (English Edition)* **11(1)** 33–38.

[19] Saravanan Rathinasamy and Raju R 2010 Sequencing and scheduling of nonuniform flow pattern in parallel hybrid flow shop *The International Journal of Advanced Manufacturing Technology* **49**(1) 213–225.

[20] Deepak Gupta, Shashi Bala and Payal Singla 2012 Optimal Two Stage Open Shop Specially Structured Scheduling To Minimize the Rental Cost, processing time Associated with Probabilities including transportation time *IOSR Journal of Mathematics* **3(3)** 01-06.

[21] Pugazhenthi R and Anthony Xavior M 2014 A Survey on Occurrence of Critical Machines in a Manufacturing Environment *Procedia Engineering* **97** 105-114.

[22] Chaouqi Mohsine, Benhra Jamal and My Ali El Oualidi 2016 Genetic Algorithm Tuning Applied to the Open Shop Scheduling Problem *International Journal of Applied Information Systems* **11(3)** 21-25.

[23] Atif Shahzad and Nasser Mebarki 2016 Learning Dispatching Rules for Scheduling: A Synergistic View Comprising Decision Trees *Tabu Search and Simulation, Computers* **5 (3)** 1-16.

[24] Dong Cao and Mingyuan Chen 2003 Parallel flowshop scheduling using Tabu search *International Journal of Production Research* **41**(13) 3059-3073.

[25] Jiabin Luo, Yue Wu and André Bergsten Mendes 2016 Modelling of integrated vehicle scheduling and container storage problems in unloading process at an automated container terminal *Computers & Industrial Engineering* **94** 32-44.

[26] Belloli D, Previdi F, Cologni A, Madaschi M and Cazzola D 2013 Gain Scheduling Control of a Gravimetric Blender with Vibrant Duct *IFAC Proceedings* **46(2)** 785-790.

[27] Jing-hua Hao, Min Liu, Jian-hua Lin and Cheng Wu 2016 A hybrid differential evolution approach based on surrogate modelling for scheduling bottleneck stages *Computers & Operations Research* **66** 215-224.

[28] Sascha Herrmann and Christoph Schwindt 2007 Planning and Scheduling Continuous Operations in the Process Industries *Management logistischer Netzwerke Physica-Verlag* 3 79-299.

[29] Lia C J and Chen W J 2004 Scheduling under machine breakdown in a continuous process industry *Computers and Operations Research* **31(3)** 415 – 428.

[30] Shi Qiang Liu and Erhan Kozan 2016 Parallel-identical-machine job-shop scheduling with different stage-dependent buffering requirements *Computers & Operations Research* **74** 31-41.

[31] Linhui Sun, Linyan Sun, Kai Cui and Ji-Bo Wang 2010 A note on flow shop scheduling problems with deteriorating jobs on no-idle dominant machines *European Journal of Operational Research* **200(1)** 309-311.

[32] Enrique Gerstl and Gur Mosheiov 2012 two-stage flow shop scheduling with a critical machine and batch availability *Foundations of Computing and Decision Sciences* **37(1)** 39–56.

[33] Omar Souissi, Rachid Benmansour and Abdelhakim Artiba 2016 An accelerated MIP model for the single machine scheduling with preventive maintenance *IFAC-PapersOnLine* **49(12)** 1945-1949.

[34] Vivek P, Saravanan R, Chandrasekaran M and Pugazhenthi R 2016 Minimizing Idle Time of Critical Machine in Permutation Flow Environment With Weighted Scheduling *ARPN Journal of Engineering and Applied Sciences* **11(5).**

[35] Pugazhenthi R and Xavior M A 2016 A heuristic to minimise the idle time of the rental/critical machine in a flowshop *International Journal of Services and Operations Management* Inderscience Publications **22(4)** 395-412.

[36] Pugazhenthi R, Anthony Xavior M and Saravanan R 2016 A case study on effect of grouping technique in a multi-stage hybrid flow shop *International Journal of Computing Science and Mathematics* **7(1)** 42-53.

Properties of Concrete partially replaced with Coconut Shell as Coarse aggregate and Steel fibres in addition to its Concrete volume

Kalyana Chakravarthy P R[1], Janani R[2], Ilango T[3], Dharani K[4]
[1,2,3,4]Assistant professors, Department of Civil Engineering, Vels University, Chennai, India.

E-mail: kalyanstructure12@gmail.com

Abstract. Cement is a binder material with various composition of Concrete but instantly it posses low tensile strength. The study deals with mechanical properties of that optimized fiber in comparison with conventional and coconut shell concrete.The accumulation of fibers arbitrarily dispersed in the composition increases the resistance to cracking, deflection and other serviceability conditions substantially. The steel fiber in extra is one of the revision in coconut shell concrete and the outcome of steel fiber in coconut shell concrete was to investigate and compare with the conventional concrete. For the given range of steel fibe from 0.5 to 2.0% , 12 beams and 36 cylindrical specimens were cast and tested to find the mechanical properties like flexural strength, split tensile, impact resistance and the modulus of elasticity of both conventional and coconut shell concrete has been studied and the test consequences are compared with the control concrete and coconut shell concrete for M25 Grade. It is fulfilled that, the steel fibers used in this venture has shown significant development in all the properties of conventional and coconut shell concrete while compared to controlled conventional and coconut shell concrete like, Flexural strength by 6.67 % for 1.0 % of steel fiber in conventional concrete and by 5.87 % for 1.5 % of steel fiber in coconut shell concrete.

1. Introduction

Concrete is a complex substance containing the multiple components in various composition. Even though, coarse aggregate constitutes the major volume and capacity of concrete, material remains a sky scraping claim. From the stand point of energy cutback and protection of natural wealth, the use of different constituents which should be light in mass to reduce the self weight of concrete in building supplies is now a worldwide fear. However, plain concrete possesses very low tensile strength, partial ductility and modest opposition to cracking. This absence has been eliminated by using fibers into the concrete. It will help in safeguarding and repairing of concrete structures.

1.1: Light weight concrete

The demand to make concrete as lighter is of quite important. LWC is made by diminishing the compactness while maintaining the strength and without unfavourably affecting the price. Introducing a new aggregate into the mix design is a familiar way to subordinate concrete mass. LWC is characteristically deals with reusing waste as aggregates towards economical growth.

1.2: Coconut shell concrete
The budding of waste coconut shell are used as a alternative for coarse aggregate in concrete. After the coconut is tattered out, the shell is regularly discharged. The bulk density of coconut shell is about 500 to 600 kg/m3, producing concrete of about less than 2000 kg/m3 in density, which makes them lightweight.The coconut shell concrete straight forwardly attains the strength around 17 N/mm2. For the past 10 years light weight concrete getting a maximum hold in the construction industry

1.3: Steel fiber reinforced concrete
Steel fibers used in concrete are of limited volume fraction. The fibers of various cross sections are used. Hooked and deformed ends of fibers improve the fatique or impact strength. For high temperature applications up to 1650 deg.-C, like in refractory structures, stainless steel fibres are used. Corrosion of steel fibers in concrete with a high water to cement ratio may cause deterioration. When fibers were exposed on a surface, they showed evidence of corrosion; however, internal fibers showed no corrosion.

1.4: Need for the project
The needs for this project are as follows:
To recover the tensile strength of the plain cement concrete with coconut shell steel fibers are to be used and to judge the results with conservative concrete with steel fibers.

1.5: Objectives of the project
The main objectives of this project are as follows:
To optimize the volume fraction of the selected aspect ratio of steel fiber in coconut shell concrete.
To study the mechanical properties of that optimized fiber in comparison with conventional and coconut shell concrete.

1.5: Scope of the project- The scope of this learning is to analyse the mechanical properties of coconut shell concrete added with steel fibres and increasing the ductile property in the manufacturing of concrete. Steel fiber is added in different percentage from 0.5 % to 2 % to coconut shell concrete and compared to conventional concrete.

2. Methodology
To study the behaviour of coconuts shell with various volume fraction of steel fibre in concrete was studied through the following methodology is shown in the figure 1 .

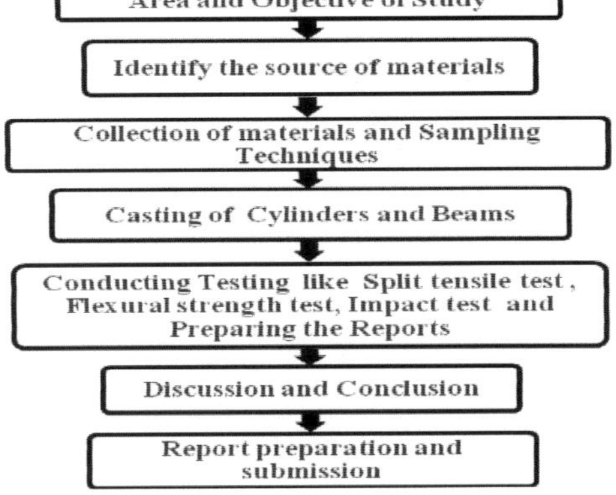

Figure 1. Methodology

3. Materials

In this experimental study the composition of concrete materials like cement, crushed granite stone, coconut shell, fine aggregate, steel fiber and water used were described as follows:

3.1 Cement

The cement used in this study is Ordinary Portland Cement (OPC) and is conforming to IS: 12269-1987. The cement properties are listed in then table1.

Table 1 Properties of cement

Description	Test Results	As per IS: 12269-1987
Fineness, %	4.12	Maximum 10%
Normal consistency, %	31	27 – 33
Initial setting time, min.	76	Not less than 30
Final setting time, min.	430	Not more than 600
Specific gravity	3.1	3.10 – 3.15

3.2 Crushed Granite Stone

The stone aggregate of size around 12.5mm is chosen as coarse aggregate for control concrete chosen by shape as per IS: 2386 (Part 1)-1963 and surface quality description of the aggregate are classified as per IS: 383-1970. The properties of crushed granite stone were shown in table 2.

Table 2 Properties of crushed granite stone

Description	Test Results	As per IS: 383-1970
Fineness modulus	6.94	5.00 – 7.00
Specific gravity	2.82	2.60 – 2.90

3.3 Coconut Shell Aggregate

The freshly unused coconut shell collected from industries was second-handed. The various size of coconut shells are used. The figure 2 shows coconut shell aggregate before crush and Figure 3 shows coconut shell after crush. The properties of coconut shell were shown in table 3.

Table 3 Properties of coconut shell

Description	Test Results
Fineness modulus	6.26
Specific gravity	1.05

Figure 2. Coconut shell before crush

Figure 3. Coconut shell after crush

3.4 Fine Aggregate

The sand used for the research programme was taken from nearby river sand and conformed to grading II as IS: 383-1970. The sand was first passed in 4.75mm sieve to eliminate any particles superior than 4.75mm. The properties of fine aggregate were shown in table 4.

Table 4 Properties of fine aggregate

Description	Test Results
Fineness modulus	2.56
Specific gravity	2.57

3.5 Steel Fiber

Steel fibers are added at various proportions from 0.5 % to 2.0 % to find the optimum level both in control concrete and coconut shell aggregate concrete by volume fraction. The properties of steel fiber used were shown in able 5.

Table 5 Properties of steel fiber

Parameters	Value
Length, mm	50
Diameter, mm	1
Aspect ratio (l/d)	50
Tensile strength, N/mm2	1225
Elastic modulus, N/mm2	210000

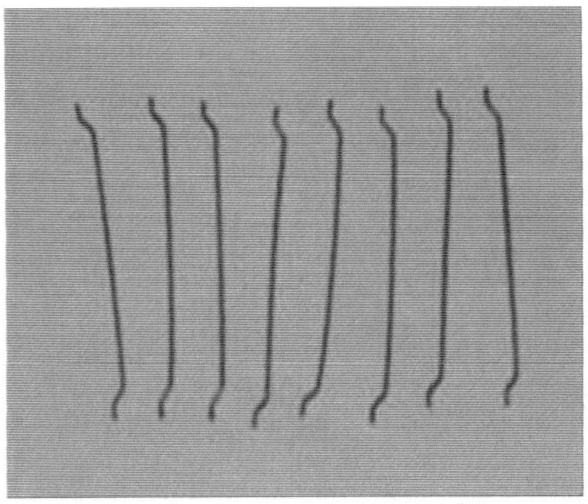

Figure 4. Hooked end steel fibers

Figure 5. Fresh concrete specimens

3.6 Water

Potable water is mostly initiated in fresh concrete and curing of specimens which remain free of charge from all impurities.

3.7 Specimen details

Rectangular cast of size 100 mm × 100 mm × 500 mm were used to prepare the concrete specimens for the determination of flexural strength. Cylindrical mould of size 200 mm height and 100 mm diameter were used to prepare the concrete specimens for the determination of split tensile strength and modulus of elasticity and cylindrical mould of size 63.5 mm height and 152.4 mm diameter were used to prepare the concrete specimens for the determination of impact resistance.

3.8 Modulus of Elasticity of Concrete

The test specimens used for determining elastic modulus were 100 mm diameter and 200 mm height. During this test, deflection at each 5000 N load was recorded. Figure 4 shows the fresh concrete cube, beam and cylindrical specimens. Figure 5 shows the fresh concrete specimens

3.9 Recent studies on Concrete replacement

Tomas and Ucol-ganiron [1] the study remains about the design a technical specification of concrete hollow block using coconut shell and fiber as aggregates that will meet the ASTM requirements in order to help contribute to the industry in saving the environment. A conventional concrete hollow block was compared to concrete hollow blocks with coconut shells and fibers of the same proportions. Observations from the tests performed were conducted in the laboratory where precise data are gathered and completely attained. Vishwas P and Kukarni [2] the three different concrete mixes with different combination of natural material content namely 0 %, 10 %, 20 %, and 30 %. Three sample specimen will be prepared for each concrete mixes. The aim behind this is to use low cost material like coconut shell and thus taking close to the concept of low cost housing. All precaution is taken to maintain serviceability, strength and durability of the members.

Sivakumar A and Manu Santhanam [3] the review shows addition of steel fibers generally contributed towards the energy absorbing mechanism (bridging action) whereas, the non-metallic fibers resulted in delaying the formation of micro-cracks. Compared to other hybrid fiber reinforced concretes, the flexural toughness of steel-polypropylene hybrid fiber concretes was comparable to steel fiber concrete. Increased fiber availability in the hybrid fiber systems (due to the lower densities of non-metallic fibers), in addition to the ability of non-metallic fibers to bridge smaller micro cracks, are suggested as the reasons for the enhancement in mechanical properties. Gunasekaran K and Annadurai R[4] investigated the effects of three types of curing on coconut shell aggregate have been studied for long term performance. The freshly discarded coconut shell collected from local oil mill was used in this study since the different species of coconut shell were processed together, the shells are found to have varying thickness of 2-8 mm. the pore structure of coconut shell has been studied through scanning electron microscope (SEM). The pore structures in coconut shell behave like a reservoir. Coconut shell specimens has very closely spaced discrete cells and also has some long continuous chain linked cells with different widths. There is no bond failure in coconut shell aggregate concrete even at the later ages and maintained good quality for the long-term performance. Song P S and Hwang S [5] investigated that the marked brittleness with low tensile strength and strain capacities of high-strength concrete (HSC) can be overcome by the addition of steel fibers. This paper investigated the mechanical properties of high-strength steel fiber-reinforced concrete. The properties included compressive and splitting tensile strengths, modulus of rupture, and toughness index. The steel fibers were added at the volume fractions of 0.5 %, 1.0 %, 1.5 %, and 2.0 %. The compressive strength of the fiber-reinforced concrete reached a maximum at 1.5 % volume fraction, being a 15.3 % improvement over the HSC.

4. Results and Discussion

The results of cube, cylinder and beam specimens are analysed for the comparison between the conventional concrete (CC) and coconut shell concrete (CSC) specimens and presented.

4.1 Fresh Concrete Properties

Coconut shell aggregate concrete M25 grade has better workability because of smooth measures. The various percentages of steel fibers are added to improve the strength. Fresh concrete properties for conventional concrete and coconut shell concrete added with various percentage of steel fiber are shown in the table 6 respectively.

Table 6 Fresh concrete properties for CC and CSC with various % of steel fiber

% of steel fiber	Conventional concrete		Coconut shell concrete	
	Slump (mm)	Density (kg/m3)	Slump (mm)	Density (kg/m3)
0	10	2503	5	1976
0.5	10	2530	5	1982
1.0	10	2538	5	2039
1.5	10	2546	5	2087
2.0	12	2558	8	2127

4.2 Hardened Concrete Properties

The mechanical properties like flexural strength, split tensile strength, impact resistance and elastic modulus of control concrete and coconut shell concrete along with optimized percentage of steel fiber was studied.

4.3 Flexural Strength

Flexural strength for conventional and coconut shell concrete with optimized percentage of steel fiber are 1.0 % and 1.5 % respectively. The flexural strength was computed from the following equation.

Modulus of rupture = $\frac{PL}{bd^2}$ N/mm^2

where,

P = maximum load applied (N)
L = supported length of the specimen (mm)
b = breath of the section (mm)
d = depth of the section (mm)

The flexural strength test results of conventional concrete and coconut shell concrete without addition of steel fiber and with addition of optimized percentage of steel fibers are shown in table 7.

Maximum flexural strength for conventional and coconut shell concrete without addition of steel fiber is 5.73 N/mm2 and 4.67 N/mm2 respectively. The maximum flexural strength for conventional and coconut shell concrete with addition of optimized percentage of steel fiber is 6.67 N/mm2 and 5.87 N/mm2 respectively. The figure 6 shows the flexural strength of conventional concrete and coconut shell concrete without addition of steel fiber and with addition of optimized percentage of steel fiber at 28 days.

Table 7 Flexural strength of CC and CSC with and without steel fiber

Type of concrete	% of steel fiber added	Flexural strength (N/mm^2)
CC	0	5.73
CSC	0	4.67
CC	1.0	6.67
CSC	1.5	5.87

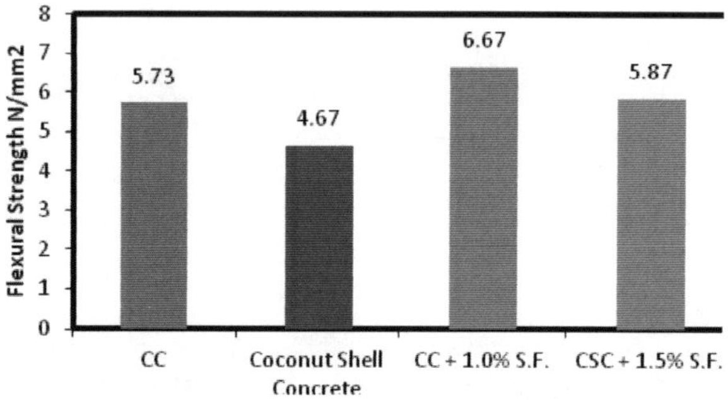

Figure 6. Flexural strength of CC and CSC with and without steel fiber

4.4 Split Tensile Strength

Split tensile strength test for conventional and coconut shell concrete with an optimized percentage of steel fiber are 1.0 % and 1.5 % respectively. Shear and bond strength between aggregate and cement paste and crack resistance are related to tensile strength of coarse aggregate and cement paste.

The splitting tensile strength was computed from the following equation.

$Fsp = \frac{2P}{\pi DL}$ N/mm^2

Where,

Fsp = Split tensile strength (N/mm^2)

P = Load at Failure (N)

D = Diameter of the specimen (mm)

L = Length of the specimen (mm)

The split tensile strength test results of conventional concrete and coconut shell concrete without addition of steel fiber and with addition of optimized percentage of steel fibers are shown in table 8.

Table 8 Split tensile strength of CC and CSC with and without steel fiber

Type of concrete	% of steel fiber added	Split tensile strength (N/mm^2)
CC	0	3.40
CSC	0	2.65
CC	1.0	4.17
CSC	1.5	3.07

Maximum split tensile strength for conventional and coconut shell concrete without addition of steel fiber is 3.40 N/mm2 and 2.65 N/mm2 respectively. Maximum split tensile strength for conventional and coconut shell concrete with addition of optimized percentage of steel fiber is 4.17 N/mm2 and 3.07 N/mm2 respectively. The figure7 shows the split tensile strength of conventional concrete and coconut shell concrete without addition of steel fiber and with addition of optimized percentage of steel fiber at 28 days.

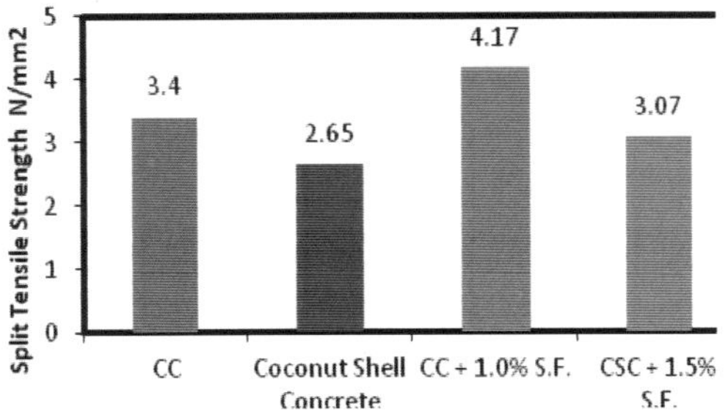

Figure 7. Split tensile strength of CC and CSC with and without steel fiber

4.5 Impact Resistance

Impact test for conventional and coconut shell concrete with optimized percentage of steel fiber is 1.0 % and 1.5 %. The test specimens used for the impact tests were 152.4 mm in diameter and 63.5 mm thick. The number o blows are listed initially or First crack and listed also for the final crack. The impact test results are presented table 9.

Table 9 Impact resistance of CC and CSC with and without steel fiber

Type of concrete	% of steel fiber added	Compressive strength (N/mm2)	Average number of blows for initial crack	Average number of blows for fractured pieces
CC	0	31.40	19	22
CSC	0	25.83	23	28
CC	1.0	37.25	38	42
CSC	1.5	29.12	45	50

The impact resistance of coconut shell concrete is more compared to conventional concrete both for initial and for failure. For conventional concrete without addition of steel fiber an average number of blows observed for making an initial crack is 19 and for making failure is 22 and with addition of optimized percentage of steel fiber for making an initial crack is 38 and for making final crack is 42. For coconut shell concrete without addition of steel fiber an average number of blows observed for making an initial crack is 23 and for making failure is 28 and with addition of optimized percentage of steel fiber for making an initial crack is 45 and for making final crack is 50. Compared to conventional concrete, coconut shell concrete offered impact resistance 20 % more approximately, this increase in impact resistance may be due to the fibrous nature of the coconut shell aggregate and its high impact resistance.

4.6 Modulus of Elasticity of Concrete

The test was carried out conforming to IS: 516-1959 to obtain the stress-strain curve and modulus of elasticity of the concrete. Stress-Strain graph for conventional concrete is shown in figure 8 and for coconut shell concrete is shown in figure 9.

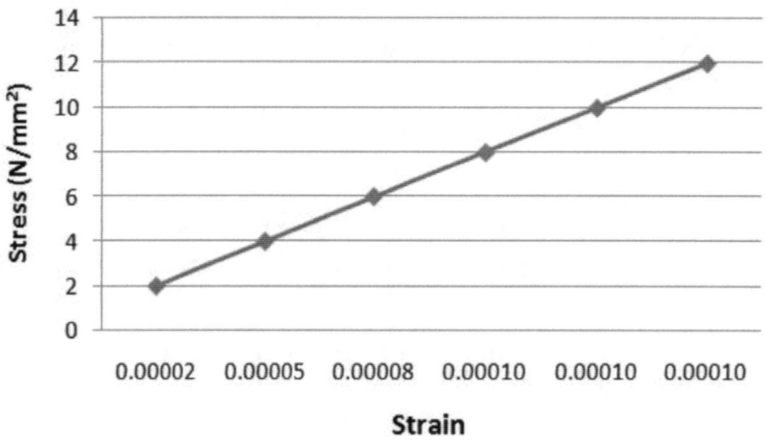

Figure 8. Stress-strain curve for CC

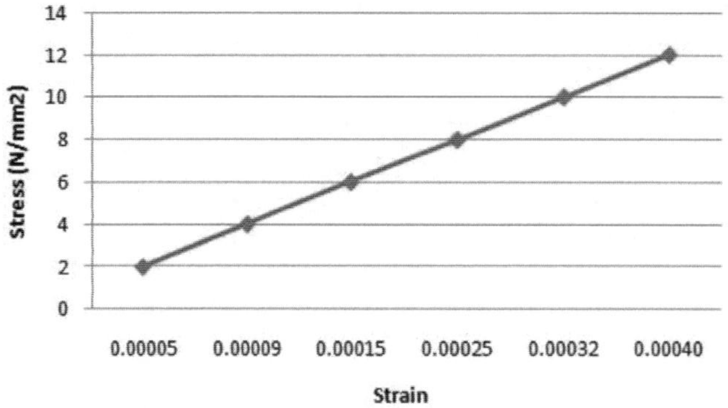

Figure 9. Stress-strain curve for CSC

Stress-Strain graph for optimized percentage of steel fiber in conventional concrete is shown in figure 10 and for coconut shell concrete is shown in figure 11.

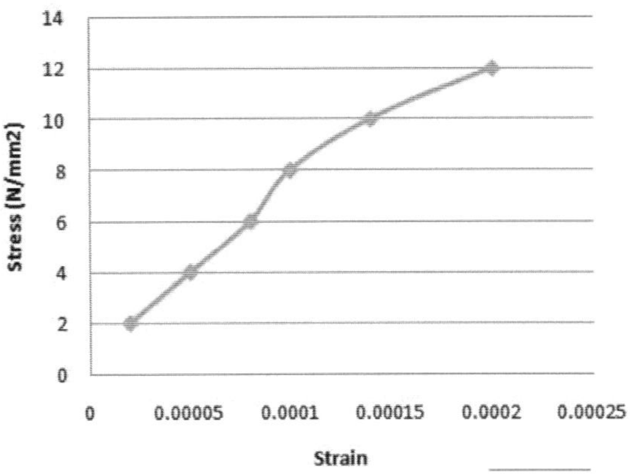

Figure 10. Stress-strain curve for CC with 1.0 % steel fiber

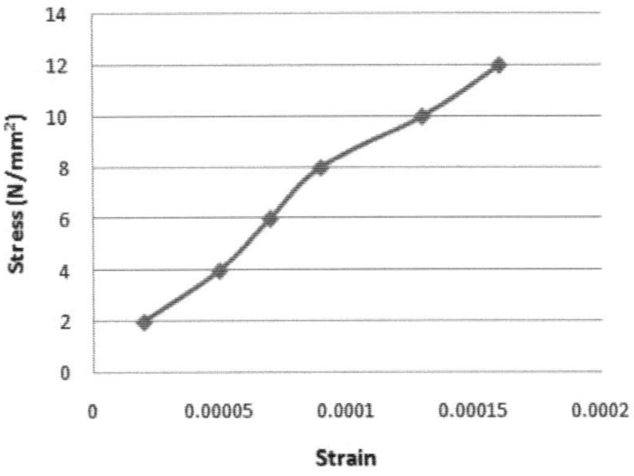

Figure 11. Stress-strain curve for CSC with 1.5 % steel fiber

The figure 12 shows comparison of stress-strain curve for conventional and coconut shell concrete without steel fiber.

From figure 12 it is noted that stress-strain diagram of coconut shell concrete is slightly below that of conventional concrete for a constant stress the strain is more in conventional concrete. Thus the young's modulus of conventional concrete will be more than that of coconut shell concrete.

The modulus of elasticity for conventional concrete and coconut shell concrete are 32608 N/mm2 and 11300 N/mm2 respectively.

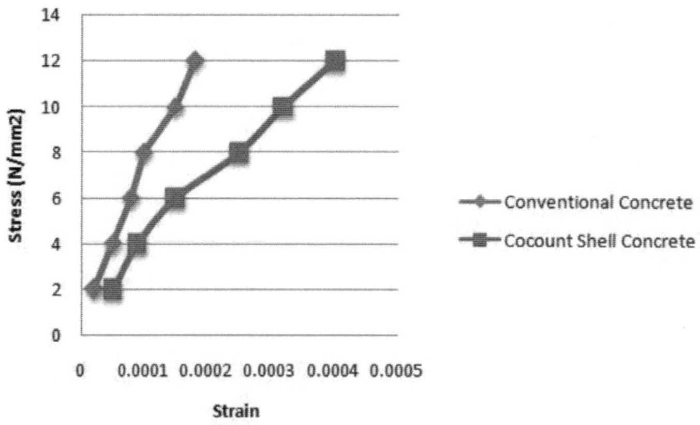

Figure 12. Comparison of stress-strain curve for CC and CSC

The figure 13 shows comparison of stress-strain curve for conventional and coconut shell concrete with optimized percentage of steel fiber.

From figure 13 it is noted that stress-strain diagram of coconut shell concrete is slightly below that of conventional concrete for a constant stress the strain is more in conventional concrete. Thus the young's modulus of conventional concrete will be more than that of coconut shell concrete.

The modulus of elasticity for conventional concrete and coconut shell concrete with optimized steel fiber are 36450 N/mm2 and 12280 N/mm2 respectively.

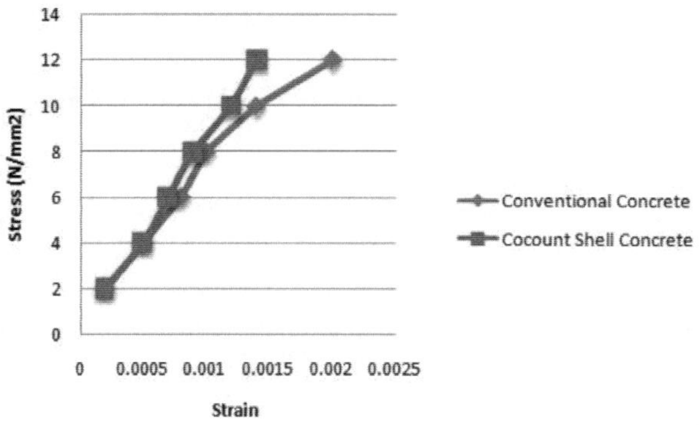

Figure 13. Comparison of stress-strain curve for CC and CSC with optimized % of steel fiber

5. Conclusions

The following conclusions have been drawn from the experimental results of this study.
The addition of steel fibers increases the strength of concrete. The optimized percentage of steel fiber used in conventional concrete is 1.0 % of the total volume of concrete and in coconut shell concrete is 1.5 % of the total volume of concrete.
For the optimized percentage of steel fiber, flexural strength of conventional concrete and coconut shell concrete is 6.67 N/mm2 and 5.87 N/mm2 at an age of 28 days. The flexural strength increased by 16.40 % in conventional concrete and 25.70 % in coconut shell concrete for optimized steel fiber

compared to without addition of steel fiber respectively

For the optimized percentage of steel fiber, split tensile strength of conventional concrete and coconut shell concrete is 4.17 N/mm2 and 3.07 N/mm2 at an age of 28 days. The split tensile strength increased by 22.64 % in conventional concrete and 15.85 % in coconut shell concrete for optimized steel fiber compared to without addition of steel fiber respectively

The impact resistance of coconut shell aggregate concrete is high when compared with conventional concrete. The impact resistance increased by 90.90 % in conventional concrete and 78.57 % in coconut shell concrete for optimized steel fiber compared to without addition of steel fiber respectively.

The modulus of elasticity for conventional concrete and coconut shell concrete with optimized steel fiber are 36450 N/mm2 and 12280 N/mm2 respectively.

5.1: Scope for future study

Effect of different fibers on coconut shell concrete can be studied

Structural behavior of coconut shell concrete with different fiber can be studied.

Making high strength of coconut shell concrete using different fibers can be studied.

6. References

[1] Tomas and Ucol-ganiron Jr 2013 , Recycling of waste coconut shells as substitute for aggregates in mix proportioning of concrete hollow blocks *Journal of WSEAS Transactions on Environment & Development* **9** 290-300.

[2] Vishwas P. Kukarni and Sanjay Kumar B.Gaikwad 2013, Comparative study on coconut shell aggregate with conventional concrete *International Journal of Engineering and Innovative Technology* **2** 67-70.

[3] Sivakumar A and Manu Santhanam 2007, Mechanical properties of high strength concrete reinforced with metallic and non-metallic fibers *Journal of Cement and Concrete Research* **29** 603-608.

[4] Gunasekaran K, Annadurai R and Kumar P S 2012, Long term study on compressive and bond strength of coconut shell aggregate concrete *Journal of Construction and Building Materials* **28** 208-215.

[5] Song P S and Hwang S 2004, Mechanical properties of high-strength steel fiber-reinforced concrete *Journal of Construction and Building Materials* **18** 669–673.

[6] Bei-xing Li, Chen Ming-xiang, Cheng Fang and Liu Lu-ping 2004, The mechanical properties of polypropylene fiber reinforced concrete *Journal of Wuhan University of Technology* **19** 68-71.

[7] Xin Luo, Wei Sun and Sammy Y.N Chan 2000, Characteristics of high-performance steel fiber reinforced concrete subject to high velocity impact *Journal of Cement and Concrete Research* **30** 907-914.

[8] Yao Wu and Jie Li, Keru Wu 2003, Mechanical properties of hybrid fiber-reinforced concrete at low fiber volume fraction *Journal of Cement and Concrete Research* **33** 27-30.

IOP Conf. Series: Materials Science and Engineering **183** (2017) 012029 doi:10.1088/1757-899X/183/1/012029

KEYWORD EXTRACTION FROM MULTIPLE WORDS FOR REPORT RECOMMENDATIONS IN MEDIA WIKI

Elakiya K, Arun Sahayadhas*

Department of CSE, Vels University, Chennai, India

*arun.se@velsuniv.ac.in

Abstract:

This paper addresses the problem of multiple words search, with the goal of using these multiple word search to retrieve, relevant wiki page which will be recommended to end user. However, the existing system provides a link to wiki page for only a single keyword only which is available in Wikipedia. Therefore it is difficult to get the correct result when search input has multiple keywords or a sentence. We have introduced a **'FastStringSearch'** technique which will provide option for efficient search with multiple key words and which will increase the flexibility for the end user to get his expected content easily.

Key Terms: mediawiki , Fast String Search , Cirrus Search

1. INTRODUCTION

In many information retrieval applications it is necessary to locate quickly some, or all the occurrence of a user specified word or phrases in one or several arbitrary text strings [1]. Specifically, the retrieval of unformatted data takes time. Therefore secondary index has been introduced which contains the keyword fragments [2]. Searching the index with user specified keywords yields the superset of pages required. In case of larger alphabets the Boyer-Moore algorithm takes larger time to search for all the occurrence of keyword in the document. Moreover, if the length of the search phrase is longer, then the performance of this technique gets even worse.

To overcome this problem, we are implementing FastStringSearch (FSS) Technique [5] [3].This technique starts comparing the text pattern from left to right with multiple keywords.The main idea of this technique is to improve the performance of the search. FSS will start to compare the pattern against some portion of the text, where a possible match occurs and will list all the pages which has any one of text specified in the input. This gives the benefit to the user by specified set of keywords instead of single keyword. Specifying multiple keywords will always give better results which is not possible with the existing system. The performance of the search is calculated by **using Squid caching** [4] **and Varnish caching** [5]. Squid stores the copies of webserver pages and when the same page is requested again, Squid will provide the page from its storage. Because of Squid, web server need not generate the page again and again which boosts the performance of webserver. Varnish caching is also same like Squid. A varnish store the page of webserver in Apache server and it retrieves back when the same page is requested again. This inturn boosts the performance of MediaWiki [6].

2. Methodology:

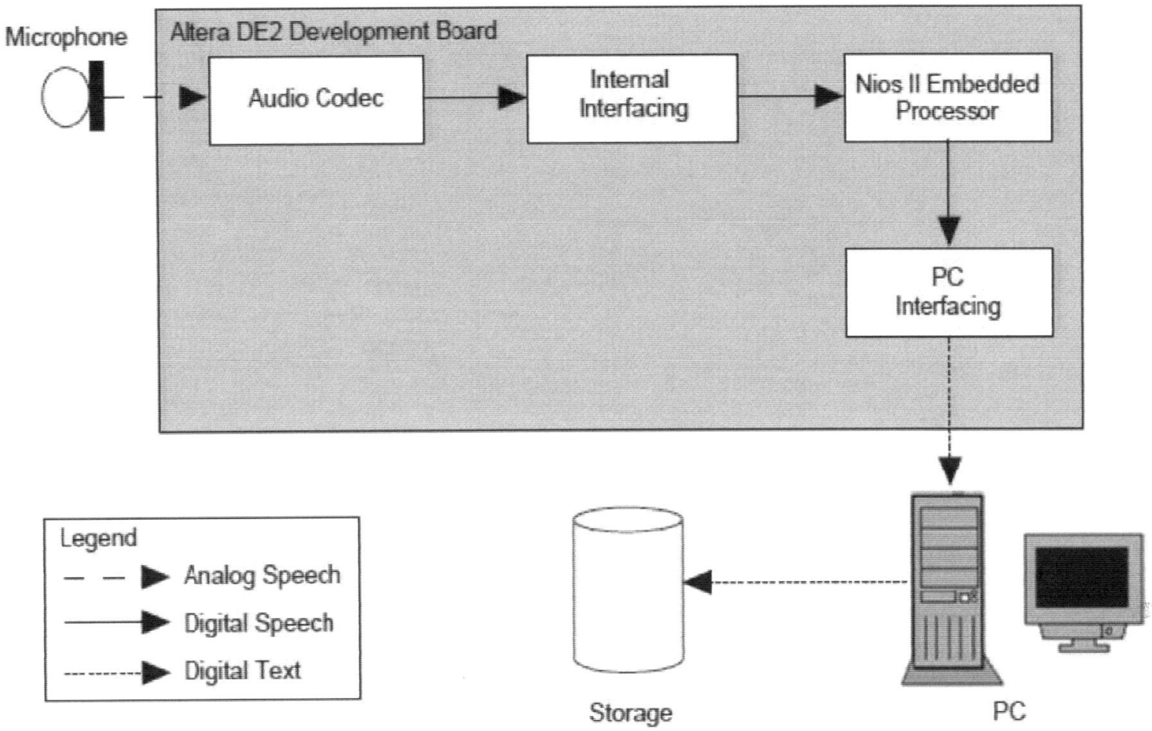
Microphone
Altera DE2 Development Board
Audio Codec
Internal Interfacing
Nios II Embedded Processor
PC Interfacing
Legend
Analog Speech
Digital Speech
Digital Text
Storage
PC

IOP Conf. Series: Materials Science and Engineering **183** (2017) 012029 doi:10.1088/1757-899X/183/1/012029

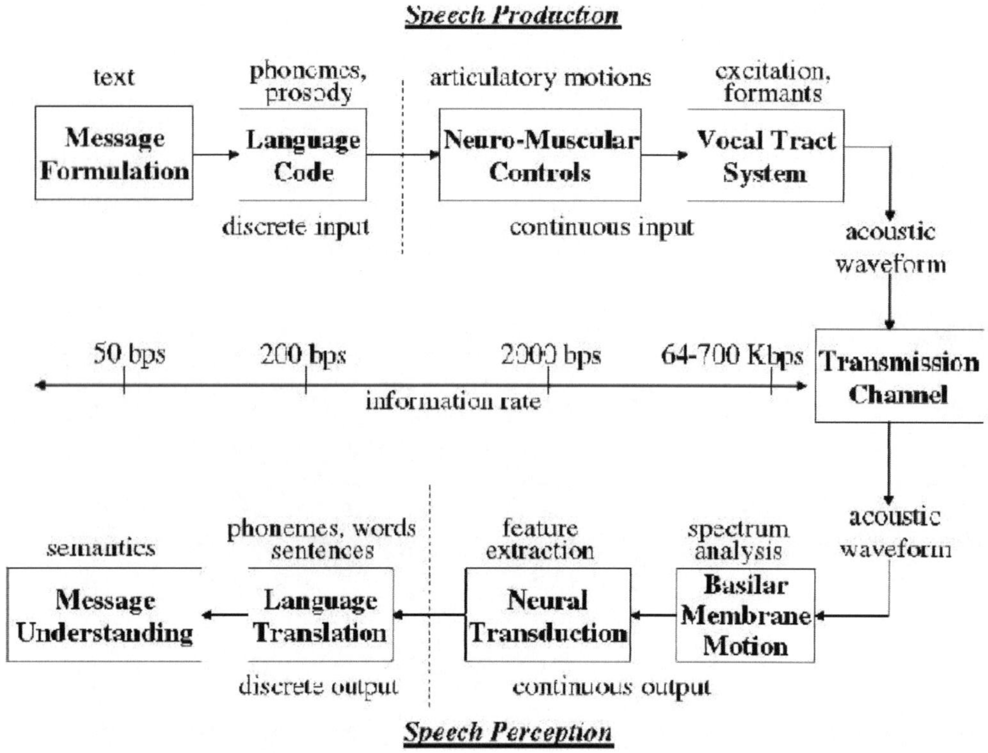

Fig 3:Audio listener from user end

User has been provided with the option to choose the preferred language. In Fig 4, 'English' has been chosen as the preferred language. Once the user chooses his preferred language, on clicking the button 'Voice', listener gets invoked and user provides his speech input via microphone.

Fig 4: conversion message from audio to text

Converter converts the recognized speech to text and fills the text box as shown in Fig 4. User also has been provided with the option to save the input text into the data base. When the user clicks on 'Talk It!' button, system converts the text into audio and reads the same to the user.

3.2. Keyword & relevant document extraction

As shown in Fig 5 and Fig 6, the media wiki takes either a single word or Multiple words search.

Fig 5: The media wiki search takes as a keyword and fetches the result.

Once speech is converted in to either single word or multiple keywords, those are given as input into MediaWiki. The system extracts the relevant documents based on either single keyword or mulitple keywords.

Fig 6: MediaWiki takes multiple word as search input and retrieves the document

3.3. Displaying the documents to the user

The system displays the extracted relevant documents to user based on the search input for both multiple keywords and single keyword.

3.4. Performance comparison of proposed system with existing system

The Proposed system gives an option to do efficient search with a sentence instead of a word which gives better results. The search is much faster when compared to the existing system. Users have an option to look the information directly in MediaWiki, and in Wikidata by reading Help:**NavigatingWikidata**. Search can be done with whole words, with spaces or punctuation marks. So if the search has the word 'book', the resulting pages with the word 'books' or 'booklet' will not be displayed. Fig 7 to Fig 9 shows the audio conversion loading time, performance and audio tuning timeline respectively for the proposed system.

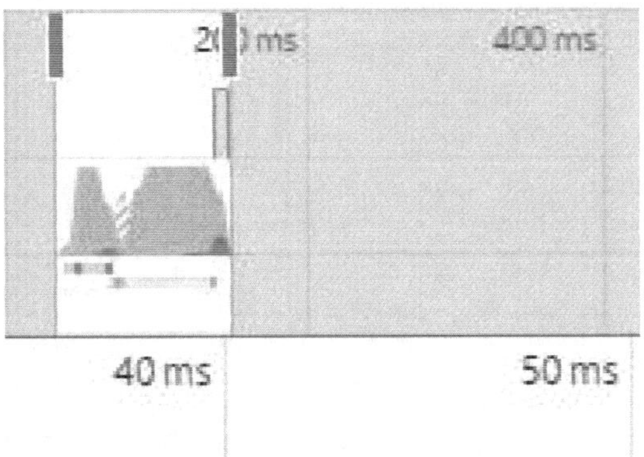

Fig 7: Audio conversion loading time for proposed system

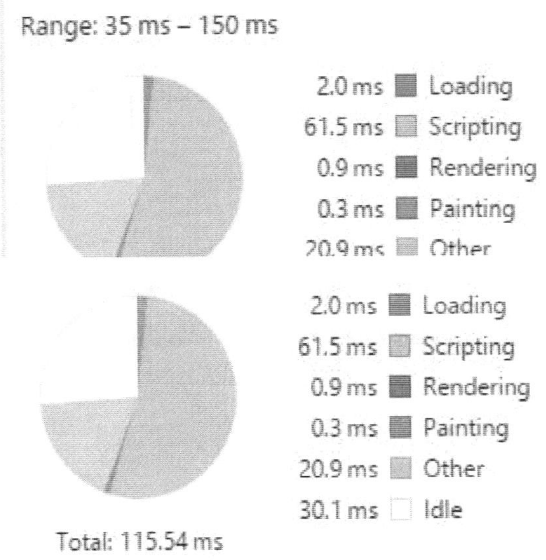

Fig 8: Audio conversion performance for proposed system

Self Time		Total Time		Activity
60.9 ms	94.5 %	60.9 ms	94.5 %	▶ Scripting
2.4 ms	3.7 %	2.4 ms	3.7 %	▶ Loading
0.9 ms	1.4 %	0.9 ms	1.4 %	▶ Rendering
0.3 ms	0.4 %	0.3 ms	0.4 %	▶ Painting

Fig 9: Audio tuning timeline for proposed system

4. Discussions:

The existing system takes an input as a word and retrieve the result by a **Automatic Content Linking Device (ACLD)** [13]. Fig 10 shows the time taken to convert a audio file into text message in the existing system and Fig 11 depicts the frequency timeline for existing system.

Some of the limitations of the Existing system are:

1. Does not searching in an efficient way (i.e) Searching using a sentence will result in incorrect results.

2. The input can only be a word and multiple words cannot be processed. For example, if the word "mobile computing" is given as the input in the search box then it will show relevant document in 60.0 milliseconds. However, if a sentence "What is mobile computing?" is given as input; it will throw an error "Incorrect sentence".

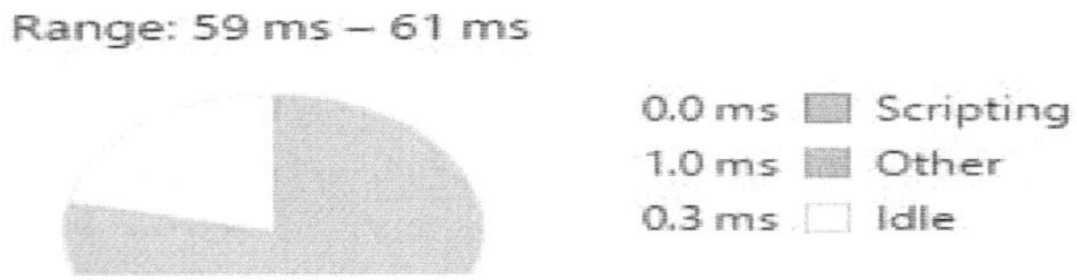

Fig 10: Converting a audio file into a text messages with a speed of 59ms-61ms for existing system

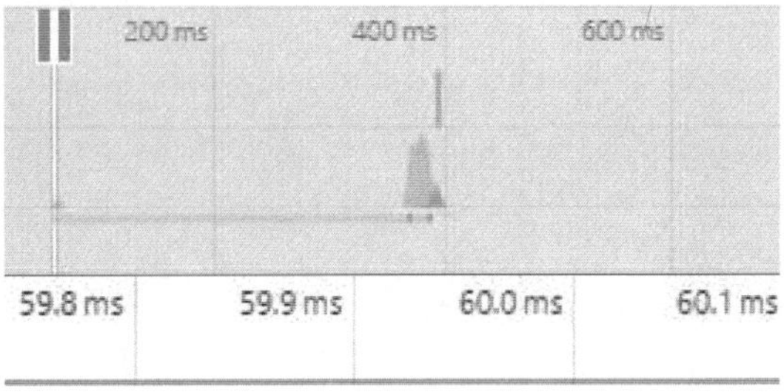

Fig 11: Audio to text conversion frequency timeline for existing system

So, we have overcome this problem by using **FastStringSearch (FSS)** [3]. Using FSS, we can search either a word or multiple words and retrieve multiple relevant documents in minimum time (40.0 milliseconds) compared to the existing systems (60 ms). Our system, takes an input as speech (audio), and then retrieves relevant document in text format from the database.

5. CONCLUSION

This fastsearchstring technique improves the search time for the user and the performance is much better than the existing systems. In the proposed system, the single word search or multiple word search retrieves the relevant document in much lesser time than the current system which retrieves only through single word search. In future, real time index updates can be added and the overall stability and performance of search can be improved by upgrading to Elastic based search 2.x.

REFERENCES

[1] C. Charras, T. Lecroq, and J. D. Pehoushek, "A very fast string matching algorithm for small alphabets and long patterns," *Comb. Pattern Matching*, pp. 55–64, 1998.

[2] D. Brodie, A. Gupta, and W. Shi, "Keyword-based fragment detection for dynamic Web content delivery," *Thirteen. Int. World Wide Web Conf. Proceedings, WWW2004*, pp. 1030–1031, 2004.

[3] A. Hume and D. Sunday, "Fast string searching," *Softw. Pract. Exp.*, vol. 21, no. 11, pp. 1221–1248, 1991.
[4] B. J. Cooper, "High Performance Web Caching With Squid Table of Contents," 2002.
[5] K. Lyngstøl, "Varnish Cache□ : What it is and why it ' s cool . Quick Glance," 2013.
[6] D. Filipiak and A. Ławrynowicz, "Generating Semantic Media Wiki Content from Domain Ontologies," *Third Int. Work. Semant. Web Collab. Spaces*, pp. 49–58, 2014.
[7] B. Venkataramani, "SOPC-Based Speech-to-Text Conversion," pp. 83–108, 2006.
[8] N. Everett, S. Engineer, C. Horohoe, S. Engineer, D. Garry, and P. Manager, "CirrusSearch."
[9] K. Kluge, "Elasticsearch in 10 seconds," 2014.
[10] D. Bijl and H. Hyde-Thomson, "Speech to text conversion," 2001.
[11] K. Saini, "Squid Proxy Server 3 . 1 Beginner ' s Guide About the Author," no. 3.
[12] M. Habibi and A. Popescu-Belis, "Diverse keyword extraction from conversations," *Proc. ACL 2013 (51th …*, no. 2010, pp. 651–657, 2013.
[13] A. Popescu-Belis, M. Yazdani, A. Nanchen, and P. N. Garner, "A Speech-based Just-in-Time Retrieval System using Semantic Search," *Acl2011*, no. June, pp. 80–85, 2011.

Heuristic for Critical Machine Based a Lot Streaming for Two-Stage Hybrid Production Environment

P Vivek[1], R Saravanan[2], M Chandrasekaran[3] and R Pugazhenthi[4]
[1]Assistant Professor, Department of Mechanical Engineering, Vels University, Chennai, Tamilnadu, India.
[2]Professor, Department of Mechanical Engineering, Ellenki College of Engineering and Technology, Hyderabad, Telangana, India.
[3]Professor, Department of Mechanical Engineering, Vels University, Chennai, Tamilnadu, India.
[4]Associat Professor, Department of Mechanical Engineering, Vels University, Chennai, Tamilnadu, India.

E-mail: vivek.acpalanisamy@gmail.com

Abstract. Lot streaming in Hybrid flowshop [HFS] is encountered in many real world problems. This paper deals with a heuristic approach for Lot streaming based on critical machine consideration for a two stage Hybrid Flowshop. The first stage has two identical parallel machines and the second stage has only one machine. In the second stage machine is considered as a critical by valid reasons these kind of problems is known as NP hard. A mathematical model developed for the selected problem. The simulation modelling and analysis were carried out in Extend V6 software. The heuristic developed for obtaining optimal lot streaming schedule. The eleven cases of lot streaming were considered. The proposed heuristic was verified and validated by real time simulation experiments. All possible lot streaming strategies and possible sequence under each lot streaming strategy were simulated and examined. The heuristic consistently yielded optimal schedule consistently in all eleven cases. The identification procedure for select best lot streaming strategy was suggested.

1. Introduction

Hybrid Flow type production environment or Hybrid Flowshop [HFS] is commonly encountered in many manufacturing environments in which two stage HFS is prominent. Lot streaming is an effective tool for time-based manufacturing strategy. Lot streaming is a practice of subdividing the production lots into smaller sub-lots in a multi-stage production systems. By use of this technique the operations of a given lot can be overlapped and minimized the manufacturing makespan [2]. However, this kind of problems received less attention to the researchers [1]. Tsubone et al. [3] investigated the impact of sequencing rules, scheduling scenarios and lot size, on makespan, work-in-process inventory and resource utilization by using simulation in a two stage HFS with lot streaming. Zhang et al. [4] considered m-1 HFS lot streaming problem with special cases of the equal-sublot version and proposed two heuristics which enumerated the number of sublots and scheduling them. This work proposed only one heuristics for both lot streaming and scheduling for all cases of HFS considered.

Chao-Tang Tseng and Ching-Jong Liao [5] dealt a single-job lot streaming problem in a two-stage hybrid Flowshop in which '*m*' identical machines at the first stage and only one machine in second stage. They used a rotation method for sequencing and scheduling then lot sizes were optimized by using linear programming. Ming Cheng et.al., [6] developed a comprehensive mathematical model for scheduling the hybrid flowshop under lot streaming. They also studied a two-stage hybrid flow shop in which only one machine at first stage and the second stage has two identical parallel machines. For solving this problem they developed mathematical programming-based heuristic methods. Mohsen Nejati et al. [7] addressed a two-stage assembly scheduling problem under lot sizing environment in which they considered 'm' machines at the first stage and 'n' assembly machines on the second stage. Their objective was to minimize the sum of weighted completion times with better machine utilization and they proposed simulated annealing and genetic algorithm for computing the optimal sequence for scheduling. Biao Zhang et al [8] proposed a mathematical model and an effective modified migrating bird's optimization (EMBO) to solve the hybrid flowshop; hybridizing with lot streaming problems within an acceptable computational time.

The shortest waiting time rule was preferred to schedule jobs which arrived concurrently for processing. Quan-Ke Pan et al., [9] discussed a rare case of hybrid flowshops with due windows. The authors aimed to minimize the weighted earliness and tardiness from the due window and presented a comprehensive computational campaign for solving the problem. This work is unique. Here the work order is initially divided into six equal groups. In each group containing similar parts and having same sequence of processing and processing times at the respective stages. Then the sub-lots are prepared by combining the groups or considering single group to process them in the two-stage HFS environment. This paper investigates lot streaming strategies in two-stage HFS to optimize lot streaming by developing a heuristic with local optima and globalize the solution based of set out successful results.

2. Two-Stage hybrid Flowshop scheduling Problem

Ruiz et al., [10] discused in detailed about the scheduling for the flow shops with multiple parallel machines per stage usually referred to as the Hybrid Flow Shop. It is a complex combinatorial problem. Djellab and Djellab [11] hybrid flowshop encountered in many real world scenarios like the industries of paper, Electronics, photographic film, Concrete production, Container handling, textile, and internet service architectures etc.. These kinds of flowshops are common manufacturing environments in which 'J' number of jobs to be processed in serial order at 2 stages. Ether stages may have more than one parallel processor, but at least one stage must have only one processor. Job may skip the stages, but it must be processed at least only one stage. The configuration of two-stage hybrid flowshop is furnished in figure1. According to this problem the jobs provided for processing of sub-lot as per group schedule. The problem is to find a group schedule which optimizes a given objective function. The hybrid flowshop problem is, in most cases, NP-hard [12]. For instance, the two stage hybrid flow shop containing two machines at one stage and the other one a single machine, However, the problem might be polynomially solvable with some special properties and precedence relationships [13].

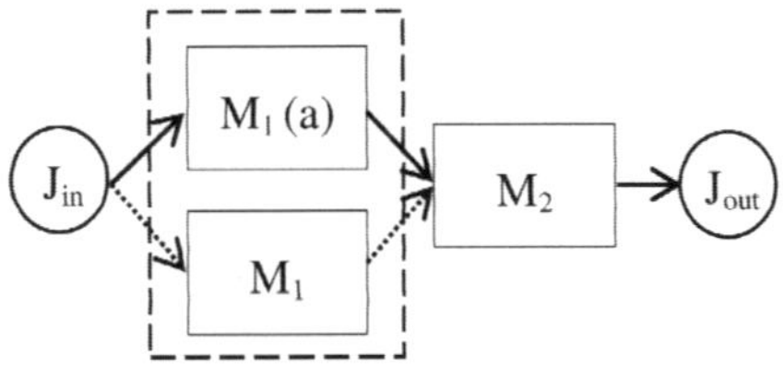

Figure 1. Two stage hybrid Flowshop

3. Mathematical Modelling

3.1. Parameters

J_g = The job J_g (J_g=**1, 2,…, 200**) of part family type, i.e., group type g (**g = 1, 2,…, 6**), where ***200*** total number of jobs to be processed in a group i.e., group size; Where ***6*** is total number group types are to be scheduled. $J_g \in J$; Where J=1200 is total Number of jobs to be scheduled.

$(P_i)_{Jg}$ = Processing times of job J_g of group type g at i^{th} stage, $i(i=1,2,\ldots s)$ where **s =2**; that is the total number of stages.

m_i = Number of parallel identical processor available in stage i.

T = Total number of time units in scheduling jobs

$(C_i)_{Jg}$= Completion time of job J of type g at the stage i.

It is a time point $(C_i)_{Jg} = t$ means that the operation completes at the end of the time unit t.

3.2. Decision variables

$(\delta_i)_{Jg}\, t$ = {1 if Jg is processed at the stage i in the time unit t;
0 other wise }

With the above notation the hybrid flow shop problem (HFSP) under the consideration can be formulated as follows.

Note that $(C_2)_{Jg}$ in the model is the final completion time of the male job J of type g, C_k after processed at require **2** stages, in other words it is the completion time of a job of group type g.

3.3. Mathematical Model

Usually in the batch processing the makespan is the completion time of the last job of the group. If there are '**g**' numbers of group means, then the makespan will be the completion time of the last job of the last group. Hence it is clear that the objective is to minimize the completion time of all jobs of all group types. Mathematically, the objective function can be defined as

$$\textbf{Minimize} \sum_{g=1}^{6}\sum_{J_g=1}^{200}(C_2)_{Jg} \tag{1}$$

Subject to

$$(C_i)_{Jg} \le (C_i)_{Jg} + 1 - (P_i)_{Jg}$$

$i\ (i = 1,2);\ g\ (g = 1,2,\ldots,6);\quad J_g\ (J_g=1,2,\ldots,200);\qquad J_g \in J;\ J=1200;$ (2)

$$\sum_{t=1}^{T_m}(\delta_i)_{Jg}\, t = (P_i)_{Jg}$$

$i\ (i = 1,2);\ g\ (g = 1,2,\ldots,6);\ J_g\ (J_g=1,2,\ldots,200);\ J_g \in J;\ J=1200;\ t = 1,2,\ldots,T;$ (3)

$$t * (\delta_i)_{Jg}\, t \le (C_i)_{Jg},$$

$i(i = 1,2);\ g(g = 1,2,\ldots,6);\ J_g(J_g=1,2,\ldots,200);\ J_g \in J;\ t=1,2,\ldots,T;\ J=1200;$ (4)

$$(C_i)_{Jg} - (P_i)_{Jg} + 1 \le t + T(1-(\delta_i)_{Jg}\, t),$$

$i\ (i = 1,2,\ldots,s);\ g\ (g = 1,2,\ldots,x);\ J_g\ (J_g=1,2,\ldots,200);\ J_g \in J;\ t = 1,2,\ldots,T;\ J=1200$ (5)

$$\sum_{Jg=1}^{e}(\delta_i)_{J_g}\, t \le m_i,$$

$i\ (i = 1,2);\ g\ (g = 1,2,\ldots,6);\ J_g\ (J_g=1,2,\ldots,200);\ J_g \in J;\ t = 1,2,\ldots,T;\ J=1200$ (6)

$(\delta_i)_{Jg}\, t \in \{0,1\}$,

$i\ (i = 1,2);\ g\ (g = 1,2,\ldots,6);\ J_g\ (J_g=1,2,\ldots,200);\ J_g \in J;\ t = 1,2,\ldots,T;\quad J=1200$ (7)

$(C_i)_{Jg} \in \{1,2,\ldots,T\}$, $\quad i\ (i = 1,2);\ J_g\ (J_g=1,2,\ldots,200);\quad J_g \in J;\ J=1200$ (8)

In the above formulation, the objective is to determine a schedule that minimizes the total completion time, while satisfying all the constraints as listed above. The objective function also helps to achieve reduced work-in- process inventory, avoid delays group submission [tardy job reduction] and improved customer services, etc. The constraint [2] represents an operation precedence among the stages of a job and ensures that an operation cannot be started until the operation of the same job at its preceding stage is finished. Constrains [3] – [5] define the time intervals for which a job is processed on a machine at a stage. Constraint [6] indicates that all these machine requirements was satisfied with the number of available machines at that time. Constraints [7] & [8] define value ranges of the variables.

4. Solution Procedure

4.1. Assumption

1. Total job can be classified as equal sized g number of groups based on similar processing time as sequence.
2. First sub-lot in the group schedule available at zero time. Subsequent sub-lots will be dispatched to the shop floor as soon as the completion of the existing sub-lot.
3. The sub-lots may or may not be equal in size [containing number of groups].
4. The pre-emption not permitted.

4.2. Heuristic Approach

Step 1: Calculate the cumulative processing time of each job Ji.

$$J_i = \sum_{j=1}^{M} P_{i,j} \forall N; i \in N$$

Step 2: Calculate the cumulative processing time of all the jobs Tp.

$$Tp = \sum_{t=1}^{g} \sum_{i=1}^{N} J_{ti}$$

Where ***t*** is no of job types, ***i*** is no of jobs in each type.

Step 3: Compute the group mean m where Tp is total processing time

$$m = \frac{Tp}{k}$$

Where '*k*' is desirable number of sub-lots

Step 4: Sub-lot formation

To find the optimal group schedule to form the sub-lots as such their total processing time deviate from the mean 'm' on both directions [increase as well as decrease] must be equal as well as minimal and hence algebraic sum of the deviations of the sub-lot processing times zero. The sum of deviations of sub-lots from it their mean value nearly equal to zero will offer near optimal solution.

The heuristic needs to be validated and verified with well known problem. The two-stage hybrid flow shop problem is discussed in Numerical example.

4.3. Simulation Model

The simulation modeling and analysis give real time value and economical to verify and validate the strategies, schedules etc. [Saravanan and Raju 2010]. The authors used such modeling analysis to verify and validate their heuristic solution for dynamic parallel hybrid flowshop scheduling problems. However, in the majority of the situations, simulation is used as a performance evaluation tool. Simulation offers the flexibility to model the complexities adequately, while the gradient computation helps in identifying a good solution quickly. Besides using simulation as a tool for optimization, this work also makes use of it to compare the performance of different capacity management schemes, providing clear conclusions about their relative performances. Simulation is a very powerful tool for the analysis and evaluation of complex systems. In many circumstances, it is used when analytical models are neither available nor easy to obtain. Traditionally, simulation is used to compare configurations, different policies, validate models, and many other qualitative features with the purpose of answering what if questions. Hence, for verifying and validating the proposed heuristic for a two stage hybrid Flowshop, the simulations modeling and analysis technique was employed. The modeling was done in extend v6. The model was verified and validated properly.

4.4. Makespan Computation

Makespan computation procedure for the example group schedule Gs = *{(l_1& l_2) - (l_3& l_4) - (l_5& l_6)}* which contain three sub-lots with equal group size of two groups per sub-lot.

$$C_j = \max\{\sum_{j1=1}^{200}(C_2)_{j1},\sum_{j2=1}^{200}(C_2)_{j2}\}+,\max\{\sum_{j3=1}^{200}(C_2)_{j3},\sum_{j4=1}^{200}(C_2)_{j4}\}+\max\{\sum_{j5=1}^{200}(C_5)_{j5},\sum_{j6=1}^{200}(C_5)_{j6}\}$$

Makespan computation for the example group schedule *{(l_1, l_2 & l_3) - (l_4, l_5 & l_6)}*, which contain two sub-lot with equal sub-lot size of three groups per sub-lot, is:

$$C_j = \max\{\sum_{j1=1}^{200}(C_2)_{j1},\sum_{j2=1}^{200}(C_5)_{j2},\sum_{j3=1}^{200}(C_2)_{j3}\}+\max\{\sum_{j4=1}^{200}(C_2)_{j4},\sum_{j5=1}^{200}(C_2)_{j5},\sum_{j6=1}^{200}(C_2)_{j6}\}$$

.

Makespan computation for the example group schedule {(l_1, l_2, l_3 & l_4) - (l_5 & l_6)} which contain two sub-lot with equal sub-lot sizes of four and two groups per sub-lot respectively, is:

$$C_j = \max\{\sum_{j1=1}^{200}(C_2)_{j1},\sum_{j2=1}^{200}(C_5)_{j2},\sum_{j3=1}^{200}(C_2)_{j3},\sum_{j4=1}^{200}(C_2)_{j4}\}+\max\{\sum_{j5=1}^{200}(C_2)_{j5},\sum_{j6=1}^{200}(C_2)_{j6}\}$$

Makespan computation for the example group schedule {(l_1, l_2, l_3, l_4 & l_5) - (l_6)}which contain two sub-lot with sub-lot sizes of the five groups and one group per sub-lot respectively, is:

$$C_j = \max\{\sum_{j1=1}^{200}(C_2)_{j1},\sum_{j2=1}^{200}(C_2)_{j2},\sum_{j3=1}^{200}(C_2)_{j3},\sum_{j4=1}^{200}(C_2)_{j4},\sum_{j5=1}^{200}(C_2)_{j5}\}+\{\sum_{j6=1}^{200}(C_2)_{j6}\}$$

Makespan computation for the example group schedule {(l_1, l_2, l_3, l_4, l_5 & l_6)} which contain only one lot of sub-lot size of six groups is:

$$C_j = \max\{\sum_{j1=1}^{200}(C_5)_{j1},\sum_{j2=1}^{200}(C_5)_{j2},\sum_{j3=1}^{200}(C_5)_{j3},\sum_{j4=1}^{200}(C_5)_{j4},\sum_{j5=1}^{200}(C_5)_{j5},\sum_{j6=1}^{200}(C_5)_{j6}\}$$

5. Numerical Example

The two stage hybrid flowshop which contains two machines in the first stage and only one machine in second stage. There are 1200 jobs to be processed. The jobs can be classified as uniform groups as 6 types with each group has equal amount of jobs 200. The processing time of each job type are presented in the Table 1. The maximum possible eleven types of lot streaming strategy were furnished in Table 2 and the number of lot streaming schedules available are also shown against the each case.

Let's discuss some example cases. If the group schedule is {(g_1, g_2, g_3, g_4, g_5 & g_6)} means the sub-lot sizes are {1,1,1,1,1&1} and number of sub lots 6 similarly for group schedule {(g_1&g_2) - (g_3&g_4) -

$(g_5\&g_6)$ is {2, 2, & 2} and number of sub-lots is 3, for group schedule $(g_1, g_2 \& g_3)$ - $(g_4, g_5 \& g_6)$ is {3 & 3} and number of sub-lots is 2, for group schedule $(g_1, g_2, g_3 \& g_4)$ - $(g_5 \& g_6)$ is {4 & 3} and number of sub-lots is 2, and for group schedule $\{(g_1, g_2, g_3, g_4\& g_5)$ - $(g_6)\}$ is {5 & 1} and number of sub-lots is 2. The makespan is the completion time of the last job in the group schedule. The makespan depends on factors such as processing time, processing sequence, job availability, machine availability and intermediate waiting time. The scheduling is the task of allocating resources by the way optimizing above parameter to achieve the minimum makespan. Hence the problem is to find optimal the group schedule. Group scheduling involves two levels. The scheduling the sub-lot is Level-II, which includes sub-lot formation and order of processing. The scheduling jobs within the sub-lot is called as Level-I.

5.1. Heuristic Approach for Lot Streaming

The Lot streaming is processes of forming sub-lots and group scheduling. The heuristic is applied here to form the sub-lots. There are 11 possible strategies of lot streaming cases were derived. In those strategies all possible lot streaming schedules (***Gs***) were simulated and makespan computed as discussed below. The optimal schedules were found and listed in the Table 2 for strategy wise. The mean ***m*** deviation of each sub lot processing times ***'d'*** and algebraic sum of those deviation $\sum d$ were tabulated in Table 2. The Sub-lot sizes were possibly varied from 1 group [200 jobs] to 6 groups [1200 jobs]. The optimal schedules posses sub-lots which deviate uniformly up and down and hence their cumulative deviations are zero and nearly zero. Hence it is proved that the heuristic solution provides optimal schedule in all 11 possible strategies by yielding minimal makespan. Such schedules are verified as validated by simulating all possible cases in under each strategy.

Table 1 Processing Time of Job types

Jobs	**Stage 1**	**Stage 2**	**Total**
J_1	5.42	1.50	6.92
J_2	2.70	4.94	7.64
J_3	2.86	2.90	5.76
J_4	4.22	3.94	8.16
J_5	5.86	1.94	7.80
J_6	4.50	3.66	8.16

6. Results & Discussion

From the illustrative case it is clear that the optimal or near optimal makespan can be achieved by using heuristic for any strategy of lot streaming. The makespan variation influence of the stage wise processing time, the number of sub-lots, groups in the sub-lot etc. The increase of the number of sub lot may increases the makespan. The optimal makespan performance based on strategies was shown in figure 2. The queue status at the critical machine with respect to strategy was illustrated in the figure 3 and figure 4. It is noticed that the lot streams case 5 and case 8 experienced lesser average queue lengths and average job waiting time than other strategies at the critical machine.

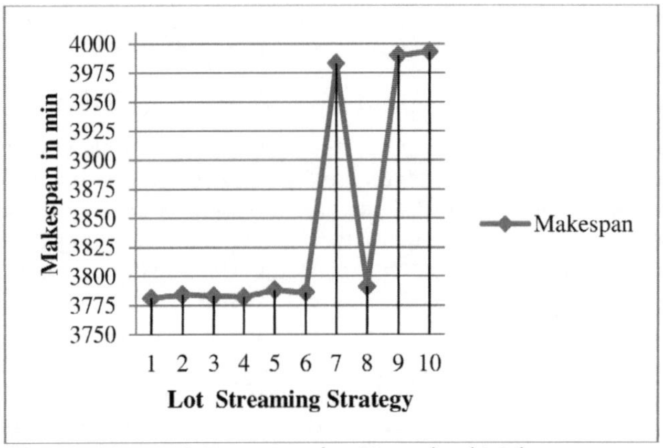

Figure 2. Lot streaming Vs optimal Makespans

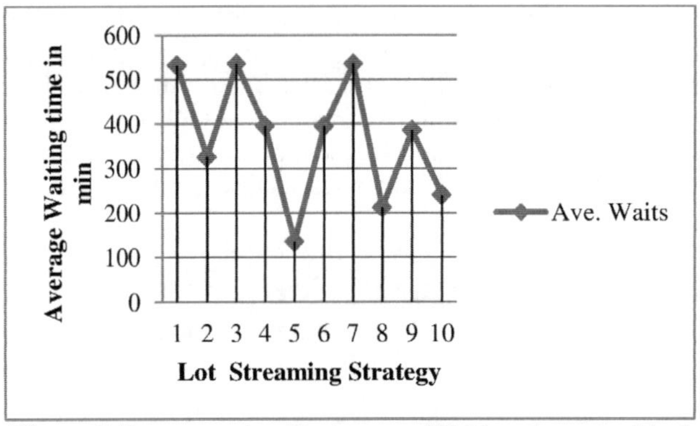

Figure 3. Lot streaming Vs Average Waiting time at critical Machine

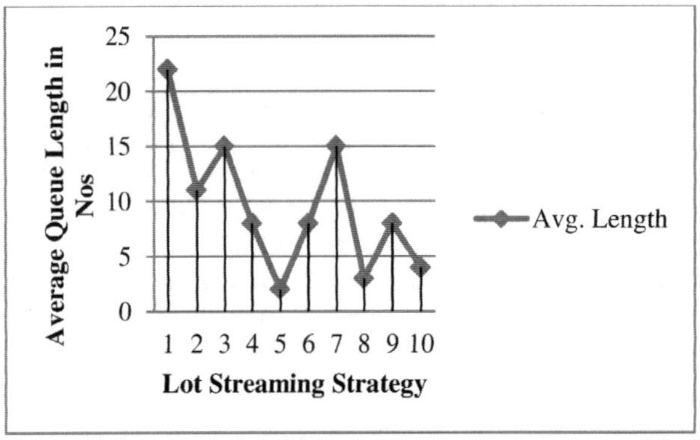

Figure 4. Lot streaming Vs Average Queue length at critical Machine

But case 5 experiences minimal queue status because the sub-lots formed in case 5 has 2 groups per sub lot equally and their processing times deviations are minimized and a number of sub-lots minimum [Three sub lots] compared case 11. Hence the best lot streaming strategy has evenly distributed groups in the sub-lots and the sub-lots deviation from its mean is minimized and possibly a less number of sub-lots.

7. Conclusion

The proposed heuristic for lot streaming type group scheduling for two stage hybrid flow shop problem was discussed. The heuristic yielded minimal makespan among all possible group schedules under that strategy. It works well, irrespective of strategy of lot streaming. The best of best strategy selection also recommended. This heuristic solution may suit for similar problems irrespective of number of groups

8. References

[1] Liu J 2008 Single-job lot streaming in m-1 two-stage hybrid flowshops European Journal of Operational Research **187(3)** 1171-83.

[2] Defersha F M and Chen M 2010 A hybrid genetic algorithm for flowshop lot streaming with setups and variable sublots International Journal of Production Research **48(6)** 1705-26.

[3] Tsubone H, Ohba H and Uetake T 1996 The impact of lot sizing and sequencing in manufacturing performance in a two-stage hybrid flow shop International Journal of Production Research **34** 3037–3053.

[4] Zhang W, Liu J and Linn R 2003 Model and heuristics for lot streaming of one job in M-1 hybrid flowshops International Journal of Operations and Quantitative Management **9** 49–64.

[5] Chao-Tang Tseng and Ching-Jong Liao 2008 A discrete particle swarm optimization for lot-streaming flowshop scheduling problem European Journal of Operational Research **191(2)** 360-373.

[6] Ming Cheng, Subhash C and Sarin 2013 Two-stage Multiple-lot Lot Streaming Problem for a 1 + 2 Hybrid Flow Shop IFAC Proceedings Volumes **46(9)** 448-453.

[7] Mohsen Nejati, Iraj Mahdavi, Reza Hassanzadeh and Nezam Mahdavi-Amiri 2016 Lot streaming in a two-stage assembly hybrid flow shop scheduling problem with a work shift constraint Journal of Industrial and Production Engineering **33(7)** 459-47.

[8] Biao Zhang, Quan-ke Pan, Liang Gao, Xin-li Zhang, Hong-yan Sang and Jun-qing Li 2017 An effective modified migrating birds optimization for hybrid flowshop scheduling problem with lot streaming Applied Soft Computing **52** 14-27.

[9] Quan-Ke Pan, Rubén Ruiz and Pedro Alfaro-Fernández 2017 Iterated search methods for earliness and tardiness minimization in hybrid flow shops with due windows Computers & Operations Research **80** 50-60.

[10] Rubén Ruiz and José Antonio Vázquez – Rodríguez 2010 The Hybrid Flow Shop Scheduling Problem European Journal of Operational Research **205** 1–18.

[11] Djellab H and Djellab K 2002 Preemptive hybrid flowshop scheduling problem of interval orders European Journal of Operational Research **137(1)** 37–49.

[12] Gupta J N D 1988 Two-stage hybrid flow shop scheduling problem Journal of the Operational Research Society **39(4)** 359–364.

[13] Saravanan R and Raju R 2010 Sequencing and scheduling of non-uniform flow pattern in parallel hybrid flow shop International Journal of Advanced Manufacturing Technology Springer **49** 213-225.

Analysis of Biomethanation Processfrom market waste to generate bio energy

S Sathish[1], A Parthiban[2], T Vinod kumar[3] and M Chandrasekaran[4]
[1]Head of the Department Automobile Engineering, Vels University, Chennai,
[2,3]Assistant Professor, Department of Mechanical Engineering,
Vels University,Chennai,
[4]Professor, Department of Mechanical Engineering, Vels University, Chennai,

E-mail: sathishamg88@gmail.com1,

Abstract. In this study was to incur that the biogas production from traditional market wastes which were represented by cabbage stem and carrot peeling, white mustard were under taken in a laboratory experiment. To produce biogas, the raw material such as cabbage stem and carrot peeling, white mustard and carrot peeling were mixed until C/N ratio close 30:1. Inoculums starter cow dung is put into digester then water is added until 500 liters. The initial pH is measured at throughout the experiments. The anaerobic digestion process is conducted at temperature of 30°C and the volumes, pH of the biogas yield were observed daily. Biogas yield and cumulative biogas, total solids were analyzed 35 days. The cumulative biogas yield at the 32th day of digestion for cabbage stem and carrot peeling (exp1), white mustard and carrot peeling (exp2) were 2140 liters and 2421 liters respectively. The highest daily biogas yield is achieved on the 22st day of digestion which is found 123 liters and 141 liters respectively. In the first 10 days, the pH level is observed decrease and increase after the day of 21. Although at the end of digestion period the pH will fall down.

1. Introduction

An anaerobic of cow dung and solid waste for treatment and generation of biogas yield can be used as bio energy. Generally this process is commonly used in bio digester. In this research the floating dome type digester is used. The large amount of market waste has solid organic waste, 70 % of which are vegetables and fruits. Although, these waste are high water content, low C/N ratio and acid level. On the other hand, the vegetable waste produces bio energy by arranging the C/N ratio. Biogas yield and sustainable environmental source if bio energy is the result of a complex biological process that requires the inclusion of a different type of microorganisms in presence and absence of air. In the other words, biogas yield is frequently produced by the anaerobic digestion or co-fermentation of organic material intimated by [1]. Vegetable wastes organic material having the high calorific value and nutritive value to the microbes, that's why efficiency of CH_4 generation can be enhanced by several orders of magnitude. It means higher efficiency and size of digester and cost of biogas yield is decreased and also, in most of the cites and town, vegetable waste is disposed in landfill for discarded which causes the public health hazards and diseases like malaria, cholera, typhoid. It emits unpleasant Odours & CH_4 which is a major greenhouse gas contributing to global warming reported Bouallagui [2]. Deressa intimated [3] the biogas production is a favorable dual – purpose technology at present, the biogas production can be used to meet energy requirements while organic waste is a useful fertilizer.

Biogas is a type of bio energy that can be generate from decomposition of organic materials and its mainly composed of methane, carbon dioxide and traces of some other gases. An interesting option for enhancing biogas yields of anaerobic digestion of vegetable wastes are co-digestion, which employs a co-substrate that has the benefit of enhancing the biogas yield due to positive synergism established in the digestion medium and the supply of missing nutrients for microorganisms suggested by Lin [4]. Vivekanandan [5] is reported that the lignocelluloses is one of the plant biomass,

It is primarily consists of three major elements such as cellulose, hemicelluloses and lignin. The other constituents such as water and proteins do not participate in organizing the structure of the organic waste materials. Generally the methanogenic bacteria and microorganisms growth is mainly depends on different parameters like pH, Carbon/Nitrogen ratio, temperature, digester design, inoculum of the digester and hydro retention time.Now a days ananerobic digestion of organic waste and sludge waste are using merging of heating process leading to lower percentage and higher for organic sludge waste to biogas yield in end of the digestion by sathish [6]. It is one of the complex process with a number of synergistic controlled factors, at various vegitable mixed waste, even a small development in the process, gives greater biogas yields which may be better for commerical one [7]. Therefore, the aim of this study is to generate biogas from tradational market waste. The experiments conducted one cubic meter digester.

Experimental setup

2.1 Inoculum

Initially the digester feed with cowdung in the ratio of 10 % for improve the microorganisms if the floating drum anaerobic digester. It is also enhance the methane yield in the digestion periods.

2.2 Materials and Methods

The figure 1shows the schematic cross sectionals view of the floating drum anaerobic digester. Biomass is chosen for these analyses are experiment 1is cabbage steam and carrot peeling; experiment 2 is white mustard and carrot peeling. These vegetable wastes from market are causes ecological and economic problem in the country. It is collected from the traditional market from Chidambaram town. This study is carried out between 1 to 35 days. Experiment 1, added 40 % of cabbage stem and 60 % of carrot peeling mixed with the same ratio of water added in the digester. Then Experiment 2, 40 % of white mustard and 60 % of carrot peeling mixed with same ratio of water feeded in the digester.

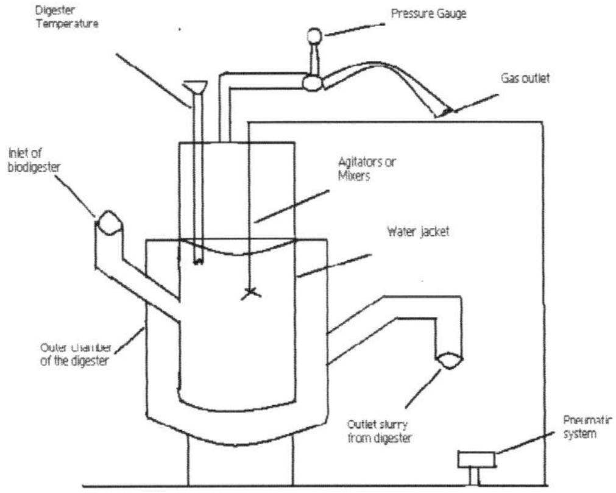

Figure 1. Schematic view of the experimental setup

The pH is measured from the digester using pH redox meter. Digester temperature is measured using thermometer (0 to 50°C). The gas flow is measured using Alborg gas flow meter. The moisture content of the wastes are determined the waste by the waste by water methods used for the digester charging. Volatile solids (Vs) are the part of organic waste matter that undergoes biodegradation to biomethanation process. Generally more volatile solids present in the organic matter to the more biogas are generated in the end of the digestion period.

Result and Discussions

Daily biogas production from Exp1 and Exp2 were shown graphically in figure 2. The biogas yield is flammable at took place at various lag periods. In this case the Exp2 having higher gas production at 22nd day with 141 liters compare to Exp2 values. Why because white mustard and carrot peeling have higher C/N ratio about 30:5 so, the biogas yield is high. Then the figure 3 shows the cumulative biogas yield with hydro retention time.

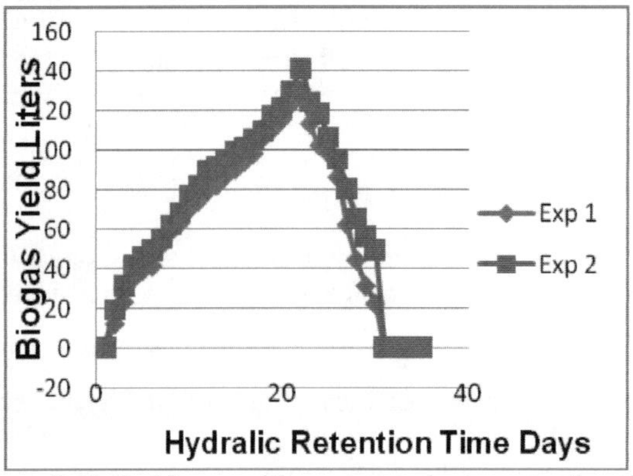

Figure 2. Daily biogas with respect to hydro retention time

Here also Exp2 achieved at 2421 liters of biogas at the end of digestion, In Exp1 achieved only 2140 liters of biogas, so white mustard and carrot peeling is the good combination of co-digestion.

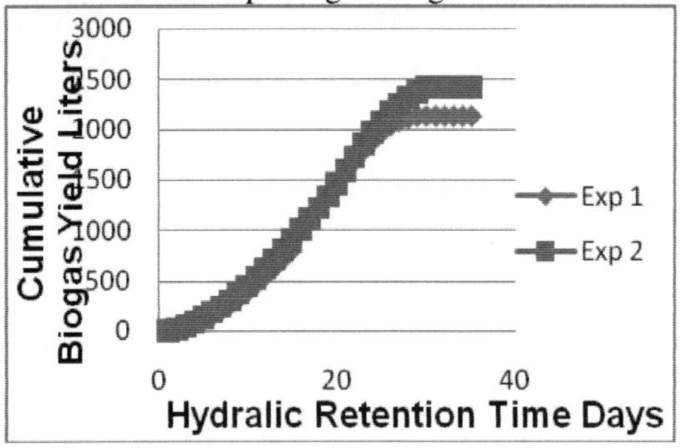

Figure 3. Cumulative biogas with respect to hdro retention time

These wastes are high in C/N ratio. These amounts of Carbon/Nitrogen nutrient source effects on the growth of microorganisms' it presence in the waste materials it causes of Enhance the biogas yield. Quantitative analysis of biogas components for the white mustard and carrot peeling indicates that CH_4 content is high for Exp1 feedstock and whereas CO_2,H_2S was found to be high in Exp2 according to the source of the organic waste material.

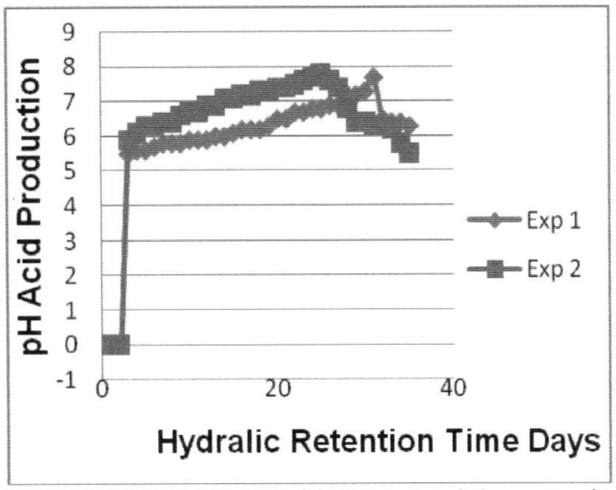

Figure 4. Cumulative biogas with respect to hdro retention time

The figure 4 shows the Acid Production (pH) with hydro retention time. Generally acid concentration greatly affects the biogas yield. Methanogenic reaction is very high as bacterias utilize the waste more readily so biogas generation decreasing. Simplication of process is methanogenic reaction that is Ts % reduced. In the biogas yield and methane form microorganisms is very sensitive to temperature changes during the digestion period. The figure 5 shows the digester temperature with respect to hydro retention time.

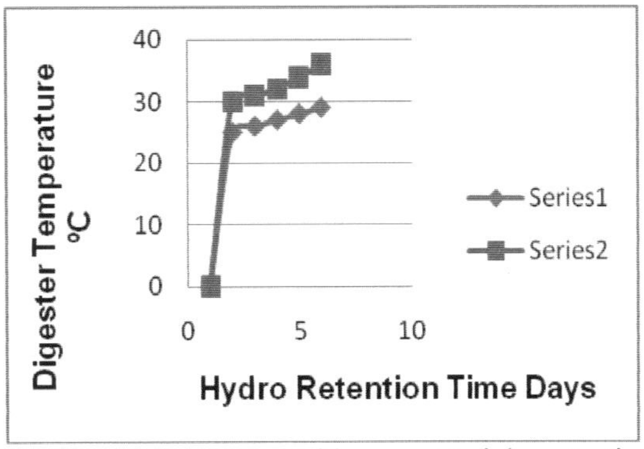

Figure 5. Cumulative biogas with respect to hdro retention time

A sudden change exceeding the temperature in Exp1 below 30ºC temperature the biogas production is very slow so temperature is one of the relative stability of biogas generation. In Exp2 above 30 to 35ºC temperature kills the pathogenic bacteria in the digester during digestion period. Then it is enhance the methanogenic bacteria in the digester. The fact that the pH of the Exp1, dropped to 6.8 in the last day of further supports the research a slow process startup. The acidifying bacteria produced acids at a much higher rate, than the methane forming bacteria were able to cope with. Another reason could be that the substrate material is decomposed to easily and therefore very susceptiable to acidification. That the methane content Exp2 constantly indicates that the certain areas possibly where a lot inoculum material is present, worked at a higher pH and produce gas while other Exp1 did not produce higher gas yield. Generally, Proper Carbon/Nitrogen ratio does the play an important role in the biogas production. Carbon/Nitrogen ratio ranging from 21:1 to 31:1 is always considered in suitable for anaerobic digestion by weiland [8]. During [9] the anaerobic digestion of beef manure in mixed and unmixed reactors. It is very clear that cow manure is an very effective feed materials for anaerobic digestion. It is significantly increase the cumulative biogas yield

4. Conclusion

These results results obtained from this research suggested that co-digestion of tradational market waste with inoculum can help to improve the biogas yield. Out of the three market wastes used (cabbage stem, white mustard, carrot peel) white mustard and carrot peel is the most effective in enhance the biogas yield. In this case the feedstock have higher C/N ratio and microorganisms. Further minimum time required for biogas yield from the biomasses tradational market wastes with increase in temperature of digestate. Thus temperature increase the production as well as the digester efficiecy.

5.References

[1] Jiang, Xinyuan and Sven G. Sommer, and Knud V. Christensen, 2011, A review of the biogas industry in China. *Energy Policy,* **39** (10), 6073-6081.

[2] Bouallagui, H., Touhami, Y and Cheikh, R. B., & Hamdi, M, 2005,Bioreactor performance in anaerobic digestion of fruit and vegetablewastes. *Process,biochemistry*, **40**(3), 989-995.

[3] Deressa, Leta, Solomon Libsu, R. B. Chavan, Daniel Manaye, and Anbessa Dabassa, 2015,Production of Biogas from Fruit and Vegetable Wastes Mixed with Different Wastes, *Environment and Ecology Research* **3** (3), 65-71.

[4] Lin, Jia, Jiane Zuo, Lili Gan, Peng Li, Fenglin Liu, Kaijun Wang, Lei Chen, and Hainan Gan. 2011,Effects of mixture ratio on anaerobic co-digestion with fruit and vegetable waste and food waste of China, *Journal of Environmental Sciences*, **23** (8), 1403-1408.

[5] Vivekanandan, S., and S. Sathish, 2015,To study the effect of temperature using pretreated rice straw to generate biogas. *American journal of engineering research,* **4** (9), 40-46.

[6] Sathish, S., and S. Vivekanandan, 2015, Experimental Investigation on Biogas Production Using Industrial Waste (Press Mud) To Generate Renewable Energy. *Int. J. Innov. Res. Sci. Eng. Technol,* **4**, 388-392.

[7] Sathish S andVivekanandan S. 2016, Parametric optimization for floating drum anaerobic bio-digester using Response Surface Methodology and Artificial Neural Network. *Alexandria Engineering Journal*. Sep 28.

[8] Weiland and Peter, 2010, Biogas production: current state and perspectives, *Applied microbiology and biotechnology* **85** (4), 849-860.

[9] Rasi, S., A. Veijanen, and J. Rintala, 2007, Trace compounds of biogas from different biogas production plants, *Energy* **32** (8), 1375-1380.

Effect on air quality and flow rate of fresh water production in humidification and dehumidification system

K Rajasekar[1], R Pugazhenthi[2], A Selvaraju[3], T Manikandan[4] and R Saravanan[5]
[1]Assistant professor, Department of Mechanical Engineering, Vels University, Chennai, India.
[2]Associate professor, Department of Mechanical Engineering, Vels University, Chennai, India.
[3]Professor, Department of Mechanical Engineering, Pondicherry Engineering College, Pondicherry, India.
[4]Assistant professor, Department of Mechanical Engineering, Thangavellu Engg. College, Chennai, India.
[5]Professor Department of Mechanical Engineering, Ellenki College of Engineering and Technology, Hyderabad, Telangana, India.

E-mail: rajmech90@hotmail.com

Abstract. Water is the vital need of any living organisms of the world when water fails, functions of nature cease the world. The water scarcity is one of the major problems to be faced by the developing world, which indicates a critical need to develop inexpensive small-scale desalination technologies. The cost of the desalination process takes more, so the world expecting the desalination plants with minimum operating cost, so the utilization of renewable energy source is a preferable one. This research article provides a glimpse of an overview of the humidification-dehumidification (HDH) based desalination method which uses the solar energy. The HDH based desalination method monitored and evaluated the performance parameters, i.e. mass flow rates of water and air.

1. Introduction

Fresh water is indispensable for the every survival of all beings on this mother planet of the Earth. There are many regions of the world which are blessed with an abundance of fresh water. But, in thick-populated areas burgeoning with rapid industrialization are experiencing more water stress, particularly when located in arid regions [1]. A survey on water resource reveals that more than a billion of people belonging to the mostly in lower-income groups, lack of access to safe drinking water in worldwide. According to the United Nations statement; in this millennium, one of the important goals is to highlight the critical need for impoverished and developing regions of the world to achieve self-sustenance in the potable water supply [2 and 3]. Especially in India the contamination content is very high fluoride and bacterial contamination is high in brackish water.

2. Desalination: A solution to scarcity of fresh water

Desalination is a technology to remove all contaminants including almost all dissolved ions, microorganisms and other pollutants [4 - 7]. The various popular desalination technologies are

1. Solar stills
2. Multi-stage flash (MSF)

3. Multi-effect distillation (MED)
4. Vapor compression (VC)
5. Reverse osmosis (RO)
6. Electrodialysis (ED)

The challenge is; How to implement the desalination technology into low-cost? and How to provide them in a community-scale and How to make them relatively maintenance-free (at least maintainable by non-technical laborers)?. Over the past five decades through the research and development in desalination has resulted that various advanced techniques have been made for desalination more efficient and cost-effective [5, 8 and 9]. In future, the desalination will be a viable technique for generating fresh water, in this research work made an attempt for desalination processes for getting fresh.

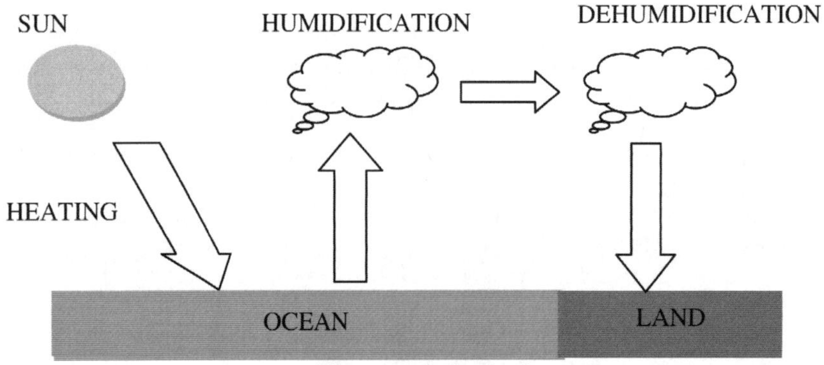

Figure 1. Rain Cycle

3. Principle of HDH system

Humidification-dehumidification (HDH) technology is one of the known techniques that offer the desalination systems of different scale depending on the demand and requirements. The operation of the rain cycle system is exemplified in Figure 1. In nature the solar energy used as a desalination resource the sea water gets heated (by solar radiation) and humidifies by the air and which acts as a carrier [10 and 11]. The humidified air rises and forms clouds and after some time the clouds dehumidified as rain. Humidification Dehumidification Desalination (HDH) system is the most promising in the recent development of the desalination processes [12]. The HDH process is based on the fact that the air can be mixed with large amounts of water vapor when the flowing air in contact with salt water [14]. A certain amount of vapor is extracted by air, which provokes cooling.

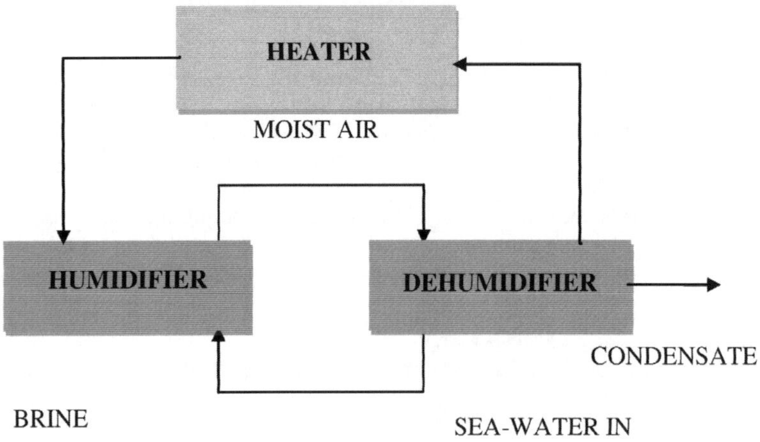

Figure 2. Simple CAOW HDH cycle

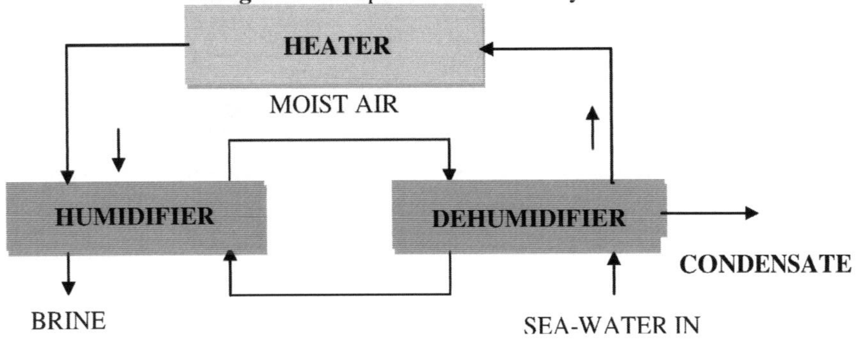

Figure 3. Simple CAOW HDH cycle

The purified water recovered from the humid, bringing air in which contact with cold surfaces. It causes the condensation of a part of the vapor-air, this occurs at the dehumidifier in which salt water is preheated by the latent heat of condensation [13].

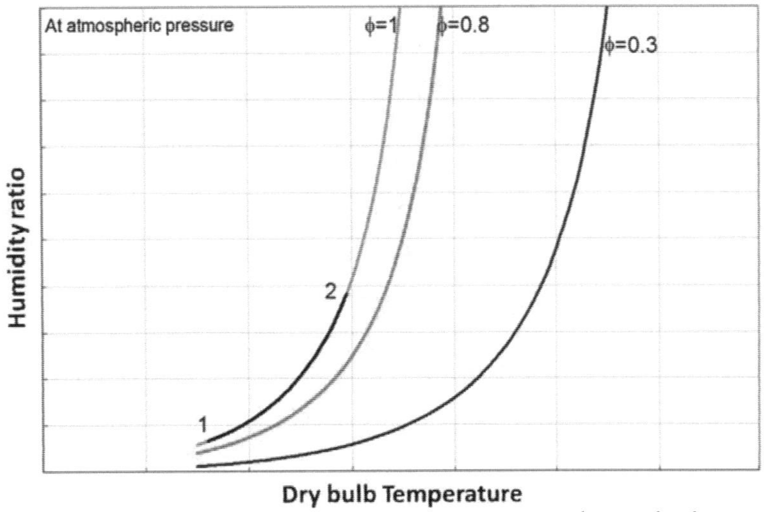

Figure 4. Water-heated CAOW HDH process on psychometric chart

4. RESULT AND DISCUSSION

The thermophysical properties required for air and sea water are computed by using thermodynamic relations; similarly, specific humidity, relative humidity, vapor pressure, and enthalpy are obtained through psychometric relations. A Program was developed in 'C' language for simulation for analyzing the HDH Desalination system, it involves input parameters such as inlet sea water temperature, inlet air temperature, inlet sea water mass flow rate, inlet air mass flow rate, heater input, the salinity of sea water spray water temperature. The 'C' program was developed to calculate and provides the information based on the various input parameters, the effect of input parameters, the condition of fluid at various stages like the temperature of fluids at each stage, relative humidity of air at each stage, specific humidity of air at various stages.

4.1: The effect of relative humidity of inlet air on the productivity of different heat input to the heater

Figure 5 provides the increased humidity of the inlet air is reduces the productivity of the system, due to the amount of water vapor required to reach the outlet state is reduced, so the amount of water evaporated in the humidifier reduced, the productivity of the system which depends on

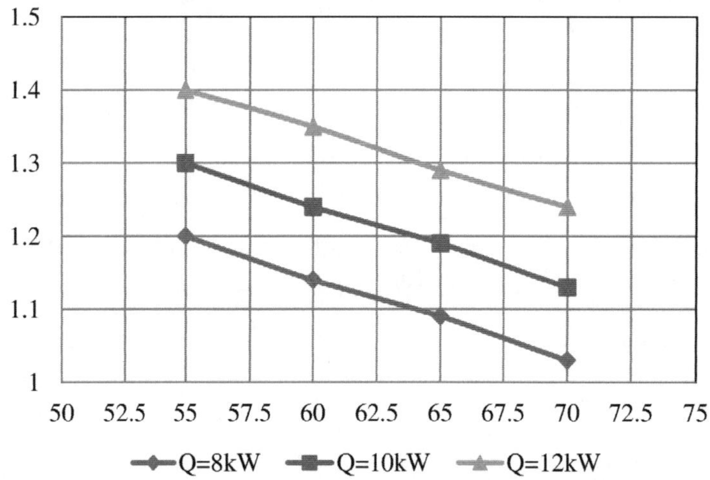

Figure 5. Effect of relative humidity of inlet air on the productivity at different heat input to the heater

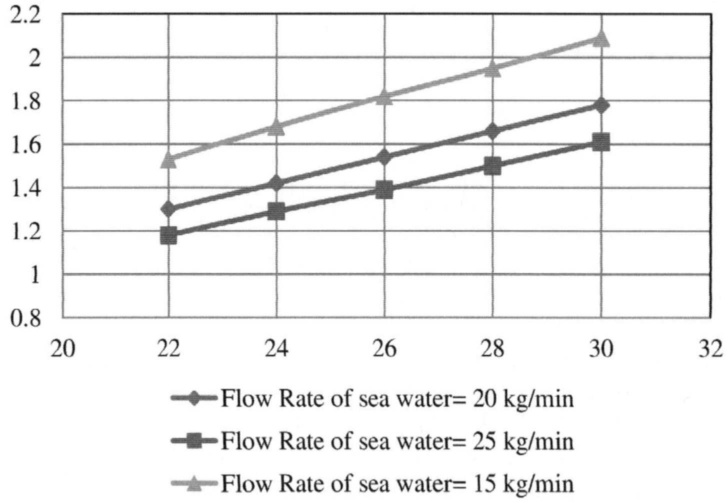

Figure 6. Effect of mass flow rate of air on productivity at different mass flow rate of sea water

Figure 6 represents the theoretical yield of fresh water production at a different mass flow rate of air and mass flow rate of water. It indicates that an increase of mass flow rate of air will increase the total amount of air carried out by the air if the sea water temperature increases then the productivity of the system to be reduced. Because it is due to the fact that for the same heat input into the outlet temperature of the heater outlet water temperature; it reduced with an increase in the mass flow rate of air, thus the productivity of the reduced system.

4.3: Effect of mass flow rate of water on productivity at different mass flow rate of air

Figure 7 shows the effect of mass flow rate of water on productivity at a different mass flow rate of air, from this it was inferred that the desalination rate decreases when the increase of the sea water flow rate and it has two opposite effects on the system performance. At the time of starting it decrease the system temperature due to the increase in the water flow rate and the associated thermal load. It effects to reduce the driving force for evaporation and, in turn, the amount of producing water. On the other hand, increasing the flow rate of air stream increases the distillate flow rate.

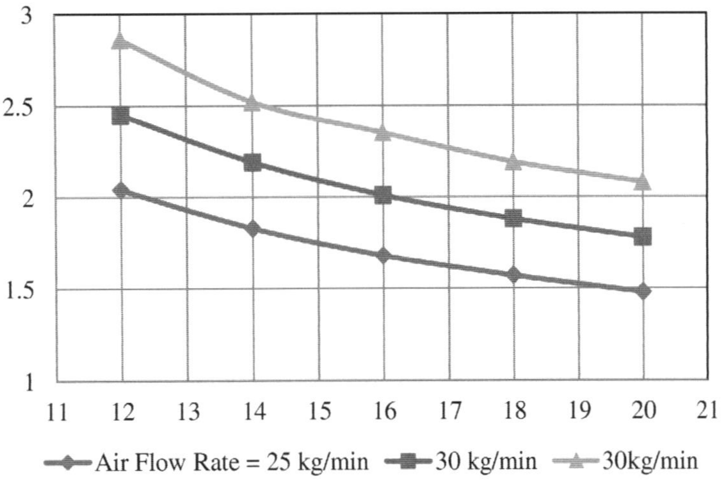

Figure 7. Effect of mass flow rate of water on productivity at a different mass flow rate of air of the increase in the total amount of water vapor carried by the air stream.

4.4: Effect of heat input to the heater on productivity at different mass flow rate of air

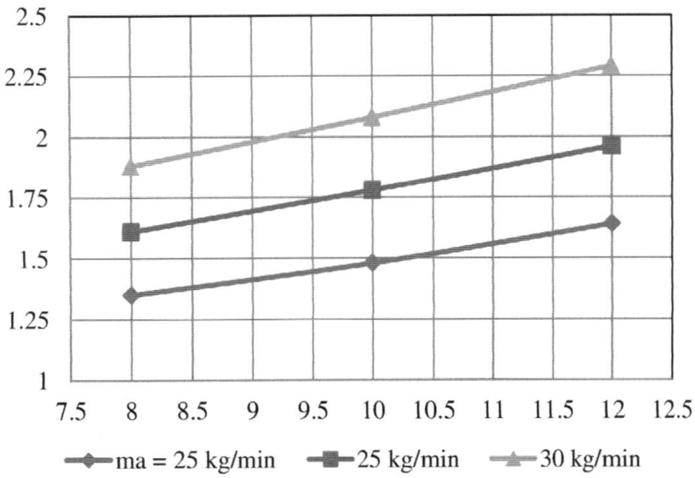

Figure 8. Effect of heat input to the heater on productivity at different mass flow rate of air

Figure 8 shows the effect of heat input and the mass flow rate of air on the productivity, the graph represents the increased heat input causes the increase the productivity of the system, due to increase of heater input temperature causes the higher heater outlet water temperature, an increase in the water temperature leads to the increased evaporation of water through the flowing air. So the amount of water evaporated increased per kilogram of dry air, so the productivity increased; on the other hand, if the mass flow rate of air increases the total amount of water carried by air is increased.

4.4: Effect of relative humidity of humidified air on productivity at different mass flow rate of air

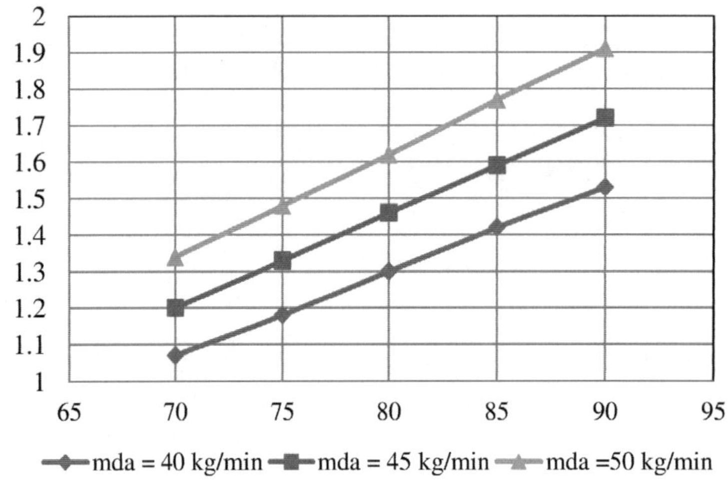

Figure 9. Effect of Relative Humidity of humidified air on productivity at a different mass flow rate of air.

The Figure 9 shows that the effect of relative humidity of the humidified air and the mass flow rate of dry air on the productivity of the total system, the relative humidity of the air increases, then the productivity also increased, this is due to increase relative humidity will lead to the increased water carrying capability of air. So the amount of water vapor per kilogram of dry air has been increased, due to this the productivity of the system also increased.

4.5: Effect of mass flow rate of water on productivity of different heat input to the heater

Figure 10 shows the effect of mass flow rate of sea water entering into the system and the heater input rate on the productivity. The mass flow rate of inlet sea water will decrease the system productivity, due to the temperature of sea water at the outlet of heater decreases with an increase in the mass flow rate of inlet sea water at constant heater input. So the humidification efficiency of the air will be reduced due to the decrease in temperature, which leads to the reduction of the water evaporation amount and reduces the productivity of the system.

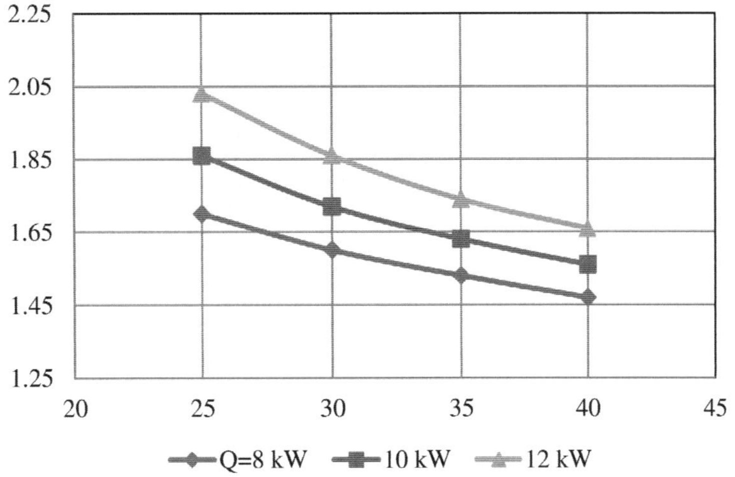

Figure 10. Effect of mass flow rate of water on productivity of different heat input to the heater

4.6: Effect of flow rate of air on recovery ratio of different heat input to the heater

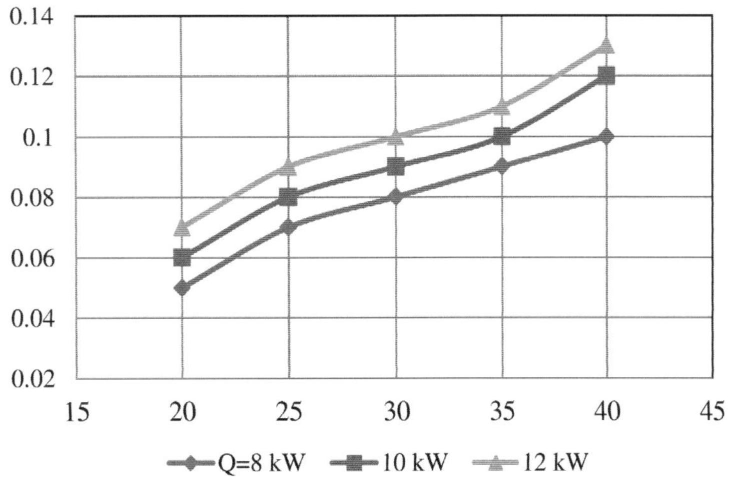

Figure 11. Effect of flow rate of air on recovery ratio of different heat input to the heater

The Figure 11 shows the effect of mass flow rate and the heater heat input rate on the recovery ratio, which represents the flow rate of air will increase the productivity of the system with constant inlet sea water mass flow rate into the system. So the amount of pure water recovered per kilogram of feed seawater increased, this shows the increase in the recovery ratio with an increase in the mass flow rate of air flowing into the system. The increased heater input leads to the higher humidification efficiency of the humidifier so the amount of water extracted in the humidifier is also increased. So the increase in heater input caused the increase of extraction efficiency or Recovery Ratio.

4.7: Effect of inlet water temperature to the heater on recovery ratio of different seawater mass flow rate

The effect of heater inlet sea water temperature and mass flow rate of water is shown in Figure 12 from the graph the increased water inlet temperature leads to the increased outlet temperature of the sea water from the sea water, so increase in the spray water temperature leads to the increase of humidification efficiency of the air so that the amount water extracted per kilogram of sea water also increased, due to this the Extraction efficiency or the Recovery Ratio increased with increase in inlet sea water temperature to the heater. On the other side increase of sea water mass flow rate decrease the productivity of the system as discussed earlier, so the amount of water evaporated per kilogram of feed water also reduced it caused the reduction of the recovery ratio with an increase in the mass flow rate of inlet sea water.

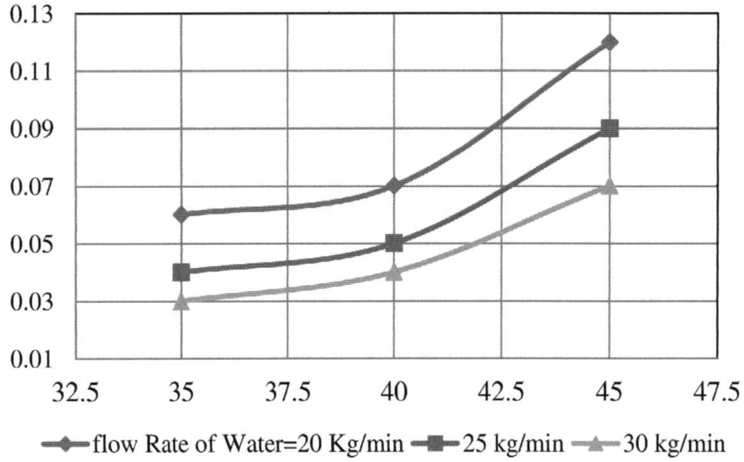

Figure 12. Effect of inlet water temperature of the heater on recovery ratio of different seawater mass flow rate

4.8: Effect of heat input to the heater on recovery ratio of different heater inlet sea water temperature

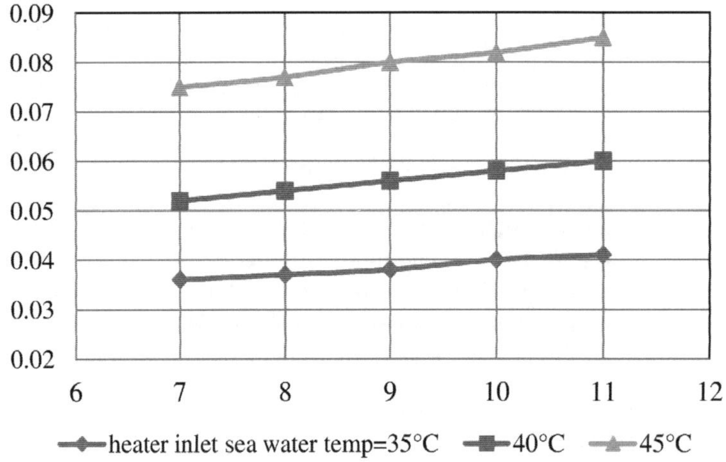

Figure 13. Effect of Heat input to the heater on recovery ratio of different heater inlet sea water temperature

The fig.13 shows the effect of heater heat input and temperature of sea water at outlet of the dehumidifier on the recovery ratio of the system, the increase in the heater input with constant mass flow rate of inlet sea water to the system will increases the productivity of the system as discussed earlier the increased heater input will increase the humidification efficiency of the air, so that the amount of water evaporated per kilogram of feed water increased, due to this the recovery ratio increases with increase in heater input. The fig.13 shows the effect of heater heat input and temperature of sea water at outlet of the dehumidifier on the recovery ratio of the system, the increase in the heater input with constant mass flow rate of inlet sea water to the system will increases the productivity of the system as discussed earlier the increased heater input will increase the humidification efficiency of the air, so that the amount of water evaporated per kilogram of feed water increased, due to this the recovery ratio increases with increase in heater input. In another view the inlet sea water temperature in the heater also causes the increase in the spray water temperature so the productivity of the system also increased, this is caused the system to increase the extraction efficiency.

4.9: Effect of RH of humid air on recovery ratio of different mass flow rate of sea water

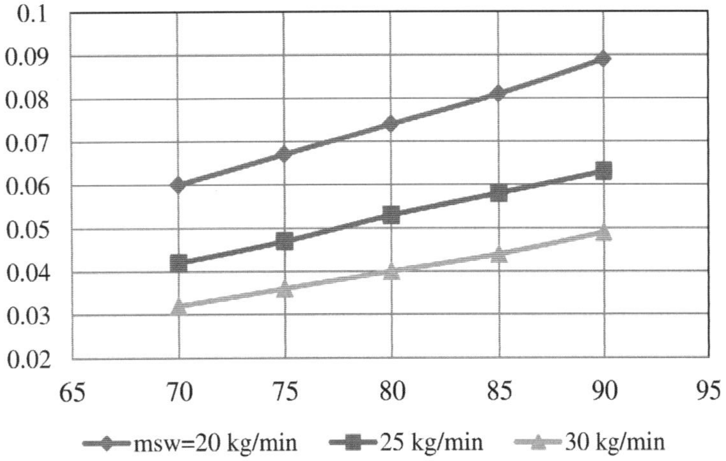

Figure 14. Effect of RH of Humid air on recovery ratio of different mass flow rate of sea water

The figure 14 shows that the effect of RH of humidified air and the mass flow rate of sea water on the recovery ratio, when the relative humidity of the humidified air increased the amount of water vapor mixed with the air also increased so the productivity of the system also increased as discussed earlier in this chapter, due to the increased productivity extraction efficiency of the system has been improved. With the increased mass flow rate of sea water with constant heater input reduces the extraction efficiency, this is due to that the increase of mass flow rate of sea water reduces the evaporative efficiency so that the recovery ratio reduced to increase a mass flow rate of inlet sea water.

5. Conclusions

The experimental study of humidification-dehumidification (HDH) system, works based on thermodynamic principles. It reveals that the increased flow rate of fresh water produced by increases the relative humidity of humid air from the humidifier. The flow rate of the dry air increases, then the evaporation efficiency of the humidifier is increased, hence the flow rate of distillation also increased. In case the flow rate of seawater increases, then the flow rate of distillate is decreased.

6. References

[1] Al-Hallaj, Mohammed Mehdi Farid, and Abdul Rahman Tamimi 1998 *Desalination* **120(3)** 273-280.

[2] Mohamed A M I and El-Minshawy N A 2011 *Energy Conversion and Management* **52(10)** 3112-3119.

[3] Mistry Karan H, Alexander Mitsos and John Lienhard H 2011 *International Journal of Thermal Sciences* **50(5)** 779-789.

[4] Narayan et al., 2010 *Renewable and Sustainable Energy Reviews* **14(4)** 1187-1201.

[5] Soufari, Zamen S M and Amidpour M 2009 *Desalination* **237(1)** 305-317.

[6] Prakash et al., 2010 *Desalination and water treatment* **16** 339-353.

[7] Ettouney and Hisham 2005 *Desalination* **183(1)** 341-352.

[8] Yamalı, Cemil, and İsmail Solmus 2007 *Desalination* **205(1)** 163-177.

[9] Guofeng et al., 2011 *Desalination* **277(1)** 92-98.

[10] Orfi J et al., 2004 *Desalination* **168** 151-159.

[11] Yamalı, Cemil and İsmail Solmus 2008 *Desalination* **220(1)** 538-551.

[12] Kabeel A E et al., 2014 *Energy* **68** 218-228.

[13] Dayem, Adel M Abdel, and M Fatouh 2009 *Desalination* **247(1)** 594-609.

[14] Sharqawy, Mostafa H, John H Lienhard, and Syed M Zubair 2010 *Desalination and Water Treatment* **16** 354-380.

IOP Conf. Series: Materials Science and Engineering **183** (2017) 012033 doi:10.1088/1757-899X/183/1/012033

Performance and Analysis of Floating dome Anaerobic Digester with Wet and Dry Feedstock

S.Sathish[1], A.Parthiban[2], S.Venugopal[2], V.M.Jothi Prakash[3]
[1, 2, 3]Department of Automobile Engineering, Vels University, Pallavaram, Chennai, India.
E-mail: sathishamg88@gmail.com1,

Abstract. The objective of this study is to evaluate the feasibility of anaerobic digestion to generate biogas yield and it's performed using wet and dry feed stock. The laboratory experiment is conducted in a floating dome type anaerobic digester with 1m3 capacity. It is made up of fibre material at continues process. The starter cowdung used as an inoculum of the anaerobic digester. Then raw materials feeded as a wet type wheat straw and dry type wheat straw is the ratio of 1:1 waste/water in both the experiments wet and dry wheat straw. In this experiments are fermented at 30°C to 35°C temperature is maintained. The daily biogas yield, cumulative biogas yield, pH, CH4, and hydro retention time these parameters is studied and analysed. The maximum daily biogas is 25liters and 42% of methane is achieved in dry wheat straw at 15th day of digestion. The highest gas yield obtained in dry condition compare to wet condition and acid level also decreased in wet digestion.

1.Introduction

The anaerobic digestion converts organic waste materials into useful biogas yield from microbial consortia under presence of oxygen and absence of oxygen. Anaerobic digestion one of the few technologies that the both produce energy and treated waste streams. In this process involved in four stages like, hydrolysis, acidogenesis, acetogenesis, and methanogenesis, create the microorganisms of the digestion[1]. Biogas generation can be enhanced by using abundant lignocellulose materials such as agricultural and forest residues, however the complex lignocellulose structure limits the accessibility of the sugar in cellulose material and hemicellulose. It means pretreated is Necessary for wheat straw why because it's one of the lignocellulose materials [2]. Basically floating drum digesters can be operated on the three temperature ranges (i) psycrophilies below 28°C (ii) Mesophilic range at 30°C to 40°C (iii) thermophilic range at 40°C to 60°C [3]. Generally the methanogenic bacteria and microorganisms growth is mainly depends on different parameters like pH range, digester design, digester temperature, Carbon/Nitrogen ratio, organic loading rate, chemical oxygen demand, biological oxygen demand and hydro retention time [4]. The anaerobic digestion process (AD) of wheat straw or waste/energy crops, paddy is considered as an important role in the renewable energy source. It is important in contribution of climate change mitigation [5]. The pH is the important factor affecting the growth of microorganisms during the anaerobic digestion process. The most anaerobic bacteria, including methane formation bacteria, exclude well within a pH range of 6.8-7.5. It is generally considered suitable for optimum biogas generation [6]. Biomass can be pretreated by physical and chemical methods. Some of the physical treatments are sizing the biomass particles and drying and removing inorganic from the feedstock etc. There are many chemical pre-treatment methods. The main aim of this method is to make the cellulose surface accessible to cellulose enzymes during the subsequent hydrolysis process. Pre-treatment prior to anaerobic digestion proves to be one of the simple and effective methods to improve biodegradability and increase biogas production of lignocellulosic materials [7]. In the agricultural wastes, particularly rice straws and rice straw have high lignocellulosic, hemicelluloses and cellulose is penetrating with the rigidifying binding material

IOP Conf. Series: Materials Science and Engineering **183** (2017) 012033 doi:10.1088/1757-899X/183/1/012033

is known as lignin. The polysaccharides are not usable for bioconversion process. Hence pre-treatment is involved to overcome the physical block of lignin and case sugar available for the microorganisms [8]. The dry anaerobic digestion process is have high ammonium concentration, it is one of the serious problem would be occurred by the low carbon/nitrogen ratio because of inherent higher ammonium concentration in a substrate such as straws, which leads to ammonia inhibition to methanogenic bacteria [9]. Recently, large amount of straws produce from agricultural form, feedlot farming high in annually, most of which are disposed into landfills or are applied to the land without treatment. AD process provides an alternative option for waste to use full energy. In this research, wheat straw is assessed for the use of anaerobic digestion with objectives of treating the wheat straw waste in wet and dry condition to decrease disposal costs and to produce biogas yield. The biogas yields mainly consist of methane and carbon dioxide and the methane is used as an energy source. The main aim of this paper is to investigate the effectiveness and the performance characteristics of floating drum anaerobic digestion process of wet and dry wheat straw for biogas generation in continuous operation. The wheat is an alkaline waste material so it pretreated with some chemical solutions. The lignocelluloses (rice straw) are a plant biomass, primarily consisting of three major elements such as cellulose, hemicelluloses and lignin. The other constituents such as water and protein are not organizing the structure of the material [10]. The conversion of agricultural wastes to energy and application of biogas had been widely accepted by household digestion process. The biogas anaerobic digestion of rice straw is converted into life fuel and also transformed to high quality of organic fertilizer [11].

2.Experimental Setup

In this experimental study a floating drum gas holder type biogas plant was used. The plant was made of fibre material, with a capacity of 1cubic meter bio-digester and the total volume of the digester was 1000 litters with an effective gas holder volume of 700 litters. Heat exchanger – hot water jacket around the digester tank by circulating warm water to raise the inside temperature of the digester in case of thermophilic condition. Gas holder – is an inverted fibre drum resting over the digester with gas holding capacity of 0.7m3. (i.e.) The drum can move up and down floating over the digester. The gas holder may float over the water jacket of the digester. This special arrangement prevents the leakage of biogas from the digester which may occur in other types of digesters. The gas valve of the gas outlet is opened to get a supply of biogas. Once the production of biogas begins, a continuous supply of gas can be ensured by regular removal of spent slurry and introduction of fresh slurry. The retention period represents the time period for which the fermentable remains inside the digester. The retention time ranges from 1 to 30 days. Longer retention periods need larger size plants and it allows more complete digestion of feed material. The figure 1 shows the schematic view of experimental setup. The cowdung were used as an inoculum of the digester, the wheat straw wet and dry condition is pretreated with 8% of NaOH which is dissolved in water to prepare NaOH solution and then added with wheat straw and mixed completely. In this study the digester feed with wet and dry wheat straw waste by volume 40:60, waste by water 30:70 in both the wet and dry condition. The digester temperature is maintained at the 30ºC to 35ºC. The experiment is started and the gas volume monitored daily with help of gas flow meter from 30 days of hydro retention time. The pH is measured using pH redox meter. The CH4 is measured using gas chromatography.

IOP Conf. Series: Materials Science and Engineering **183** (2017) 012033 doi:10.1088/1757-899X/183/1/012033

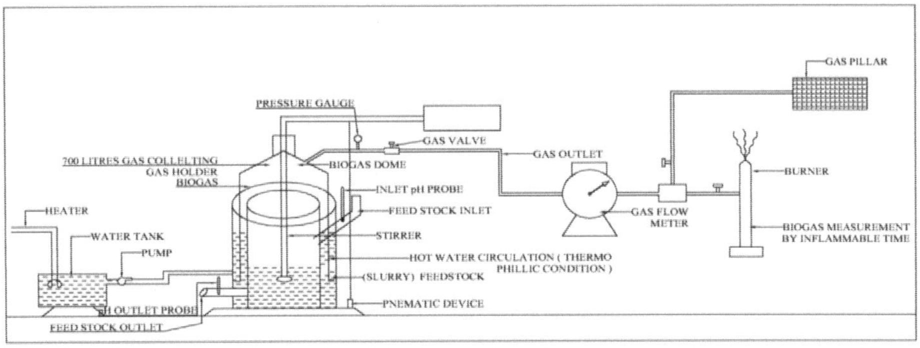

Figure 1. Schematic view of the Experimental setup

3.Result and Discussion

The daily yield during the study period or hydro retention time (HRT) is shown in figure 2. It is observed that biogas production is actually slow at starting and the end of observation in both the conditions. During the 13th day biogas yield is achieved at 39 liters in dry condition. There is less biogas yield obtained from wet condition and mainly due to the lag phase of microbial,

Growth whereas, from the range at 15 to 18 days of observation; biogas yield enhanced substantially due to the exponential growth of methanogenic bacteria. The similarly highest cumulative biogas yield from dry feedstock (wheat straw) compare to wet wheat straws.

This is probably due to high concentration of ammonia nitrogen is toxic to anaerobes, which will reduce to biogas yield from the digestion period. The cumulative biogas yield shown in Figure 3. During the anaerobic digestion of dry wheat straw is achieved at highest methane content of 42%.

The acidifying bacteria produced acids at a much higher rate, than the methane forming bacteria were able to cope with. Another reason could be that the substrate material is decomposed to easily and therefore very susceptible to acidification. That the methane content Exp2 constantly indicates that the certain areas possibly where a lot inoculum material is present, worked at a higher pH and produce gas while other Exp1 did not produce higher gas yield. Generally, Proper Carbon/Nitrogen ratio does the play an important role in the biogas production. Carbon/Nitrogen ratio ranging from 21:1 to 31:1 is always considered in suitable for anaerobic digestion by weiland [8]. During [9] the anaerobic digestion of beef manure in mixed and unmixed reactors. It is very clear that cow manure is an very effective feed materials for anaerobic digestion. It is significantly increase the cumulative biogas yield

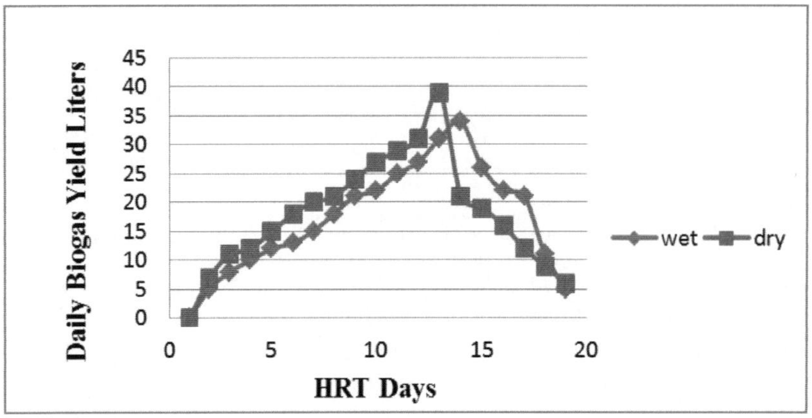

Figure 2. Daily biogas yields with HRT

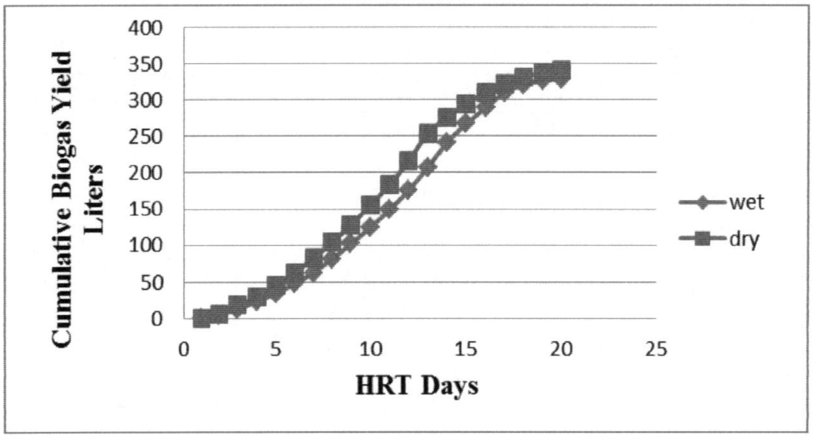

Figure 3. Cumulative biogas yields with HRT

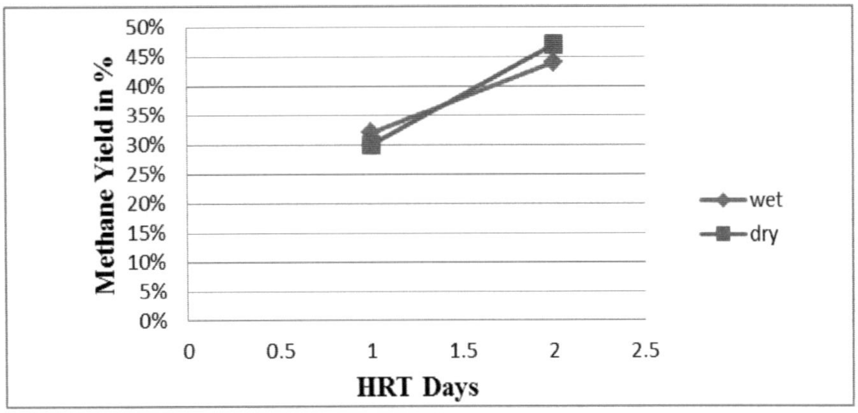

Figure 4. Methane percentage with Retention Time

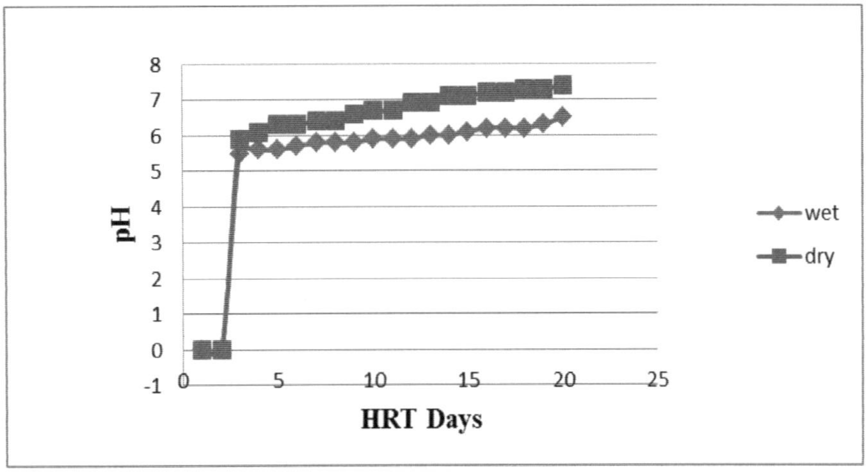

Figure 5. pH range with HRT

4.Conclusion

The result of this research, two types of feedstock's wet and dry wheat straw were attempted for biogas and methane production using floating dome anaerobic digester. Results revealed that the maximum methane and biogas production is found in dry wheat straw compared to wet wheat straw. The methane from dry feedstock is ignited the cumulative gas generation is found to be more qualitatively and quantitatively. The results showed that the dry wheat straw might be one of the feedstock for efficient biogas generation and waste treatment.

5.References

[1] Niu, Qigui, "Microbial community shifts and biogas conversion computation during steady, inhibited and recovered stages of thermophilic methane fermentation on chicken manure with a wide variation of ammonia." *Bioresource technology* **146** (2013): 223-233.

[2] Amon, Thomas, "Biogas production from maize and dairy cattle manure—influence of biomass composition on the methane yield."Agriculture, *Ecosystems & Environment* **118.**1 (2007): 173-182.

[3] Sathish S, Vivekanandan S. Experimental Investigation on Biogas Production Using Industrial Waste (Press Mud) To Generate Renewable Energy. *Int. J. Innov. Res. Sci. Eng. Technol.* 2015; **4**:388-92.

[4] Deepanraj A, Vijayalakshmi S, Ranjitha J. Production of Bio Gas from Vegetable and Flowers Wastes Using Anaerobic Digestion. In *Applied Mechanics and Materials* 2015 (Vol. **787**, pp. 803-808). Trans Tech Publications.

[5] Vivekanandan, S. Sathish. S "To study the effect of temperature using pretreated rice straw to generate biogas." *International of Environmental Science and Engineering*.2015 (Vol **10**) No 2, pp631-636.

[6] Kuruti K, Rao AG, Gandu B, Kiran G, Mohammad S, Sailaja S, Swamy YV. Generation of bioethanol and VFA through anaerobic acidogenic fermentation route with press mud obtained from sugar mill as a feedstock. *Bioresource technology.* 2015 Sep 30; **192**:646-53.

[7] Sathish S, Vivekanandan S. Parametric optimization for floating drum anaerobic bio-digester using Response Surface Methodology and Artificial Neural Network. *Alexandria Engineering Journal*. 2016 Sep 28.

[8] Kaparaju P, Serrano M and Thomsen AB and Kongjan and Angelidaki I. Bioethanol, biohydrogen and biogas production from wheat straw in a biorefinery concept. *Bioresource Technology*. 2009 May **31**; 100(9):2562-8.

[9] Riya S, Suzuki K and Terada A and Hosomi M and Zhou S. Influence of C/N Ratio on Performance and Microbial Community Structure of Dry-Thermophilic Anaerobic Co-Digestion of Swine Manure and Rice Straw. *Journal of Medical and Bioengineering* Vol. 2016 Feb; 5(1).

[10] Zheng, Yi, "Pre-treatment of lignocellulosic biomass for enhanced biogas production." *Progress in Energy and Combustion Science* **42** 2014: 35-53.

[11] Sathish, S., and S. Vivekanandan. "Optimization of Different Parameters Affecting Biogas Production from Rice Straw: An Analytical Approach." *International journal of simulation science and technology*, 2015Feb, pp: 78-84.

IOP Conf. Series: Materials Science and Engineering **183** (2017) 012034 doi:10.1088/1757-899X/183/1/012034

A Simple Control Strategy Using Speedy Transient Response for Multiphase Buck Converter

K Sasikala[1], D Ravikumar[2] and B Rubini[3]
[1, 3]Assistant Professor, Department of Electrical and Electronics Engineering, Vels University, Chennai, India.
[2] Assistant Professor, Department of Electronics and Communication Engineering, Vels University, Chennai, India

E-mail: sasikala1410@gmail.com

Abstract. The high performance of recent originating digital processors requires power supply for transient output, firm voltage regulation and high current. Industry has taken up interleaved multiphase synchronous buck converters for meeting such demands. Application of the voltage mode (ripple) hysteretic strategy to multiphase Voltage Regulator Units (VRU) would help satisfying many of the upcoming digital processors and IC's powering demands. This arises from the use of the various advantages that can be obtained from this technique. This current division method is known for several benefits it provides: easy accessibility in application to all control methods, with freedom from the prime control technique in voltage regulation loop, absence of any necessity for current reference, better accuracy in current division of equal magnitude at almost every load conditions in standstill and transients. As against this, there are several challenges that involve distribution of multiple control signals, current sharing and sensitivity to noise, high speed comparators, hysteretic band accuracy and stable at large load transients. In addition, application of the current division technique presented was made to a multiphase voltage mode ripple controlled Buck Converter that produced a novel and easy control strategy having minimum output voltage ripples and speedy transient output.

1. Introduction

There is a need for DC-DC converters and Voltage Regulator Units (VRU) for the future origination of IC's for digital processors that are inclusive of large current capability ,small output voltage variation in standstill and transient conditions, better power and current densities , light weight, miniature in size and greater efficiency[1],[2],[3]. There is an imperative need to get all these requirements satisfied at very low output which may drop below IV in the following years.

The interleaving (multiphase) and the single phase voltage mode (ripple) hysteretic control strategy have the ability to bring down the output voltage ripple to a greater value. But, there is still the need for investigation of the overshoot and undershoot demands during transients of higher magnitude [4]. Obtaining low output overshoot involves the requirements of the design of better operation feedback control from the multiphase VRM's. Such feedback control should have the ability to provide current sharing [5]. This is the reason why a vigilant control design in the output capacitor size is required for satisfying the maximum overshoot as well as undershoots limits during heavy load transients, despite the reduction of ripples in the output voltage and assistance in attaining speedy current transient output coming from interleaving techniques[6]. It is possible for the hysteretic window to be placed to aelucidated level in the single phase ripple voltage mode control strategy. This is for ensuring speedy response from the controller to transitory load and correcting the voltage prior to variation in the undershoot and overshoot values. Despite the output capacitor still playing a significant role, there is a noticeable reduction in its size arising from hysteretic control.

There are many advantages seen in voltage mode hysteretic strategy over many other techniques. Those include easy accessibility, absence of necessity for closed loop compensation, an almost quick reaction to load changes and absence of any constraints on large frequency output voltage ripple having small switching frequency, identical current sharing (division) among the phases gives greater current carrying capacity and speedy transient output.Besides voltage regulation, high performance and current sharing functionality is needed from multiphase converters for ensuring almost identical division of load among the phases at almost every load conditions. The paper presents a multiphase voltage mode (ripple) hysteretic controlled Buck converter with current division technique.

2. Interleaved Buck Converter Using Voltage Mode (Ripple) Hysteretic Strategy

The below mentioning points require careful attention at the time of application of the hysteretic (ripple) voltage mode strategy to interleaved converters

(i) Frequent division has to be made for the control signal derived from the hysteretic (ripple) control, simultaneously maintaining the original control signal (interleaving).

(ii) Multiple control operation has to be disabled during transients to enable simultaneous switch ON andswitch OFF for the switches of all the phases. The converting frequency is a minimum to achieve the speedy transient output and identification among the phases at the time of transients.

Figure 1 is the functional diagram of N interleaved DC-DC converter using voltage mode (ripple) control strategy. Generating of the control signal (Cm) is to calculate the output voltage at a maximum value (VH) and minimum value (VL) to create a periodic window. The starting pulses between the phases is done to ensure that one phase upper sided switch is in the ON state at a particular time, but with all the remaining switches are being in the OFF state during the same time which generates the multipurpose control signals C1, C2, C3……CN.

As comparator 1 in a conduction state, the other comparator has the threshold limit of VLT<VL switches ON and all the upper side of a switches is (MOSFETs) and switches OFF of all the lower side of a switch is at a lower to higher load transients (thus the voltage undershoot).The new Threshold (VLT) is involved to ensure the transient output to produce speed at a maximum identify current sharing at ephemeral state. With no addition of this threshold, only one switch is in ON state at lower to higher ephemeral state when Vo<VL with the result of a slower transient output. On the other hand, every upper side switches are in the OFF state at a higher to lower load ephemeral state (those create potential overshoot) by the hysteretic comparator with new threshold which is larger than VH.

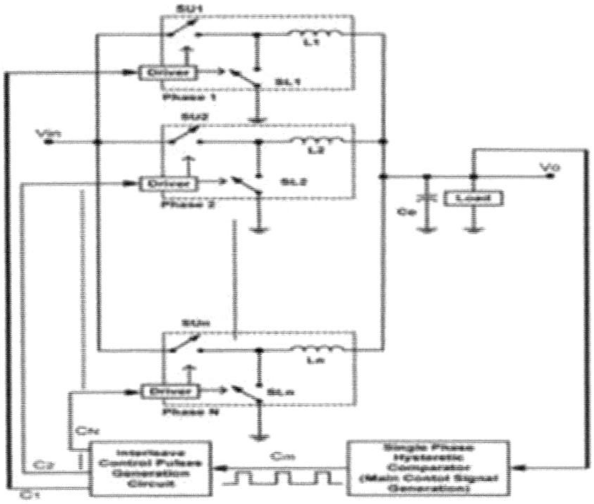

Figure 1 Functional Diagram of N interleaved converter using voltage mode (ripple) hysteretic control

3. Multiphase Voltage Mode (Ripple) Hysteretic strategy using Current Sharing (Division) Method

Identical distribution of the current between the phases at a multiphase converter is an important point to note. Unfortunately, there is a contingency of the components, connections and lay out deviations that cause inequality at a current distribution (sharing), more particularly at a large load transients.

Achievement of current division between the interleaved converters generally becomes a fast control over ON time at a converter phase. The ON time of particular phase carrying larger current is lesser compared to the ON time of other phases. It is also a way that the phase carrying the lowest current has the highest ON time. This implies that the current sharing (division) regulation is achieved by abridging the ON time of (phase) switches. There is the dependence of the current sharing (division) task on the control procedure that finds to use in the voltage regulation. Additionally, complexity is found for the current sharing functionality over and above the direct loop to the closed loop design. The ON time of the switches is a direct derivative from the ripples in the output voltage through fastening of the switches OFF/ON's with the output voltage hitting the upper as well as the lower constraints of the hysteretic comparator.

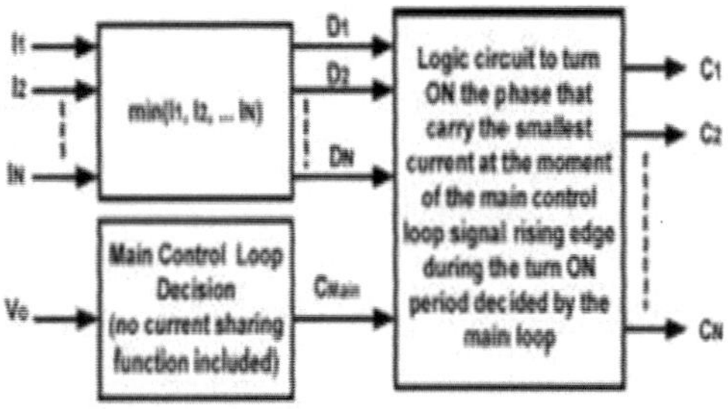

Figure 2 Basic functional diagram of the Proposed Method

Therefore any change in ON duration of a phase switch involves change in the hysteretic window of other switch that is turned ON indirectly. When the other switch is ON, numerous turn ON's are happened inside the comparator whose switching time period resulting in a higher switching frequency for each and every phase.

Figure 2 is the basic functional diagram of the proposed current sharing (division) method where the instantaneous currents at every phase are senses (I_1, I_2, I_3....I_N) and each compared to the another phases of instantaneous currents to produce the comparison digital code (D_1, D2, D3,....D_N), without any necessity for a allusion to identify the phase that carries the minimumcurrent. As the main regulating control loop having zero current division function determines the time to switch ON a switch, the upper sided switch in the converter for instance, furnish energy to the output of the converter. The phase that has the lowest current is switched ON through the introduction of a rising edge signal (C_{main}), while the remaining phases are switched OFF generating the terminal control signals (C_1, C2, C3......C_N).Stated in elementary languages, one switch is switched ON at any point of time. The choice of this switch is done through monitoring of the specific phase that is seen having the lowest instantaneous current during when a switch needs turned ON.

Despite the instantaneous comparison of current in every phase in relation to other phases on a continuous basis, decision on the comparison takes the effect at an increasing effect of the original control signal (C_{main}) by switching ON to the suitable phase. Therefore, the instantaneous position switch is the only effect on the current sharing process. Several methods are used for sensing instantaneous currents. One of the standard methods is through the inductors placing resistors in series having a small value. this is known for the simplicity and relative accuracy, downgrades is a result of the drop in the current sensing resistor, more particular applications involves more current. A second method is realizing the potential across the lower side switches.

i) Availability of the instantaneous current data for a DC-DC converter where the lower sided switch is in the ON state and the upper sided switch is in the OFF state.
ii) All the lower sided switches are ON period to turning ON one upper sided switch, resulting in the availability of the current information.

Figure 3 is the fundamental diagram meant for the multiphase voltage mode (ripple) hysteretic controller using the proposed current division technique. Figure.4 shows the conventional waveform samples meant for current division (sharing) process in a steady state. The basic working principle of the circuit is given in Figure.3 which can be elucidated with the related waveforms depicted in Figure.4. Generation of the main control signal (C_{main}) is done through comparison of the voltage at the output with a least voltage V_L and highest voltage $V_{H,}$ approximating to a reference voltage V_{Ref} with the use of hysteretic comparator Comp2. The rising edge of the turn ON signal of $C_{main}$is utilized as a clock for D Flip flop causing activation of current sharing (division) decision of Comp1. A lower limit $V_{LT}<V_L$ is shown for Comp 3. The output of the comparator is logic 0 (low) in the standstill operation when$V_0>V_{LT}$ and logic 1 (high) during start up and also during the lower to higher load transients. $V_0<V_{LT}$ is the cause of the current sharing decision of Comp1.

Comp 3 has a lower limit $V_{LT}<V_{L.}$ The output of the comparator is logic 0 at the time of standstill working when $V_0>V_{LT}$ and a logic 1 at the time of startup when $V_0<V_{LT}$ that is responsible for the phases to switch simultaneously at the time of transients. This process generates the required control signals C_1 and C_2 for achieving voltage regulation in addition to current sharing

IOP Conf. Series: Materials Science and Engineering **183** (2017) 012034 doi:10.1088/1757-899X/183/1/012034

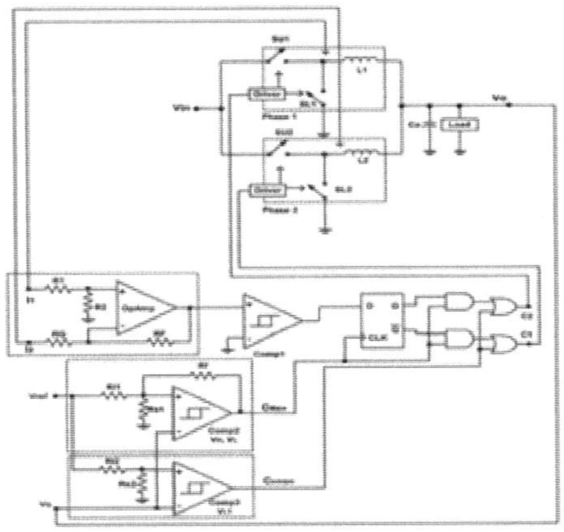

Figure 3 Proposed hysteretic Controller method

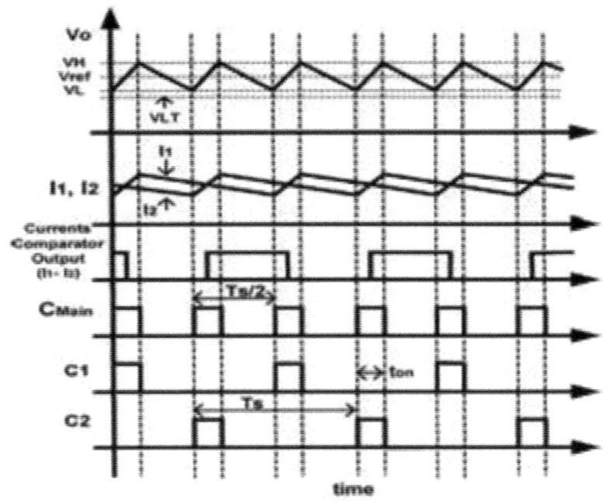

Figure 4 General waveform samples for the current division method

4. New control technique for n interleaved phases

Figure.5 is the basic widely used controller diagram for N multiphase voltage-mode (ripple) hysteretic strategic current division converters. Figure.5 depicts the outputs D_1,D_2……,D_N of the circuit using current comparison as a logic high at a time, making the phase carrying the lowest instantaneous current to switch ON at the rising edge of the voltage mode control loop as elucidated in the preceding part of the two-phase method[8].

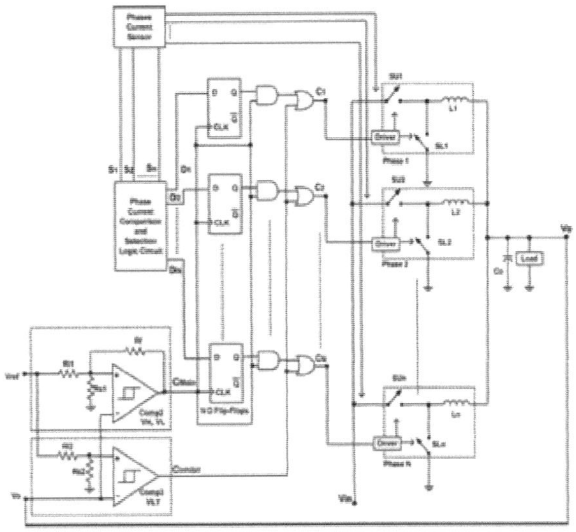

Figure 5 Fundamental controller diagram of N multiples of voltage-mode (ripple) hysteretic strategic current sharing (division) converters.

5. Theoretical analysis

This section carries an analysis of hysteretic controlled buck converter in a multiphase voltage-mode. The frequency equation is also obtained. There are certain assumptions made during the analysis. One is the buck converter is in standstill condition. Another is that all components are perfect with the exception of the following non idealities: The output capacitor consists of an identical Inductance (ESL) and identical Resistance (ESR), the switches have ON inductor and the resistance. There is an assessment that the voltage at the input is stable having zero ripples and the control loop having limited time duration [9].

Figure.6 depicts the voltage across the output capacitor in a stand still state considering ESL and ESR of the capacitor. In standstill state, the ripple voltage across the capacitor considering ESL and ESR is written by

$$v_{ripple}(t) = v_C(t) + v_{ESR}(t) + v_{ESL}(t) \qquad (1)$$

Where

$v_{ripple}(t)$ = Capacitor output voltage ripple

$v_C(t)$ = capacitor output voltage;

$v_{ESR}(t)$ = capacitor output voltage having Equivalent Resistance

$v_{ESL}(t)$ = capacitor output voltage having Equivalent Inductance

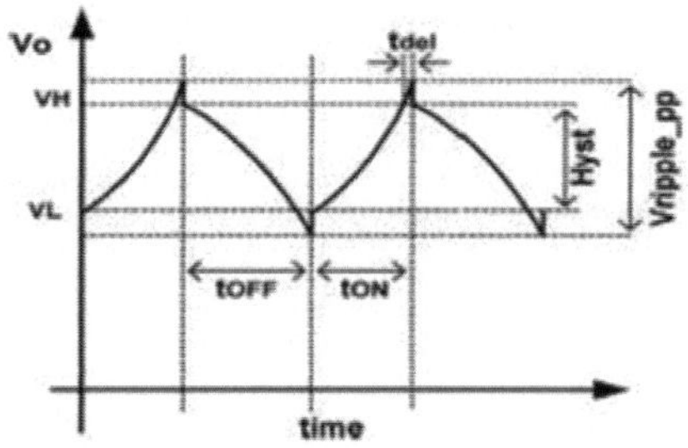

Figure 6 Output Voltage with its Ripple and delay time

In standstill condition two modes of operation are possible. At the time of Mode 1, as one of the upper -side switches in the ON condition, (1) turns

$$v_{ripple}\_ON(t) = \left(\frac{\Delta I \times t^2}{2\times C_O \times D \times T_S} - \frac{\Delta I \times t}{2\times C_O}\right) + \left(ESR \times \left(\frac{\Delta I \times t}{D \times T_S} - \frac{\Delta I}{2}\right)\right) + \left(ESL \times \frac{\Delta I}{D \times T_S}\right)$$

$$t_0 \le t \le DT_S. \qquad (2)$$

At the time of Mode 2, as all upper-side switches are in the OFF condition, (1) is written by

$$v_{ripple}\_OFF(t) = \left(\frac{\Delta I \times t^2}{2\times C_O \times (I-D) \times T_S} + \frac{\Delta I \times t}{2\times C_O}\right) + \left(ESR \times \left(\frac{\Delta I}{2} - \frac{\Delta I \times t}{(1-D)\times T_S}\right)\right) + \left(-\frac{ESL \times \Delta I}{(1-D)\times T_S}\right)$$

$$DT_S \le t \le T_S \qquad (3)$$

Where

ΔI = N phase overall inductor's current ripple;

T_S = N phase switches period;

D =duty cycle

C_O = N phase total output capacitance

The hysteresis window is then given by

$$H_{yst} = v_{ripple\_}ON\ (t_{ON\_}t_{del}) - v_{ripple\_}OFF\ (t_{OFF}\text{-}t_{del}) \tag{4}$$

Where

H_{yst} =hysteresis window;

t_{ON} =upper side switches ON time, where $t_{ON}=DT_S$;

t_{OFF} =upper side switches OFF time, where $t_{OFF}=(1\text{-}D)\ T_S$;

t_{del} = feedback loop's overall time delay.

Assume all the output inductors are symmetrical in all the phases, now the overall ripple current of inductor(s) having N phases is written by

$$\Delta I = V_{in} - I_0 R_{eq} - NV_O \times D \times T_S/L \tag{5}$$

Where

V_{in} = input voltage; V_O = output voltage;

R_{eq} =total path resistance;

L =output inductor of single phase

$$D = \frac{tON}{Ts} = \frac{V0 + I0/Req}{Vin} \tag{6}$$

The switching (changing) frequency relies on the current flowing through the load; this leaning is constrained, considering the same equivalent circuit of synchronous rectifiers over the switch period[7]. However there is a decrease in switching frequency with increase in the phases for the symmetrical hysteresis window. As a matter of fact, the output voltage has a smaller ripple when each phase's switching frequency maintains the same as that of N phase.

6. Simulation Results

Simulation of two-phase using the proposed current sharing method for DC-DC buck converter and hysteretic control was obtained. The design specifications include: Vin = 12V, V_0 = 1.5 V, 50A full load, L_{phase1} =L_{phase2} = 1 mH, and C_o = 20μF, transient hysteretic band of 300μV and hysteretic band of 100μV.. There is an assumption of three capacitors being e paralleled with ESR = 8mΩ, ESL = 4.8nH. Another simulation parameters were t_{del}=100 ns and R_{eq}=10mΩ.

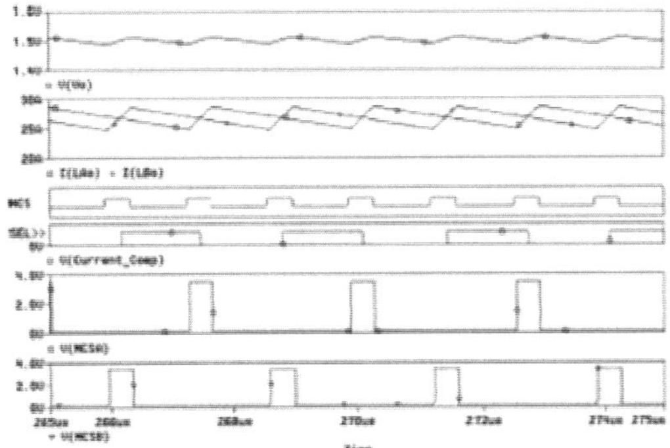

Figure 7 Simulation Results during the Steady-State condition

Figures 7-9 gives the outputs at different states. a) V (V_o) be the voltage output wave. b) I ($L1_o$) and I ($L2_o$) be the inductor output current wave for Phase 1 as well as Phase 2. c) MCS is the wave for output hysteretic comparator. d) V (Current Comp) be the current comparison waveform i) logic high when Phase 1< Phase 2 , ii) logic low when Phase 2<Phase 1. e) V (MCS 1) and V (MCS 2) is the upper -side switches driving waves of Phase 1 as well as Phase 2.

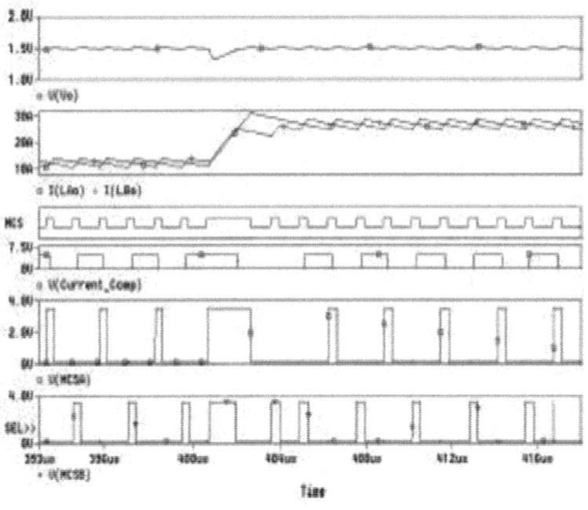

Figure 8 Simulation outputs in logic 1 to Logic 0 Load Transient State

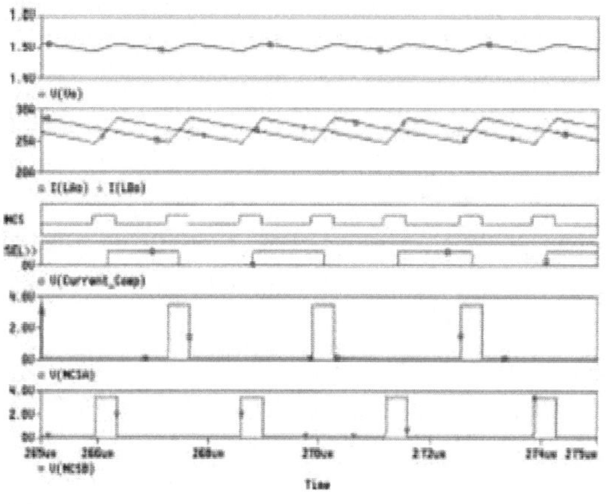

Figure 9 Simulation outputs in logic 0 to Logic 1 Load Transient State

7. Conclusion

This paper presents the current sharing method for multiphase voltage-mode (ripple) hysteretic strategy for DC-DC converters which has several advantages. These include simplicity in application, being free of the original control technique applied in the voltage regulation loop, absence of any necessity of current reference, better current sharing precision of equal magnitude at almost every standstill and transients load states, as a consequence not affecting the speed of the controller. In addition to the above conditions, there is another possibility of this sharing approach to any control method including events with no loop compensation like multiphase (ripple) hysteretic control. The implementation of the above method has been widely applied for N interleaved phases.

There is also some applications of this method to a multiphase voltage mode (ripple) hysteretic controller dc-dc converter, with the result of a recent and easy control strategy with low ripples in the output voltage and speedy transient output. The frequency equation has been found for N interleaved phases by including ESL, ESR as well as the path resistance with the output current effect.

8. References

[1] Muhammad H Rashid *Power electronics circuits devices and applications second edition*, prentice hall publication.

[2] Panov Y and Jovanovic M M Stability and dynamic performance of current sharing control for paralleled voltage regulator modules *IEEE transactions Power Electronics* **17**.

[3] Abu-Qahouq J A Mao H and Batarseh I 2002 Novel control method for multiphase low-voltage high current fast transient VRMs Proceedings in *IEEE 33rd Annual Power Electronics Specialists Conference* **4** 1576-1581.

[4] Huang W 2001 A new control for multi-phase buck converter with fast transient response in *Proceedings 16th Annu. IEEE Applied power electronics conf.* **1**.273-279.

[5] Xunwei Zhou Peng xu Lee F C 2002 A novel current sharing control technique for low voltage regulator module applications *IEEE Transactions on Power Electronics* **15(6)** 1153-1162.

[6] Osama Abdel-Rahman Issa Batarseh 2009 Transient Response Improvement in DC-DC Converters Using Output Capacitor Current for Faster Transient Detection *IEEE Power Electronics Specialists Conference.*

[7] Mini P Varghese Manjunath A Raghu C N 2012 An inventive approach of stability for Voltage Regulator Modules with high frequency multiphase buck converters *IEEE International Conference on Power Electronics, Drives and Energy Systems (PEDES.)*

[8] Nagaraja H N Deba prasad Kastha Amit Petra 2011 Design Principles of a Symmetrically Coupled Inductor Structure for Multiphase Synchronous Buck Converters *Industrial Electronics* **58(3)** 988 – 997.

[9] Santhos Ario Wibowo Zhang Ting Tetsuya Taura Yasunori Kobori Haruo Kobayashi 2008 analysis of coupled inductors for low-ripple fast-response buck Converter Circuits and Systems *IEEE Asia Pacific Conference* 1860 – 1863.

Investigation of solar water heater by using flat plate collector and evacuated tubes

J J Jayakanth[1], S Ramasubramanian[2], M Chandrasekaran[3] and S Manavalan[4]

[1]Assistant Professor, Department of Mechanical Engineering, Vels University, Chennai, India
[2]Assistant Professor, Department of Mechanical Engineering, Vels University, Chennai, India
[3] Professor, Department of Mechanical Engineering, Vels University, Chennai, India
[4]Assistant Professor, Department of Mechanical Engineering, Bharath Institute of Science and Technology, Chennai, India

E-mail: jkanth.se@velsuniv.ac.in

Abstract. Collection, storage and utilization of solar energy by solar water heating by flat plate collector and evacuated tube is cheapest and effective renewable energy technologies. This solar water heater design uses a nonconventional source of energy which can be used for house hold applications. The cost of production will be comparatively low in cost and high in capacity. The G.I. sheet collector boxes are replaced by copper tube, stainless steel water tank, thick costly PUF insulations and toughened glass etc. Pebbles are used as a medium of heat storage this increases the capacity of solar water heater.by using this method natural source of energy can be used instead of depending on electric water heaters.

1. Introduction

The solar water heater uses the incident solar radiation for generating heat and it converts the radiation energy in to thermal energy. The heat generated by the thermal energy source is used for different application. There is a rapid growth in the world of solar water heater market which accounts for an annual growth rate of 15%. Major market share for solar energy water heater are India China and Europe. Overall the cost of unit is expensive so there is a need to develop a low cost construction of this water heater. Government of India has a target of 20 million units to be produced by the end of 2022 for Solar Water Heater collectors. It is being implemented in different phases the first phase with 7 million units for the year 2010 to 2013, the second phase with a target of 15 million units for the year 2014 to 2017 and the target for 2022 is 22 million units. The solar water heating system has the following components solar thermal collectors, water storage tanks, connecting pipelines, and the water to be heated and circulated within the solar water heating system. The source of heat is used by heat falling on to the surface which generates the heat source. The heat source can be generated by either a flat plate collector or a evacuated tube collector. The sun rays falls directly over the collector tube which results in the heating of tube. The water is passed to flow through the heated tube which results in change in temperature of the water. Now the water gets heated up and the temperature starts increasing.

Ahmet Samanci[1] concluded that the efficiency of the closed system is lower than the two phase system and also by using two phase SWHS heat is transferred by the process of evaporation & condensation as a result of quick response to change in solar radiation. Sanjay Kumar Sharma et al[2] suggested that the system is designed for 5 litres water inside & outside of the tank for per hour and 30 litres water capacity tank. The research work focuses upon the process of energy conversion from the collector to the working fluid. This is accomplished by employing an aluminium sheet placed at the base within the system to induce a gradient of heat capacitance and the Flat Plate Collectors (FPC) and Evacuated Tube Collectors (ETC) are more popular devices used in the solar water heater system. This domestic V- Through flat plat collector system with a capacity of 100 litters per day is capable of achieving significant energy savings in hot climate countries. Particularly, in the present situation of acute energy short age and most suitable to supply the needs as a family of four persons.The evacuated tube collector method is more popular in the last decade which has a great impact in the solar water heater industry. There are more manufactures and suppliers for the Evacuated Tube Collectors because of this there is greater expansion in the solar water heating businesses. Due to this the cost of manufacturing came down which resulted in reduction in costs of the domestic solar water heater systems. Rajakrishnamoorthy P [3]suggested that the efficiency of the conventional collector is lesser when compared to the Integrated collector storage solar. Pankaj Kumar Mongre[4] suggested that increase in the instantaneous efficiency 55.24 % due to increase in glazing area. Morrison et al[5] has identified that during high temperature operations the evacuated tube solar collectors has a good performance than the flat plate collector. The U-tube glass evacuated solar collector is being used widely when compared to a heat pipe method. Some experimental investigation has resulted that U-tube welded inside a circular fin has a good result in thermal efficiency, which has a benefit over other different shapes of the absorber tube. The parts which are used for the fabrication of this evacuated U-tube solar water heater are discussed in the components and description section.

2. Components and Description

The components used to fabricate physical setup of this work are Acrylic Fiber, Copper Tube, Water Flow Controller, and Mild Steel.

2.1 Acrylic Fiber

The Acrylic fibres are synthetic fibres which are made from a polymer called polyacrylonitrile. This polymer has an average molecular weight of ~100,000, which about 1900 monomer units. The first acrylic fibres were created in 1941by DuPont. The DuPont then created the trademark under the name orlon. The Acrylic is called as acrilan fabric. The acrylic fibre are used as lining in boots, gloves, sweaters and tracksuits and it is also used in furnishing fabrics and carpets. The manufacturing of acrylic is made as filaments and then it is cut into short lengths as staple which is similar to wool hairs and then it is spun into yarns.

2.2 Copper Tube

Copper tubes are generally used in HVAC system as a refrigerant line and it is also used transport hot and cold water. The copper tube is of two types one is soft copper tube and other is rigid copper tube. To join the copper tubes we use flare connection or compression connection or solder method. The purpose of using copper is due to its property of having higher corrosion resistance but it is becoming expensive. The soft or ductile copper tubes can be easily bent, so that it can be easily fixed around the path which has obstacle in the path of tubing. The copper tubes can be bent for more exposure to the sun rays which improves the heating method and easy to carry water.

Figure 1. Copper tube

2.3 Water Flow Controller

The flow of water should be controlled to achieve a linear flow rate so that the water flowing through the tube will have even heating through the thermal conductivity achieved through the copper tube. A self sealing latex tube is used for the extra medication and easy flushing highly kink resistant tubing roller clamp is used for the flow controller with special beveled extra strong spike disc filter in drip chamber facilitating rapid adjustment of fluid at 23G needle for smooth flow drop size which is reduced to 60 drops / ml.

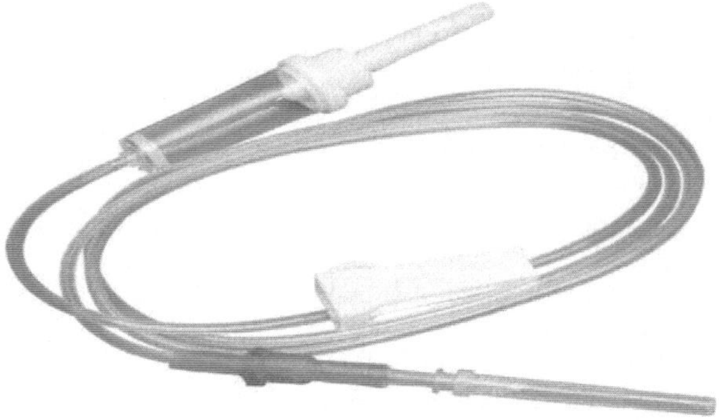

Figure 2. Water flow controller

2.4 Mild Steel

The propertied of the mild steel and its usage in different fields are compiled here. The mild steel has a density of 7861.093 kg/m3. This is the calculated value for industrial grade mild steel. The Young's modulus of mild steel and the measure of its stiffness is around 210,000 MPa. The value can be changed based on the amount of carbon that is being added to the steel, which makes this steel different from other types. Carbon atoms get affixed in the interstitial sites of the iron lattice, making it stronger and harder. But if the hardness is increased then the ductility is being decreased.

A passive system is considered as the basis for the design work. The passive systems do not need any electricity or relay to work on it. It works based on natural circulations of hot water in the system through the tube with some applied gravity feed and the pressure flow from water. The solar water heater is designed for the Community Resources centre where the product has to be made cost effectively. The design of solar water heating system is made based on the following need:

- Daily water consumption(by assumption): 25 litres/day
- Water inlet temperature = 34°C

- Design the water heater based on the parameter.
  - Atmospheric temperature = 34°C
  - Collector Inclination angle = 30^0
- Fluid to tube heat transfer coefficient is 205 W/m^2 °C

The following design parameters are considered for the solar flat plate collector.

Collector length	= 77cm
Collector breadth	= 46cm
Collector height	= 11cm
Absorber plate thickness	= 0.5cm
Storage tank length	= 59cm
Storage tank breadth	= 29.5cm
Storage tank height	= 29.5cm
Outer dia. of copper tube	= 0.9cm
Inner dia. of copper tube	= 0.7cm
Copper tube length	= 458cm

2. Experimental Setup

Figure 3. Experimental setup

The figure shows the designed solar water heater which is going to provide hot water for bathroom. The required operating temperature range of 55°C to 65°C can be achieved.

3.1 Collector setup

The collector is designed with a dimension of 77cm x 46cm. G.I. sheet is used for the fabrication of the collector. Two HDPE pipe is used in the design of 32mm diameter at the header and baser at the top and bottom, drip pipe of diameter 12mm.in the HPDE pipe 12 holes in are made. The holes are made to insert the drip pipe. Once the preparation hole in header and baser pipe is made, the drip pipe is connected between headers and baser and rest flow is connected by using T-joint as a lateral pipe.

3.2 Stand setup

Based on the collector angle the stand is fixed with the support of MS angle camp. The collector should face at the required angle so that maximum efficiency can be achieved. The considered values for the stand are as follows

Angle	= 30°C
The height of stand	= 73cm
The base length of stand	= 112cm

IOP Conf. Series: Materials Science and Engineering **183** (2017) 012035 doi:10.1088/1757-899X/183/1/012035

3.3 Storage Tank structure

The storage tank has a capacity of 25 litres in drum shape is fabricated. For providing insulation to the drum. Another drum is prepared using G.I. sheet for the outer. In between inner tank and outer tank Polystyrene is inserted and glass-wool is also used between the two drum. Then sawdust and glass wool is also used to retain the heat. Multilayer packing is given in between the two drum structures.

3.4 Experimental Connections

The construction setup has three holes in the storage tank. One is used at the bottom and the other two holes used at top of the setup. The top hole is used to connect through headers of the collector. The top hole used to extract hot because the hot water remains at the top of the storage tank. The lower hole is connected through the HDPE-pipe to the baser. The lower hole also connect inlet pipe for incoming water. One side of the collectors is joined by using T-joint and another side in closed using MT cap. This way the connections are established and ready for testing.

4. Results And Discussions

Water was allowed to flow through the SWH continuously with the help of the pump and the temperature of water in the insulated tank was noted down at regular intervals of time.

4.1 Calculation of collector efficiency (complete cycle)

The efficiency of the SWH is calculated for the trial-1 for the given period of heating time

Total duration of heating	= 100 minutes
ΔT achieved = (50-34)	= 16°C
Mass of water	= 25 kg
Cp of water	= 4187 J/kg°C
Q heater= m*Cp	= 1674.8 KJ
Wattage heater= Q/t	= 279 Watts

Average solar power in India is 5 KW/meter^2

Surface area of the collector = 0.3542 meter^2

So net solar power incident on collector surface = [5 * 0.3542 * cos 30°]

{Cos 30° gives the projection of the collector on the ground, as it is tilted at an angle of 30° with the horizontal surface}

So Efficiency of the collector = [W heater/W solar] * 100 = 18.19 %

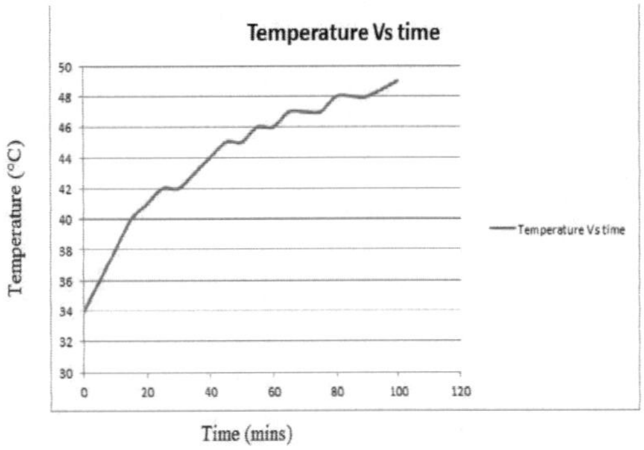

Figure 4. Temperature Vs time trial run-1

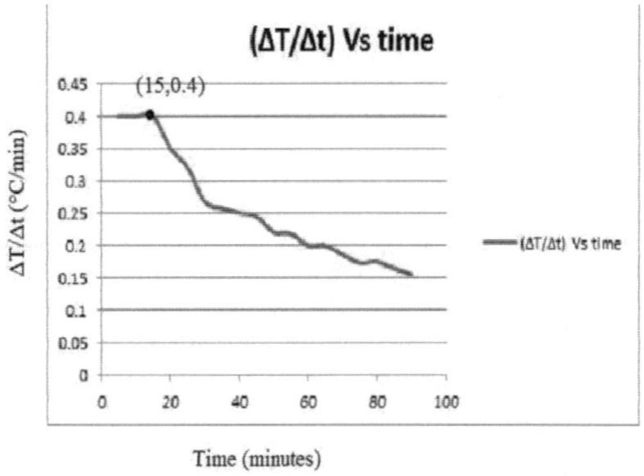

Figure 5. Graph $\Delta T/\Delta t$ Vs time trial-1

4.2 Calculation of efficiency for the first 15 minutes

The graph [Fig 5] shows that the ΔT/Δt is maximum and steady for the first 15 minutes. So the maximum efficiency of the collector comes, if it is operated for this duration only.

ΔT is 8 degrees in 15 minutes.

Calculation of efficiency in the same procedure we have

Wattage heater = (m*Cp*ΔT/Δ t)

= (25* 4187 * 8) /(15 * 60) = 930 Watts

So efficiency = (0.930/1.53)*100= 60%

5. Conclusions

The hybrid solar collector thus fabricated was put to test under the sun, while circulating water through it at different flow rates. The SWH efficiency was as high as 60 %. These types of solar water heaters can easily be used to heat pool water. A flow rate comparable to that of .2lpm is best for heating. The heaters are to be used intermittently for heating a batch of water till 15-20 minutes until the efficiency drops down and then fresh water be circulated through it. A cycle like this would give us the maximum output in minimum time.

6. References

[1] Ahmet Samanci and Adnan Berber, 2011 Experimental investigation of single-phase and two phase closed thermosyphon solar water heater systems *Scientific Research and Essays* **6(4)** 688-693.

[2] Sanjay Kumar Sharma, Dheeraj Joshi, 2013 Fabrication and Experimental Investigation of V-Through Flat Plate Collector in Hot Climatic Conditions of Rajasthan: A Case Study of Jaipur *International Journal of Emerging Technology and Advanced Engineering* **5(3)**.

[3] Rajakrishnamoorthy P, Alexander R and Suthan C, 2014 Experimental Investigation of Integrated Collector Storage Solar Water Heater *International Journal of Innovative Science Engineering & Technology* **2(1)**.

[4] Pankaj Kumar Mongre, Dr Bhupendra Gupta, 2013 Experiment study of solar water heater with circulating pump and using of aluminium tube *International Journal of Engineering Trendsand Technology* **3(4)**.

[5] Morrison G L, Budihardjo I and Behnia M, 2001 Water-in-glass evacuated tube solar water heaters *Solar Energy* **76(3)** 135-140.

[6] Raj Thundil Karuppa R, Pavan P and Reddy Rajeev D, 2012 Experimental Investigation of a New Solar Flat Plate Collector *Research Journal of Engineering Sciences* 2278 – 9472, Vol. **1(4)** 1-8.

[7] Hosni I Abu-Mulaweh, 2012 Design and development of solar water heating system experimental apparatus *Global Journal of Engineering Education* **1(14)**.

[8] Razali Thaib, Hamdani, Irwansyah and Zaini, 2014 Investigation Performance of Solar Water Heater System using Paraffin Wax *ARPN Journal of Engineering and Applied Sciences* **10(9)**.

[9] Saravanan M and Karunakaran N, 2014 Experimental Investigation on Cylindrical Heat Pipe with V-Trough Solar Collecting System *International Journal of Research in Advent Technology* **11(2)**.

IOP Conf. Series: Materials Science and Engineering **183** (2017) 012036 doi:10.1088/1757-899X/183/1/012036

Processing Of Neem And Jatropha Methyl Esters –Alternative Fuels From Vegetable Oil

S Ramasubramanian[1], S Manavalan[2], C Gnanavel[3], and G Balakrishnan[4]

[1,2,3] Asst. Professor, [4]Asso. Professor
[1]Department of Automobile Engineering, Vels University, Chennai, India
[2]Department of Mechanical Engineering, Bharath Institute of Science and Technology, Chennai
[3]Department of Automobile Engineering, Vels University, Chennai, India
[4]Centre of Excellence in Patterned Multiferroics & Nanotechnology, Bharath Institute of Science and Technology Chennai, India

E-mail: raams87@gmail.com

Abstract. Biodiesel is an alternative fuel for diesel engine. The methyl esters of vegetable oils, known as biodiesel are becoming increasingly popular because of their low environmental impact and potential as a green alternative fuel for diesel engine. This paper deals with the manufacturing process of Biodiesel from jatropha and neem oil. Biodiesel was prepared from neem oil and jatropha oil, the transestrified having kinematic viscosity of 3 & 2.6 centistokes, methanol ratio is 6:1 & 5.1respectively. The secondary solution is preheated at 65 C & 60 C and reaction temperature is maintained at 60C & 55 C and reaction time is 60 minutes approximately with NaOH catalyst and low viscosity oil is allowed to settle 24 hours. The average yield of neem and jatropha methyl esters was about 85%. These methyl esters shows excellent alternative under optimum condition for fossil fuels.

1. Introduction

There are three most extensive methods for biodiesel production they are Transesterification, Pyrolysis and Micro emulsion. Chemically, biodiesel is defined as the monoalkyl ester of long-chain fatty acids derived from renewable bio-lipids. Biodiesel is better than diesel fuel in terms of sulfur content, aromatic content and bio-degradability. Among this process physical characteristics and free fatty acid of vegetable oil are nearer to that of diesel. The transestrification process is very simple and two step process. Gaurav Paul et al [1] concluded that the usage of jatropha biodiesel decreases the torque and brake thermal efficiency in CI engines. Due to the higher oxygen content in Jatropha there is a increase in NO_x. The complete combustion of bio diesel reduces the PM and smoke emission. Kavati Venkateswarlu et al [2] suggested that increase in WGR percentage results in BTE increase at first and then start deceasing wile BSFC initially decrease in peak pressure NO_x and exhaust gas results with the increase in percentage of EGR. Nagaraj et al. [3] analysed the Honge oil methyl ester. When this oil is used directly it resulted in poor performance and reduced BTE with increased emission. But when silver nano – particle additive is added it results in improved performance and drastic reduction in emission

2. Experimental Setup

Figure 1 shows the transestrification setup, the components are three neck flask (capacity: 5 liters max), heater windings with controller, stirring apparatus, thermometer, measuring jar, conical flask, separating funnel, magnetic stirrer, hot air oven, weighing machine, pipette.

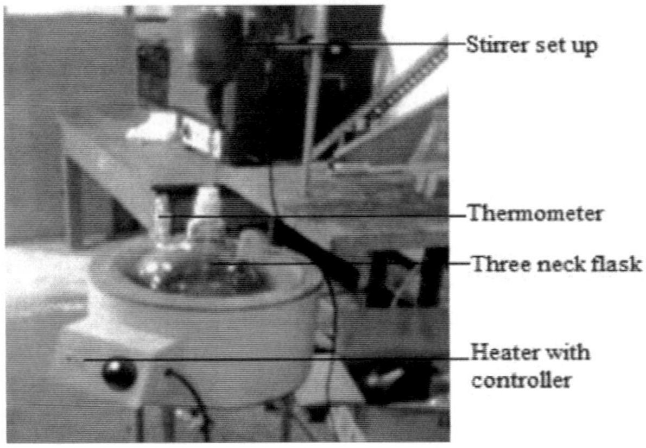

Figure 1. Photograph of Transesterification set up

3. Vegetable oil details

In particular tropical countries like India having higher chance to grow plants like Jatropha, which is an energy crop. India imports about 45-56% edible oil for its domestic requirement and therefore, it is not possible to divert the edible oil resources for biodiesel production in the country. In order to avoid food vs fuel problem, the usage of non-edible oil is gaining interest. So the non-edible oil resources like Jatropha, Pongamia, neem etc., seem to be the only possibility for biodiesel production in the country.

In this study, jatropha and neem oil was chosen for biodiesel production and key properties which decide the transesterification also tabulated. Hence two step transesterification works well. According NOVOD there are more than ten non edible oils for production of biodiesel. Out of which, Jatropha and neemlesser extend Pongamiapinnata have received much attention. Vegetable oil main constituents are Oleic and Linolenic acid, which are highly unsaturated. The fatty acid and moisture contents were tested in standard laboratory.

Table 1. Key Properties of Vegetable oil

Properties	Jatropha oil	Neem oil
Free Fatty Acid content (in %)	9.53	9.58
Moisture Content (%)	0.52	0.56

4. Acid catalysis

To reduce the viscosity of raw vegetable oil and to improve flow properties transestrification process is performed. Since jatropha oil having higher fatty acid content cannot be evolved for base catalysis, because alkali catalyst react with FFA to form soap, resulting in serious emulsification and separation problem. Due to inappropriate storage condition and improper handling the quality of jatropha oil gradually decreased. The vegetable oil is exposing to sunlight and open air for long period free fatty acid level would be increased above 1%. It is well known that improper handling of jatropha oil would

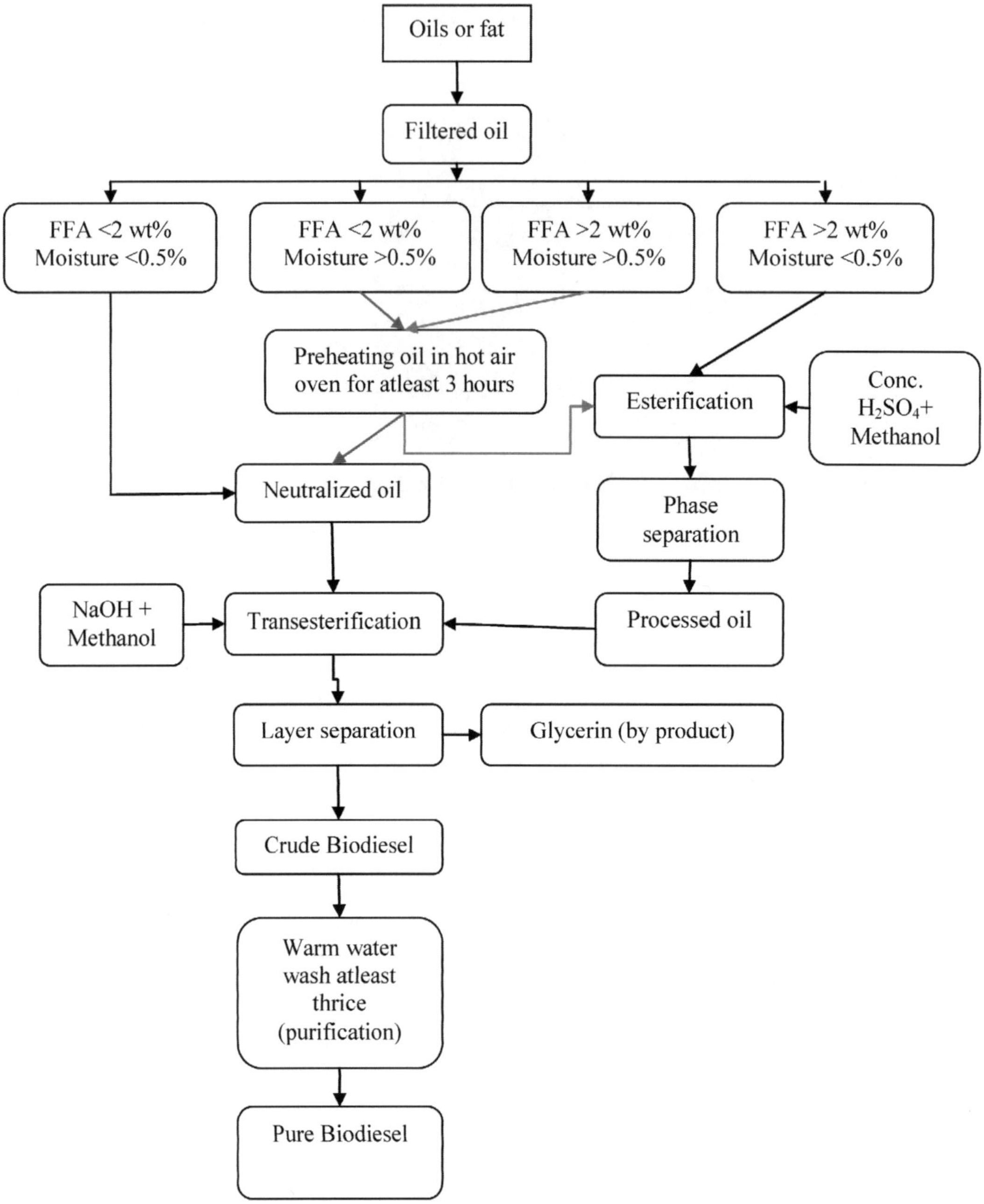

Figure 2 Optimization Flowchart

cause increased water and free fatty acid content. Free fatty acid percentage will vary depending on the feed stock quality. Removing or reducing FFA content in the oil is the pre-treatment and first step acid esterification, which is a pretreatment step, which will be reduce the FFA in the presence of optimised methanol and catalyst concentration.

5. Base Catalysis

For direct alkali transesterification requires lower FFA, hence higher FFA content oil is pre-treated and it is evolved to base catalysis using optimized amount of methanol and base catalyst. After the completion of the reaction, it forms two layers. The upper will be methyl ester and higher molecular weight glycerin settles at the bottom.

6. Process optimization

The objective of transesterification is to reduce the viscosity of higher molecular fatty acid by converting in to lower molecular weight ester. The parameter to be optimize for transesterification are Methanol to oil ratio, Catalyst to oil ratio, Temperature, Stirring speed and Reaction time. For small scale production and process optimization magnetic stirrer with heater set up was used. Two step processes (Acid-Base catalysis) was chosen for transesterification. In the first step pretreatment (acid catalysis) methanol quantity was varied from 0.2 to 0.6 w/w ratio of oil and Concentrated Sulphuric acid was used as acid catalyst and it is varied from 0.5 to 1.5 %.

In the next step (base catalysis), methanol quantity was varied from 0.1 to 0.3 w/w ratio of oil and base catalyst as sodium hydroxide in varying proportion from 0.5 to 1.5 %. For both catalysis, in order to stirr the mixture, the stirring speed was optimized to 300 rpm and temperature at which acid catalysis takes place also varied from 45 to 60°C. Reaction time also varied from 30, 60, 90, 120 minutes. Based on this result the optimized quantity was tabulated and it was utilized for mass production.

Table 2. Optimised quantity for transesterification

Parameters	Neem Oil		Jatropha Oil	
	Acid catalysis	Base catalysis	Acid Catalysis	Base Catalysis
Methanol to oil ratio (w/w)	0.7	0.4	0.6	0.24
Catalyst to oil Ratio	1% Con H_2SO_4	0.8% NaOH	1% Con H_2SO_4	0.8% NaOH
Temperature (°C)	60	65	55	62
Stirring speed (rpm)	300	300	300	300
Time (hours)	1.5	2	1.5	2

7. Transesterification Process

The raw Jatropha oil and neem oil was filtered to remove all insoluble impurities followed by heating at 100°C in hot air oven for 20 min to remove all moisture. In the first step, the higher fatty acid oil was processed using 0.6w/w ratio of methanol to oil in presence of concentrated Sulphuric acid as a catalyst (1 wt% of oil). The reaction takes place for one hour at optimum temperature of 55°C and stirred at 300rpm in a three neck flask with heater as shown in Figure. After the reaction the mixture was allowed to settle in a separating funnel for 5 hours. The methanol water mixture was separated and processed oil was removed.

In the second step, the processed oil (which posses Fatty acid lower than 0.5%) was transesterfied using 0.24 w/w ratio of methanol to oil and 0.8 % of NaOH as an alkaline catalyst to produce biodiesel at temperature of 60°C [3]. The reaction takes place for 2 hours, at optimum speed of 300 rpm. Allow the mixture to settle for 6 hours to overnight in a separating funnel for gravity separation, and then it form two layers. The higher molecular weight glycerin separates at the bottom, which looks dark brown colour and upper yellow layer is methyl ester, it was separated. The methyl ester was washed gently using water for at least 3 times at a temperature of 40°C, in order make ester to neutral (PH=7) and also remove catalyst, glycerin and un reacted fatty acids. Final yield of biodiesel was 85%.

Table 3. Comparision of properties of biodiesels with diesel

Properties	Diesel	Neem Methyl Ester	Jatropha Methyl Ester
Density (kg/m^3)	822	868	860
Calorific value (kJ/kg)	42,500	38,500	40
Viscosity (mm^2/s)	3.2	4.5	5.2
Cetane number	45-55	50	48
Autoignition temperature (°C)	280	302	286
Flash point (°C)	70	120	145

8. Result and Discussion

Physical and chemical properties of jatropha and neem biodiesel were found, Biodiesel properties are similar to conventional diesel. The approximate yield of biodiesel was obtained 80%. It was confirming to the European and American standards.

9. References

[1] Gaurav Paul, AmbarishDatta and Bijan Kumar Mandal 2014 An Experimental and Numerical Investigation of the Performance Combustion and Emission Characteristics of a Diesel Engine fueled with Jatropha Biodiesel 4th *International Conference on Advances in Energy Research* Energy Procedia **54** 455 – 467.

[2] Kavati Venkateswarlu, BhagavathulaSree Rama Chandra Murthy and VissakodetiVenkataSubbarao, 2013 The Effect of Exhaust Gas Recirculation and Di-Tertiary Butyl Peroxide on Diesel-Biodiesel Blends for Performance and Emission Studies *International Journal of Advanced Science and Technology* **54.**

[3] NagarajBanapurmath, Narasimhalu T, AnandHunshyal, RadhakrishnanSankaran, Mohammad HussainRabinal, NarasimhanAyachit and RohanKittur, 2014 Effect of silver nano-particle blended biodiesel and swirl on the performance of diesel enginecombustion *International Journal of Sustainable and Green Energy* **6(3)** 150-157.

[4] Sungyong Park, Hwanam Kim and Byungchul Choi, 2009 Emission characteristics of exhaust gases and nanoparticles from a diesel engine with biodiesel-diesel blended fuel (BD20) *Journal of Mechanical Science and Technology* **23** 2555-2564.

[5] Santhanamuthu M, Chittibabu S, Tamizharasan T and Mani T P, 2014 Evaluation of CI engine performance fuelled by Diesel-Polanga oil blends doped with iron oxide nanoparticles *International Journal of ChemTech Research* **2(6)** 1299-1308.

[6] Shashi Kumar Jain, Sunil Kumar, and AlokChaube, 2011 Technical Sustainability of Biodiesel and Its Blends with Diesel in C I Engines: A Review International *Journal of Chemical Engineering and Applications* **2(2).**

[7] HanumanthaRao Y V, Ram SudheerVoleti , SitaramaRaju A V and Nageswara Reddy P, 2009 Experimental investigations on jatropha biodiesel and additive in diesel engine *Indian Journal of Science and Technology* 4**(2).**

[8] JinlinXue, Tony Grift E, Alan Hansen C, 2011 Effect of biodiesel on engine performances and Emissions Renewable *and Sustainable Energy Reviews* **15** 1098-1116.

[9] Enrico Mattarelli, Carlo Alberto Rinaldini and TommasoSavioli, 2015 Combustion Analysis of a Diesel Engine Running on Different Biodiesel Blends *Energies* ***8*** 3047-3057.

[10] GauravDwivedi, Siddharth Jain and M. P. Sharma, 2011Impact of Biodiesel and its Blends with Diesel and Methanol on Engine Performance *International Journal of Energy Science* **2(1)** 105-109.

IOP Conf. Series: Materials Science and Engineering **183** (2017) 012037 doi:10.1088/1757-899X/183/1/012037

Investigation of Performance Analysis and Emission Characteristics of Waste Plastic Fuel

[1] M Ruban, [2]S Ramasubramanian, [3]R Pugazhenthi, [4]Sivaganesan

[1,2,4]Asst. Professor, [3]Asso. Professor
[1,2]Department of Automobile Engineering, Vels University, Chennai, TN, India.[2]
[3,4]Department of Mechanical Engineering, Vels University, Chennai, TN, India.

E-Mail: rubanmurugesan@gmail.com

Abstract. Today the world is confronted with the twin crisis of fossil fuel depletion and stringent emission norms, because of the environmental awareness. The disposal and degradation of waste plastic is a major issue and scarcities of fuel were major focus area of the researchers. In this virtue the waste plastic fuel extraction makes more attention to the researchers. In this research work focused to find the performance of the waste plastic fuel and compared to diesel. The waste plastic fuel extract from thermal cracking method this process the polymer chains were breakdown into useful lower molecular weight compounds and it becomes plastic pyrolysis it can be utilized as a fuel. The properties of the waste plastic fuel is obtained by various testing process and which is analyze and compare with the fossil fuel diesel. It is found that almost it has similar properties to the diesel and almost all properties of the pyrolysis is closer to that of diesel. The characteristics of the pyrolysis were tested in the engine test bed. The pyrolysis / waste plastic fuel can be directly used in diesel engines over the entire load spectrum smoothly without any major modification. The performance of the waste plastic fuel / pyrolysis is evidenced that it is one of the best alternative fuel as well as the waste plastic can be converted into a useful fuel

1. Introduction

Plastic is a one of the greatest innovation of this century; there are used numerous ways because of the less weight, does not oxidation or decompose, low cost, reusable and conserves natural resources It saves energy and CO2 emissions during their using as well as; it substitute in all applications with the existing mix of alternative materials. The continuous innovation and application of new things the plastics production has increased by an average of almost 10% in every year since 1950. But the waste plastic management is the biggest problem of the current scenario. Sharma et al. (2016), the disposal of waste plastic very is a serious environment challenge all over the world (1).
The researchers are concentrate to find the fuels for I.C. engines. In this circumstance the waste plastic are currently receiving transformed interest, the oil can be derived from waste plastics. The aim of this paper is to analyze the engine performance, combustion and emission characteristics of diesel engines fuelled with waste plastic oil and its blends with petroleum-based diesel fuel.

2. Literature Survey

2.1. Sources and properties of plastic wastes

The major resources of waste plastic were collected from the municipal waste and industrial wastes, dones of plastic waste arise as a by-product in industry. From the total waste plastics 78 % of the quantity is thermoplastics and the remaining to thermosets[1]. The thermosets cannot be recycled because they includes epoxy resins and polyurethanes, But thermoplastics can be recycled; they are composed of polyolefins such as polystyrene, polyethylene, polyvinyl chloride and polypropylene [2]. The waste plastics categorized according to their combustibility either burnable or unburnable. burnable wastes are used for heat production,Unburnable waste are used for recycling, or land filling, based on the material. Ampaitepin and Tetsuo , Municipality waste giving high heat value, such as (WPF) waste plastics fuel [3].

2.2. Waste plastic oil / fuel

The plastic industry had an extraordinary growth rate for the past three decades only. The various formation of plastic has enlarged more than 100 times in the last 30 years The various formation of plastic has been used in different kind of applications (e.x:car industry,robotic industry & aerospace). Now a days plastic handling is biggest issue. So that plastic wastes are safely & quickly recycleable. Mitsuhara et al.,the new trent of eco-friendly waste plastic is a alternative fuel for steel production[4]. Williams and Williams, using the waste plastic fuel in power plant it will help to increase the thermal efficiency[5]. Plastics are non-eco-friendly generally it containing carbon, hydrogen, and few other elements. In 2010 nationlwide survey conducted in India the plastic demand is 12 million tons per year. 14,000 metric tons produced every day and remaning is imported from other countries. Mani and Nagarajan says All the plastics are recycled and sometimes it is not done so due to lack of sufficient market value [6].

Hai et al., The extent of conversion of plastics or plastic derived waxes into light engine fuels can be increased by the application of stable hydrocracking catalysts. Similar to petroleum derived cracking products, the fractions from plastics processing contain appreciable quantities of aromatics and unsaturated hydrocarbons [7].

2.3. Waste Plastics in to Fuel Conversion Processes

De-Polymerization of plastic involves breaking down long chains of hydrocarbons present in plastics molecules into smaller components, such as - petrochemical gases & liquid fuels. This process of breaking down plastic into lighter components is known as 'Cracking'. Thermal cracking (or pyrolysis) requires high temperature & pressure conditions. However, in the presence of a specialized catalyst, cracking can be efficiently achieved at relatively much lower temperature & near atmospheric pressure, thus making the process commercially attractive & industrially safe. This method of cracking is known as 'Catalytic De-polymerization'. Our De-polymerization model employs a rotary reactor operating in continuous production mode to convert waste plastic into fuel. A typical output consists of about 70% liquid fuel, 20% coke & 10% petroleum gasses. Liquid fuels are give the high energy with low sulphur content. The fine coke also gives high energy so its used for industrial & domestic applications.

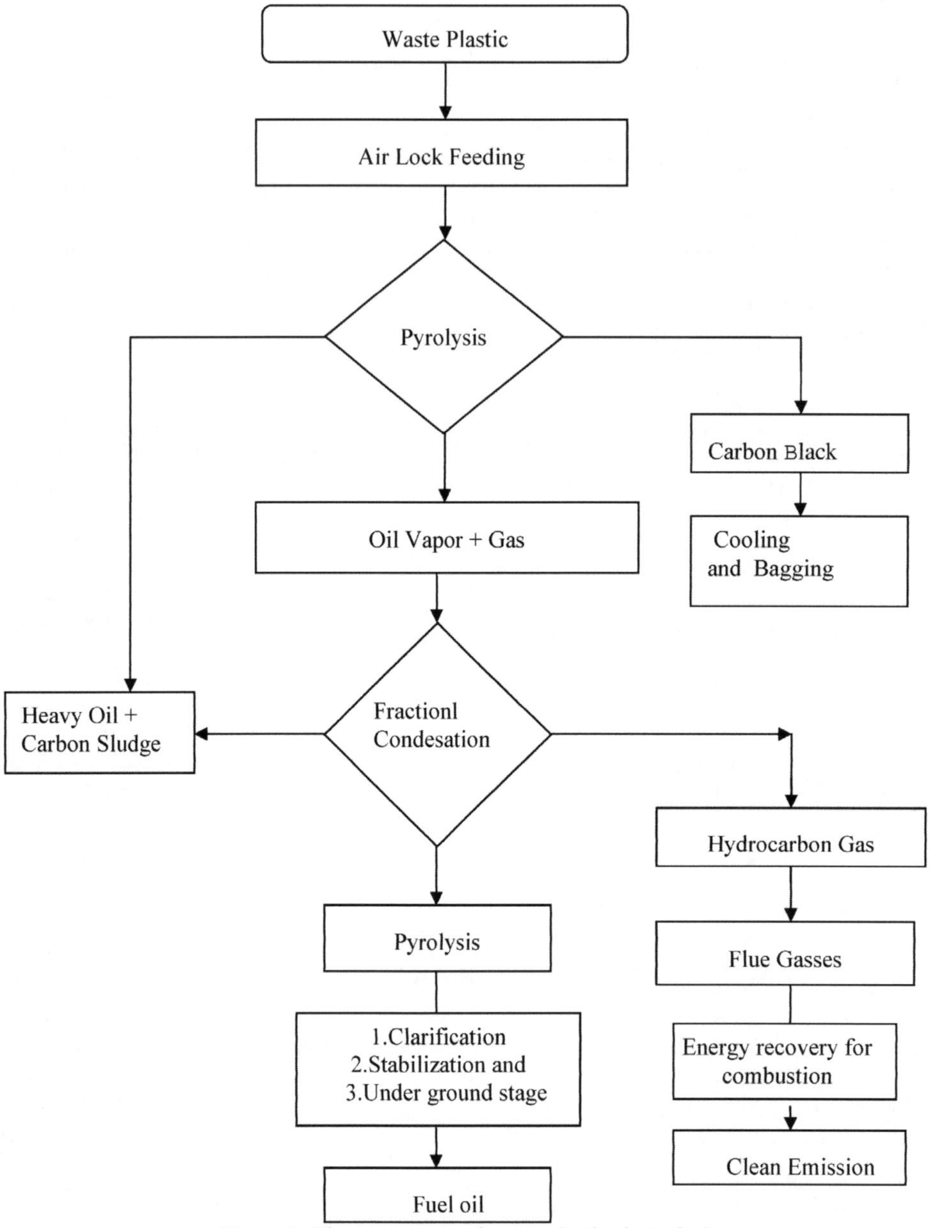

Figure 1. Flow processes of waste plastics in to fuel

The petroleum gasses produced are used internally to meet the energy requirements of the process, thereby making the process highly cost effective. Beinga closed loop process; it allows mass & enthalpy conservation exercise to be carried out so as to establish that no material is expelled in to the environment. Converted and recovered as output in the form of high calorific value solid, liquid & gaseous fuels. The material balance as well as energy balance at the output vis-à-vis that at the input make an excellent statement, and that is, the process has nothing to discharge into the atmosphere, thereby, pollutions of air, water & soil are NOT created. The output of the oil contains no chlorine, sulphur, nitrogen or heavy metals. Any of that material would remain in the ash. The Figure. 1 shows that the flowchat of the various process of waste plastics in to fuel conversion processes. This system consists of the knapper, extrusion machine, pyrolysis reactor & catalytic upgrade reactor.

3. Properties of the Waste Plastic Fuel To Diesel Fuel

Table 1 and 2 shows the Properties of the Waste Plastic Fuel to Diesel Fuel and Details of Engine Bed.

Table 1. Property of Waste Plastic Fuel vs. Diesel

S.No	Properties	WPF	Diesel
1	Density (kg/m²)	793	850
2	Ash Content (%)	<1.01%wt	0.045
3	Calorific Value(kJ/kg)	41,800	42,000
4	Kinematic Viscosity at 40^{0}C(cst)	2.149	3.05
5	Cetane Number	51	55
6	Flash Point ^{0}C	40	50
7	Fire Point ^{0}C	45	56
8	Carbon Residue (%)	0.01%wt	0.20%
9	Sulphur Content (%)	<0.002	<0.035
10	Pour Point ^{0}C	-4	3-15
11	Colour	Pale Black	Orange
12	Specific Gravity at 30^{0}C	0.8355	0.84 to 0.88
13	Atomic Content (%)	55	20
14	Existent Gum (gm/m^2 max.)	36	-

Table 2. Engine test bed detail

Make and Model	Kirloskar, AV1 make
General Details	Four stroke, compression ignition, Constant speed, Vertical, Water Cooled, Direct injection
Number of cylinder	Single cylinder
Bore	80 mm
Stroke	110 mm
Compression ratio	16.5: 1
Rated Output	3.7 kW at 1500 rpm
Rated Speed	1500 rpm
Injection pressure	240 bar
Type of combustion chamber	Hemispherical combustion Chamber
Connecting rod length	235 mm
Fuel	Diesel

4. Results and Discussions

4.1. Carbon monoxide

IOP Conf. Series: Materials Science and Engineering **183** (2017) 012037 doi:10.1088/1757-899X/183/1/012037

The CO formations shown in Figure 2 for fuels and blends with corresponding load. The emission of CO depending upon the physical and chemical properties of the fuel used. The decrease in CO emission for JB20 is an oxygenated fuel, it leads to better combustion of fuel, resulting in the decrease in CO emission.

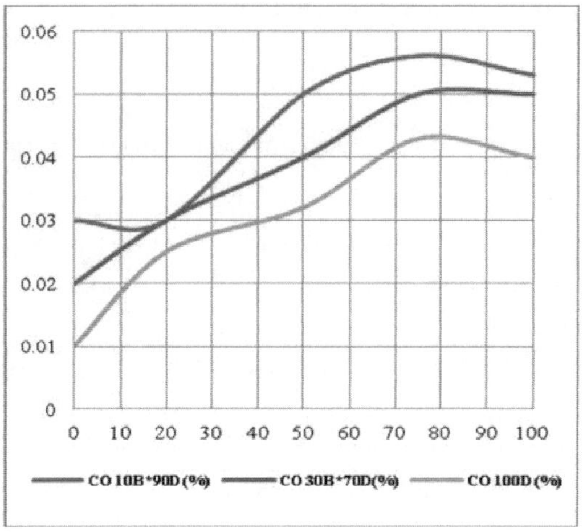

Figure.2 Variation of CO for plastic fuel and diesel blend.

4.2. Hydrocarbons

The variant of hydrocarbons with respect to load for both fuels and their blends is as shown in the Figure 3. HC emissions are observed for the blend at low load conditions.

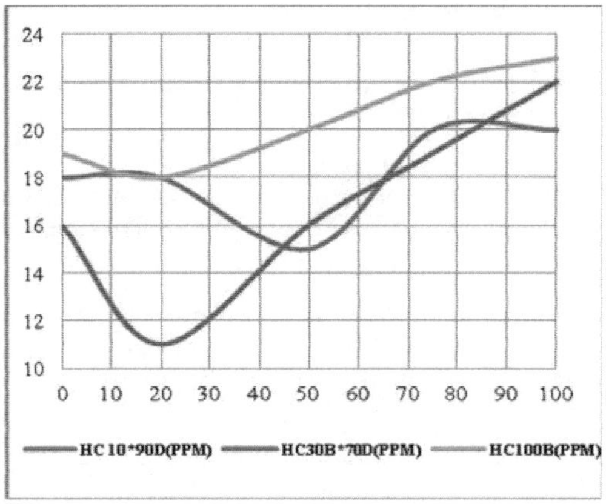

Figure 3. Variation of HC for plastic fuel and diesel blend

In low load conditions, fuel injected is leaner mixture result is high HC emission & poor performance. The HC emission of the plastic fuel is less than that of diesel fuel because of higher cetane number .

4.3. Oxides of Nitrogen

Generally plastic fuel is generates higher amount of Nox than diesel. The NOx emission is enhance the percentage of Particulate matter emission in the fuel. Nitrox of oxide enhance for JB20 may be associated with the O2 content of JB20, Causes of the O2, present in the fuel, may provide excessive oxgen for NOX formation which is shown in Figure 4. The generation of NOX emission is particularly governed by the magnitude of peak cylinder temperature and the crank angle at which it mainly occurs.

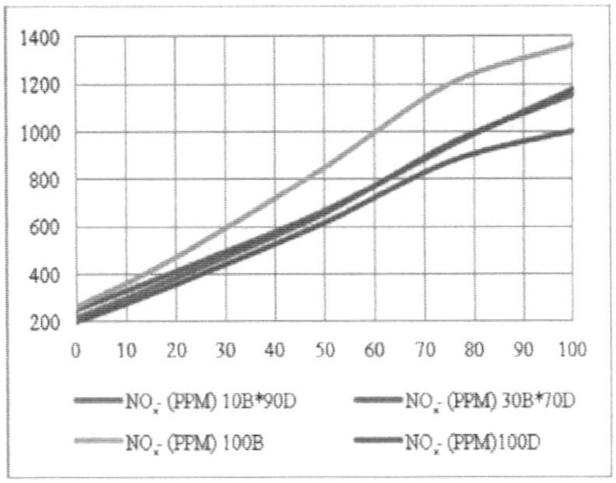

Figure 4. Variation of NOx for plastic fuel and diesel blend.

4.4. Carbon dioxide

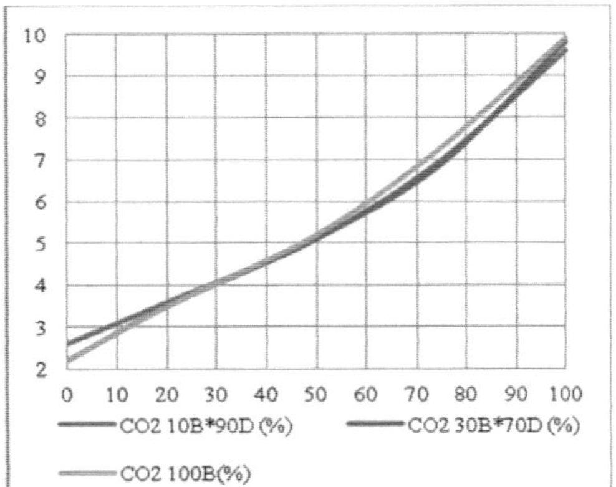

Figure 5. Variation of Carbon dioxide of plastic fuel Vs Load

The increasing volumetric fuel consumption. It is observed that the CO2 emission of plastic fuel is slightly more than the diesel which is shown in Figure 5. This is attributed to the presence of oxygen and high cetane number of the plastic fuel.

5. Conclusion

The properties of the waste plastic fuel is analyzed and compared with the fossil fuel diesel and found that almost it has similar properties to the diesel. The characteristics of the waste plastic fuel is obtained from various studied and summary of the results are the higher heat release rate in waste plastic oil compared with diesel fuel due to better combustion. The Nox increase 40% due to higher heat release rate and combustion and temperature. The CO emission increased 7% in waste plastic oil compared to diesel operation. Unburned hydrocarbon emission is higher by about 15%. Waste plastic fuel exhibits higher thermal efficiency up to 75%. Activated carbon using with this experiment fuel is clean and color is bright yellow. This fuel burns cleaner and burning time is also longer.

References

[1] Sharma, Sanjeev, Rudra Pratap Singh, and Surender Kumar 2016 Development of Nano-Structured Metals Processed by Severe Plastic Deformation *Indian Journal of Science and Technology* **9.27**

[2] Buekens AG, Huang H. 1998 Catalytic plastics cracking for recovery of gasoline range hydrocarbons from municipal plastic wastes *Resources Conservation and Recycling* **23** 163–81.

[3] Ampaitepin S, Tetsuo T 2010 The waste-to-energy framework for integrated multi-waste utilization: Waste cooking oil, waste lubricating oil, and waste plastics. *Energy* **35** 2544-2551.

[4] Mitsuhara Y, Soloiu VA, Nakaishi Y 2001 Application of new fuel produced from waste plastics and heavy oil to diesel engine. *Transact. Jpn. Soc. Mech. Eng.* **67** 2618-2624.

[5] Williams PT, Williams EA 1990 Interaction of plastics in mixed plastics pyrolysis. *J. Energy Fuels* **13** 188-196.

[6] Mani M, Nagarajan G 2009 Influence of injection timing on performance, emission and combustion characteristics of a DI diesel engine running on waste plastic oil. *Energy* **34** 1617-1623.

[7] Hai VP, Nishida O, Fujita H, Harano W, Toyoshima N, Iteya M 2001 Reduction of NOx and PM from diesel engines by WPD emulsified fuel, *SAE Technical Paper* -01-0152.

A Comparative Characteristic Study of Jatropha and Cardanol Biodiesel Blends

R Pugazhenthi[1], M Chandrasekaran[2], R K Muthuraman[3], P Vivek[4] and A Parthiban[5]

[1,5]Associate professor, Department of Mechanical Engineering, VELS University, Chennai, Tamil Nadu, India.

[2]Professor, Department of Mechanical Engineering, VELS University, Chennai, Tamil Nadu, India.

[3]Assistant professor, Department of Aerospace Engineering, Periyar Maniammai University, Thanjavur, Tamil Nadu, India.

[4]Assistant professor, Department of Mechanical Engineering, VELS University, Chennai, Tamil Nadu, India.

E-mail: pugal4@gmail.com

Abstract. The demand in fuel needs and the depleting fossil fuels raised the need towards bio-fuels. The emerging trend in research field is highly focused on biodiesel production and their characteristic analysis. Since pollution is a major threat to the environment, emission parameter analyses are much important to be concentrated. As the entire world contains plenty of biofuels, it is necessary to explore them for its efficiency and analyze their parameters. In this experimental work jatropha and cashew nut shell biodiesel (Cardanol) was extracted and they were blended with diesel. The characteristics of jatropha and cardanol biodiesel were studied in the DI diesel engine by varying the load at the same speed. In brief, this experimental analysis is carried out to compare the emission characteristics between Jatropha biodiesel at 20% (B20) and 40% (B40) and Cardanol biodiesel blends at 20% (C20) and 40% (C40).

1. Introduction

In this current era, the consumption and price of hydrocarbon fuels (Petrol, Diesel) have increased progressively. The existing hydrocarbon fuels can be utilized for the next 50 to 60 years only, to fulfill our fuel requirements. In order to meet the demand of these fuels, there is a need to develop alternative energy sources like biodiesel, solar power, natural gas, gasohol etc., which could be an alternative in the future. Dhanasekaran and Mohankumar [1] revealed the importance of the alternative fuel in the current scenario. Biodiesel is described as a fuel comprised of mono-alkyl esters of long chain fatty acids derived from animal fats or vegetable oils. The use of edible oil to produce biodiesel in India is not feasible as the demand is more due to the inadequate supply of such oil for commercial use. In Indian economic condition, the promotion of non-edible biodiesel is essential; it can be grown in even non-cropped marginal lands and waste lands in large scale. India has more than 300 species of trees, seeds, and fruits which produce the oil bearing seeds [2].

The raw biodiesel obtained cannot be used directly, because of its higher density and viscosity when it compared to hydrocarbon fuels. This biodiesel is very poor in atomization and vaporization with air, which leads to poor combustion, lower thermal efficiency, and higher emissions. So the biodiesel and their blends can be used to promote favorable environmental and performance benefits. From the literature survey, there are two main non-edible oil crops are used; among them, the first one is Jatropha curcas. The seed of Jatropha contains 26 - 40% oil, which can be processed to produce a good quality of biodiesel. The other one is Cashew Nut Shell oil (CNSL). It is also one of the most efficient and high productivity Non-edible oil crops The usage of CNSL reduces HC, CO, NOx gasses than the hydrocarbon fuels [3]. In this comparative study, 20% and 40% of the blends of Jatropha curcas and Cashew Nut Shell are used; they are mentioned here as B20, B40 (Jatropha) and C20, C40 (CNSL).

2. Literature survey

Anand et.al. [4] derived the biodiesel by using transesterification process from waste cooking oil in response surface method. The derived biodiesel characteristics were studied in a single cylinder, four-stroke, direct injection diesel engine at a constant speed of 1500 rpm to find the combustion characteristics, the effect of injection pressure, timing of performance and emission. The combined variation of injection pressure and timing increase the injection pressure about 280 bars and injection timing in advance of 25.50 BTDC, which shows the appreciable improvement in brake thermal efficiency, cylinder gas pressure, and heat release rate and it is also reduced the nitric oxide (NO) and smoke emission.

In general, the transesterification process is used to produce the biodiesel, which involves a chemical reaction between an alcohol and the triglycerides. The fatty acids are the suitable catalyst which leads to the formation of fatty acid alkyl esters and glycerol. Alcohol to oil molar ratio, catalyst concentration, reaction temperature, reaction time, free fatty acid (FFA) content and stirring rate are the major factors of affecting the transesterification process. Velmurugan (2011) tested of commercial diesel fuel and bio-fuel from CNSL with ethanol as an additive (BDEB) [5]. Experimental Investigation of CNSL as the alternative was undertaken by V. Palvannan and the first trial run with a maximum of 20% CNSL-diesel blend was successfully conducted on single cylinder CI engine. Engine power output and smoke emissions are found to be satisfactory, then it was decided to proceed further testing. Achten W.M.J., et.al., [8] studied the performance of the biodiesel blends and emission characteristics with 0.5%, 10%, 15%, 20% and 25% blends of Cardanol and Diesel. They reported that up to 20% blends of Cardanol bio-fuels may be used in combustion-ignition engines without any modifications.

Sayin Cenk et al., investigates the influence of compression ratio (CR) and injection parameters such injection timing (IT) and injection pressure (IP) on the performance and emissions of a DI diesel engine for various biodiesel blended with different CRs, ITs and Ips at the speed of 2200 rpm. The study gives the best results for BSFC, BSEC and BTE were observed at increased the CR, IP. The experiments were carried out using DI diesel engine which was operated in a range of 700 kPa brake mean effective pressures with the interval of 100 kPa. The increased IP gave the best results for BSFC, BSEC and BTE compared to the ORG and decreased IP. Finer breakup fuel droplets obtained with increased IP provides more surface area and better mixing with air and this effect improve the combustion. OP, HC, and CO emissions decreased and NOx emissions increased with the increase in IP for the all fuel blend [6].

Narayana Reddy and Ramesh [7] study the performance, emissions and combustion of Jatropha in a single cylinder, constant speed, direct injection diesel engine and results were compared with neat diesel operation. Anand. R., et.al., [4] studied the effect of diestrol fuel on the performance, emission and combustion characteristics of a DI diesel engine for varying injection pressure and timing. From the experimental investigation, the maximum brake thermal efficiency of 31.3% was obtained at an injection pressure of 240 bars and injection timing of 25.5° btdc it reduced the nitric oxide (NO) emission by 4.3%. The unburnt hydrocarbon (UHC) was observed and a minimum ignition delay was observed in 12.7° CA with diestrol fuel.

3. Jatropha curcus

The Jatropha is a shrub type plant that grows to an average height of 3-5 meters. Its fruit contains seed, which is rich in non-edible oil. It can grow in poor soils, but generally, prefer the heat of the tropics and subtropics and doesn't need irrigation. According to favorable natural conditions for the growth of this plant, it can be cultivated throughout India.

3.1: Jatropha oil extraction

Jatropha curcas seeds contain 27% to 40% of the oil that can be helped to produce a high-quality of biodiesel fuel. It can be used in the standard diesel engines, especially if the oil of the seeds is well extracted [9]. Many oil extraction methods are available based on accuracy, cost effectiveness, kind and nature of the feedstock and efficiency. The methods include mechanical pressing, solvent extraction and supercritical fluid extraction [4]. Thananchayan et al. their experiments indicate that the Jatropha gives a higher biodiesel yield than Castor oil by transesterification reaction in first order reversible reaction in this the oil was extracted in excess compared to methanol. The mechanical pressing (expulsion) process involves the use of pieces of equipment like screw press or a piston, extruder expander and mortar. In this process, oil is expelled from a dry feedstock. The main drawbacks of this process are particle size and nature of the feedstock. Also, the moisture content of the seeds has a role to play in the efficiency of the process [10]. In solvent extraction, oil can be extracted chemically using a solvent such as n-hexane, chloroform, and benzene. N-hexane is the most commonly used solvent that mixed with the feedstock paste and then, later, distilled in order to obtain pure oil. This method is not expensive, and it is efficient [12].

3.2: Biodiesel production from jatropha

The extracted oil has to undergo a chemical process known as transesterification to yield biodiesel. The process was carried out by reacting the extracted Jatropha oil with methanol in the presence of potassium hydroxide (which has been reported to give high yield and conversion of the transesterification reaction with minimal side reaction) as a catalyst to produce ester and glycerol. At the end of the reaction, glycerol and biodiesel will be formed in two layers. After settling, the glycerol was at the bottom while the biodiesel was at the top of the container used. The layers were later separated from each other by draining the glycerol from the bottom of the flask containing the mixture.

The initial triglyceride content of the oil was determined by titrimetric analysis. Keeping the reaction temperature and stirring rate constant at 65°C and 460 rpm respectively, three different biodiesel production conditions were varied. Initially, the transesterification was done by varying the reaction time and keeping the oil to methanol molar ratio of 4:1 and 2.5% catalyst. The first set of biodiesel was produced. Secondly, the reaction time was kept at 75 minutes, molar ratio was 4:1 and the catalyst load was varied. Thus yields the second set of biodiesel. Finally, the methanol to oil molar ratio was varied, keeping constant reaction time at 75 min and the catalyst concentration at 2.5%. Thereafter, third biodiesel production was carried out using the obtained optimum parameters from the investigations carried out before, and the yield of that was determined by titrimetric analysis and appropriate calculations. The general equation for transesterification of vegetable oils containing triglycerides is shown in figure 1.

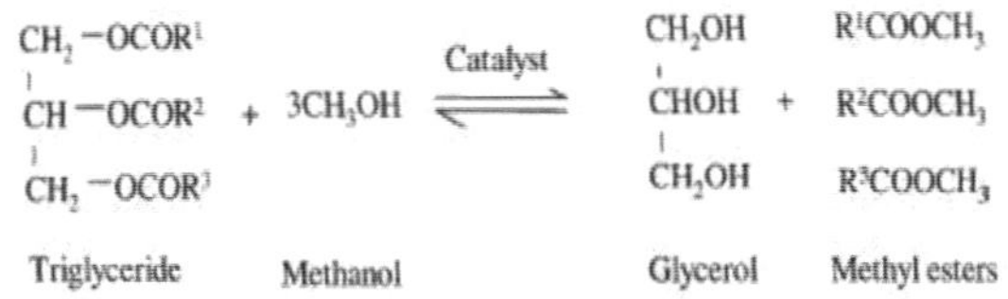

Figure1. Transesterification process

Finally, 880ml of Jatropha biodiesel (i.e. 88% yield) and 120ml of glycerol was obtained from one liter of jatropha curcas oil.

4. Cashew Nut Shell Liquid (CNSL)

The cashew processing industries produce CNSL as a byproduct and nearly 30 – 35 % of CNSL oil is available in the cashew nut shell. There are three main methods generally used in extracting cashew nut shell liquid from cashew nuts, namely thermal, mechanical and solvent extraction [11]. At first, the heating process which leads to decarboxylation of anacardic acid to form Cardanol, this comprises of 10% cardanol and 30% polymeric material and the remaining percentage is made up of substances. After the heating process, distillation is done at a reduced pressure in order to remove the polymeric material. Now, the distilled CNSL consists of 78% Cardanol, 8% cardanol, 2% polymeric material and remaining other substances. At present, the commercial extraction process is carried out by boiling or heating CNSL up to 108^0 to obtain Cardanol rich oil. The cardanol is obtained from cashew Nutshell Liquid (CNSL), the Mono hydroxyl phenol having hydrocarbon chain in the Meta position which was shown in figure 2.

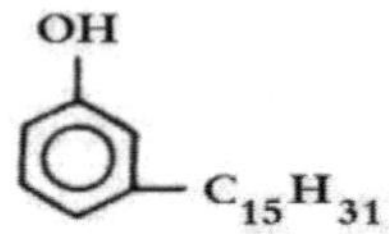

Figure 2. Structure of Cardanol

4.1. Biodiesel from CNSL

The phenomenon of the high and low volatility of non-edible oils affect the atomization of the fuel, Which leads to improper combustion and severe carbon deposits, injector choking and piston ring sticking, etc., The following methods are used to reduce the viscosity of the biodiesel,
1.Emulsification,
2. Pyrolysis,
3. Dilution and
4. Transesterification.

Among these, the transesterification is most commonly used the successful process to produce good biodiesel. The CNL biodiesel is one of the clean and environment-friendly fuel. A 50ml of extracted cashew nut oil measured and poured into a conical flask and which was preheated up to 70°C. In another conical flask, a 0.225 gram of sodium hydroxide and 20ml of methanol are mixture taken and it is mixed in gently and properly, it was mixed by stirring till the entire pellet dissolves to form sodium methoxide solution. The same procedure was followed while using potassium hydroxide as catalyst. The mixture was kept at 70°C for a period of 1.5 hours and then it was kept overnight for proper setting of the glycerin produced.

Even after the transesterification process some impurities i.e enriched methanol, potassium methoxide, and glycerol are present in the Fatty Acid Methyl Esters (FAME). The extracted biodiesel needs some purification process before it can be used in diesel engines. The water is used as a washing agent of the biodiesel; a 30ml of water is supplied with the product gently and mixed it to avoid foam formation. This biodiesel and water mixture is permitted 5 hours of settling, they were settled in two phases in the form of biodiesel and water impurities, then they separated and used [12].

IOP Conf. Series: Materials Science and Engineering **183** (2017) 012038 doi:10.1088/1757-899X/183/1/012038

4.2. Properties of diesel and various biodiesel blends

Table 1. Properties of Biodiesel vs Neat Diesel

FUEL PROPERTY	**DIESEL**	**JATROPHA**		**CARDANOL**	
		B20	**B40**	**C20**	**C40**
Density@ 15^0 C	825	836.5	846.8	910.01	932.6
Kinematic viscosity @ 40^0C	3.52	2.88	3.16	3.12	3.98
Gross calorific value (MJ/kg)	45000	44635	43119	44850	43090
Flash point (°C)	48	58	52	89	92

5. Engine Test Bed

The performance of the jatropha and Cardanol blends were tested with different loads on Kirloskar make water-cooled, direct injection system was used and compression ratio 20:1 was maintained. Tests were conducted at different loads, with diesel, Jatropha blends and Cardanol blends for comparative analysis and the experimental setup of the engine is shown in figure 3.

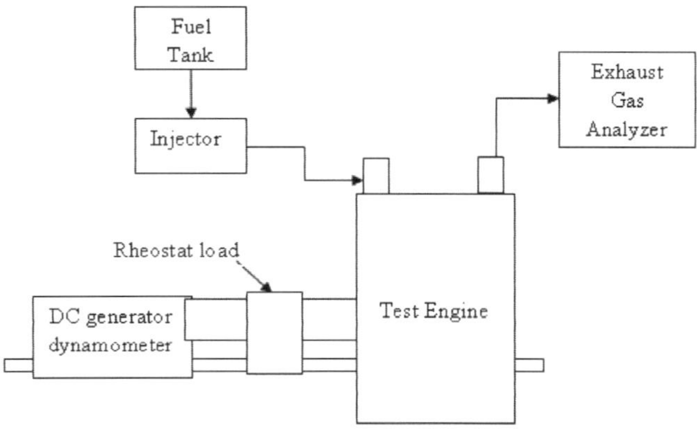

Figure 3. Experimental Setup

The dynamometer was used to find then load measurement and the engine speed was sensed, which indicates the inductive pickup on the flywheel it consists the digital output. The Chromel-alumel thermocouple was used for finding the exhaust gas temperature and a smoke meter used to find the smoke measurement. The Carbon monoxide (CO), carbon dioxide (CO_2), hydrocarbon (HC), nitrogen oxides (NOx) was measured by emission monitoring systems. To measure the accurate value of the fuel consumed a burette place with two sensors apart from the two markings of measure 20cc.

5.1: Engine specification

The engine test bed's engine specification is shown in table 2.

Table 2. Specification of the Engine

S.No.	Details	Specification
1	MAKE AND MODEL	Kirloskar oil engine TAF1
2	Bore diameter and StrokeLenth	87.5 mm X 110 mm
3	Engine Compression Ratio	17.5 : 1
4	Engine Rated Speed	1500 rpm
5	Cubic Capacity	0.661 liter
6	Injection timing	26 Btdc
7	Injector opening pressure	200 r

5.2: Operating parameter

Badal Dev Roy. et.al. (2016), discussed the various operating parameters of the engine test bed [13], Table 3 shows the best operating range of parameters which was tried in the present testing.

Table 3. Operating Parameter

S.No	Parameters	Range
1	Percentage of Load	20%, 40%, 60%, 80% and 100%
2	Engine Speed	Constant speed (1500 rpm)
3	Engine Compression ratio	17.5:1
4	Injection Timing °Btdc	26°Btdc
5	Injection Pressure	200 bars

5.3: Test procedure

1. The engine tests were conducted at 26° btdc injection timing and the injection pressures of 200bar.
2. The engine started on diesel (Hydrocarbon fuel) and it warmed up to the engine cooling water temperature to be stabilized.
3. The fuel consumption, exhaust temperature, exhaust emissions such as NO_X, CO_2, CO and exhaust gas opacity were measured and recorded in different loads.
4. The tests were repeated for 20%, 40%, 60%and 80% loading conditions.
5. To find the best value of three readings was taken from each load and an average value is calculated.

6. Results & Discussion

The engine test results of both jatropha and cardanol biodiesel's performance and emission characteristics were compared with the diesel and they are tabulated from the following table's which is also being represented in the graphs. Table 4 shows the comparison of brake specific fuel consumption.

Table 4. Comparison of brake specific fuel consumption

BMEP	BFSC Kg/kW hr				
(BAR)	DIESEL	B20	B40	C20	C40
0	0	0	0	0	0
0.89	1.08	0.484	0.512	0.673	0.646
1.78	0.6	0.39	0.468	0.385	0.373
3.57	0.45	0.338	0.421	0.252	0.247
4.46	0.39	0.303	0.357	0.231	0.225

In Figure 4, the BFSC for diesel is initially increasing with BMEP and a further increase in BMEP decreases the BSFC. This is same for the cases with B20, B40, and C20, C40 blends. The Jatropha blends have lower BFSC values than CNSL blends. In brake, specific fuel consumption C20 performs very well than diesel. B20 and C40 also perform well. The brake specific fuel consumption for C20 is around 50% relatively less of biodiesel.

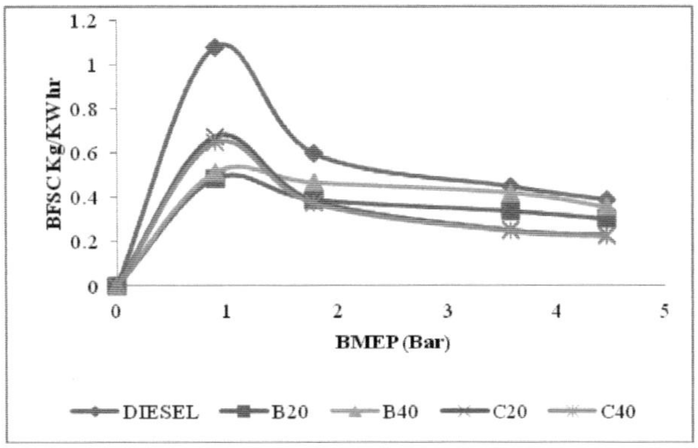

Figure 4. Comparison on brake specific fuel consumption

Table 5. Comparison on brake thermal efficiency

BMEP (BAR)	BTE (%) DIESEL	B20	B40	C20	C40
0	0	0	0.00	0	0
0.89	7.2	8.03	10.12	12.45	12.953
1.78	13.07	14.73	16.77	21.761	22.452
3.57	17.3	19.88	20.98	33.199	33.85
4.46	19.95	23.21	26.04	36.268	37.131

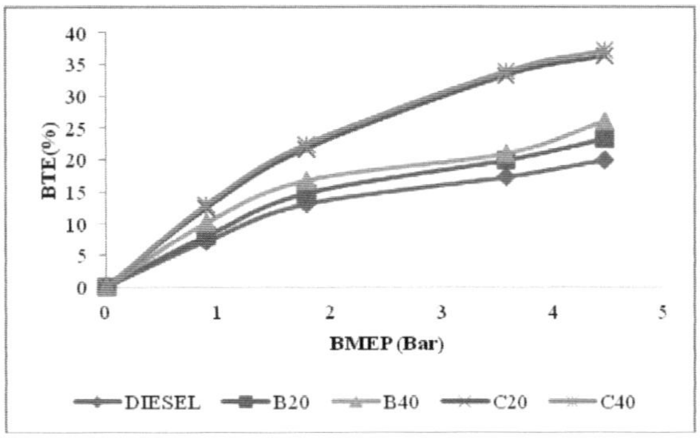

Figure 5. Comparison on brake thermal efficiency

Figure 5, shows that for CNSL blends have little higher BTE than Jatropha blends. In brake thermal efficiency C20 and C40 performs very well than diesel and B20.

Table 6. Comparison on CO emission

BMEP (BAR)	CO EMISSION (%) DIESEL	B20	B40	C20	C40
0	0.03	0.01	0.02	0.095	0.14
0.89	0.04	0.02	0.02	0.09	0.12
1.78	0.05	0.02	0.02	0.087	0.11
3.57	0.06	0.02	0.02	0.081	0.098
4.46	0.06	0.04	0.05	0.079	0.097

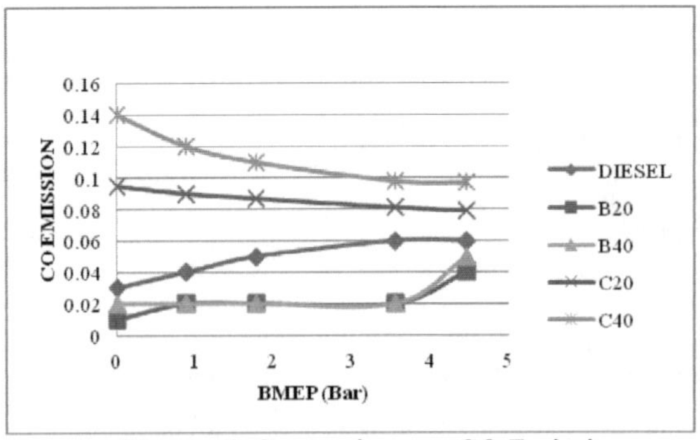

Figure 6. Comparison on CO Emission

Figure 6 illustrates that the CO emission for Jatropha blends which are lower than that of CNSL blends. In CO emission B20 and B40 performs very well than diesel, C20 and C40; C20 is relatively lesser in CO emission than C40.

Table 7. Comparison on HC emission

BMEP	HC EMISSION (%)				
(BAR)	DIESEL	B20	B40	C20	C40
0	12	10	11	8	11
0.89	10	12	11	8	9
1.78	9	9	10	6	9
3.57	11	9	10	5	10
4.46	13	10	11	5	7

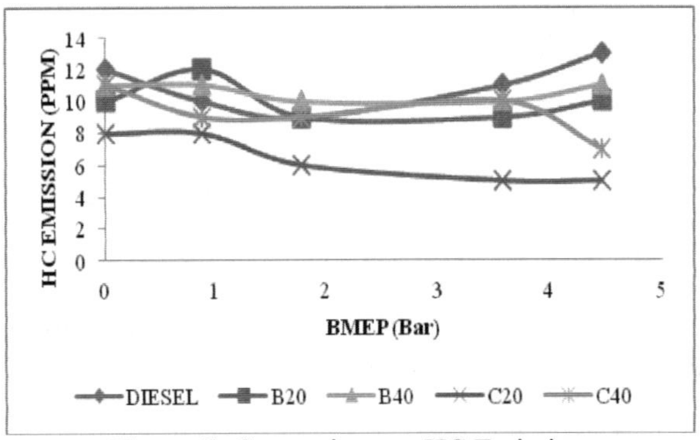

Figure 7. Comparison on HC Emission

Figure 7, Shows that the C20 blend has a low HC emission than any Jatropha blends and C40. In hydrocarbon emission, C20 is relatively very less than all other biodiesel blends as well as diesel. The hydrocarbon emission for C20 gradually reduced with an increase in pressure.

Table 8. Comparison on NO_X emission

BMEP	NO_X EMISSION (ppm)				
(BAR)	DIESEL	B20	B40	C20	C40
0	40	110	115	16	24
0.89	61	206	217	36	52
1.78	78	314	328	73	106
3.57	210	674	692	254	250
4.46	330	1075	1104	312	326

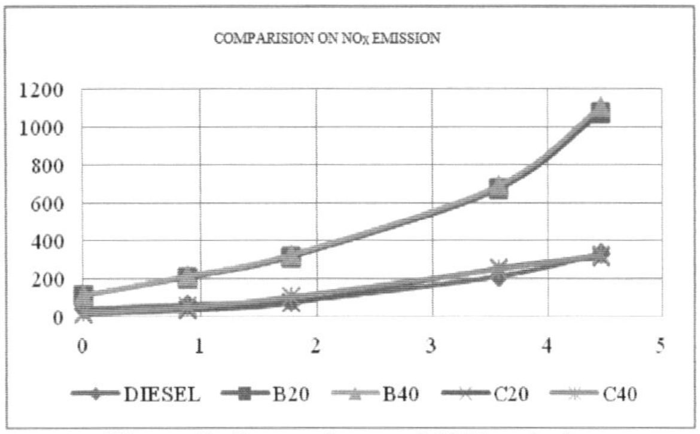

Figure 8. Comparison on NOX emission

Figure 8, illustrates that the NO_X emissions are lower for CNSL blends. In NOX emission C20 and C40 are relatively lesser than the diesel, B20 and B40.

6. Conclusion

The characteristics of Jatropha and CNSL biodiesel blends are studied and compared with diesel by varying the pressure, C20 performs very well with respect to brake specific fuel consumption. C20 and C40 perform very well with respect to brake thermal efficiency. C20 performs very well in reducing the hydrocarbon emission. C20 and C40 perform very well by reducing the NOX emission. The reduction of CO emissions was not found to be satisfactory for C20 and C40, whereas B20 and B40 were relatively efficient in reducing the CO emissions.

7. Future work

The comparative character study of the performance and emission characteristic on Jatropha and CNSL revealed that C20 and C40 are not effective in reducing the CO emission when compared to B20 and B40. Hence the future work was targeted in the study of substances to be added with C20 and it has to be effectively reducing the CO emissions, thereby reducing the greenhouse effect.

8. References

[1] Dhanasekaran C and Mohankumar G, 2016 Dual fuel mode DI diesel engine combustion with hydrogen gas and DEE as ignition source *International journal of hydrogen energy* **41** 713-721.

[2] Subramanian, K.A, Singhal S.K, Saxena M, Singhal S, 2005 Utilization of liquid biofuels in automotive diesel engines *J. Biomass and Bioenergy* **29** 65-72.

[3] Mallikappa D, Reddy R.P and Murthy C.S 2012 Performance and emission characteristics of double cylinder CI engine with Cardanol biofuel blends *Renewable energy* **38** 150-154.

[4] Anand R, Kannan G R, Nagarajan S, Velamathi S 2010 Performance emission and combustion characteristics of a diesel engine fueled with biodiesel produced from waste cooking oil. *SAE Paper* 2010-01-0478.

[5] Velmurugan A and Loganathan M 2011 Performance and emission characteristics of a DI diesel engine fuelled with cashew nut shell liquid (CNSL) – Diesel blends *World Academy of science, Engineering and technology* **58**.

[6] Sayin Cenk, Gumus Metin, and Canakci Mustafa 2010 Effect of fuel injection timing on the emissions of a direct-injection (DI) diesel engine fueled with canola oil methyl ester-diesel fuel blends *Energy Fuel* **24** 2675-82.

[7] Narayana Reddy J and Ramesh A 2006 Parametric studies for improving the performance of a Jatropha oil-fuelled compression ignition engine *Renew Energy* **31** 1994-2016.

[8] Achten, W.M.J, Mathys E, Verchot L, Singz V.P, Aert R and Muys B 2008 Jatropha bio-diesel production and use *Biomass and Bioenergy* **32(12)** 1063-1084.

[9] Salar-García M.J, et al. 2016 Analysis of optimal conditions for biodiesel production from Jatropha oil in supercritical methanol: Quantification of thermal decomposition degree and analysis of FAMEs *The Journal of Supercritical Fluids* **112** 1-6.

[10] Thananchayan T, Krishnakumar G, Pushpraj M, Ajay Avinash S P and Karunya S 2013 Biodiesel Production From Jatropha Oil And Castor Oil By Transesterification Reaction - Experimental And Kinetic Studies *International Journal of ChemTech Research* **5(3)** 1107-1112.

[11] Patel, Rajesh N, Santanu Bandyopadhyay, and Anuradda Ganesh 2006 Extraction of cashew (Anacardium occidentale) nut shell liquid using supercritical carbon dioxide *Bioresource Technology* **97.6** 847-853.

[12] Kasiraman G, Nagalingam B, and Balakrishnan M 2012 Performance, emission and combustion improvements in a direct injection diesel engine using cashew nut shell oil as fuel with camphor oil blending *Energy* **47.1** 116-124.

[13] Badal Dev Roy. et.al., 2016 Experimental investigation of turbocharger mapped by datalogger in I.C. engine *ARPN Journal of Engineering and Applied Sciences* **11(7)** 4587 – 4595.

IOP Conf. Series: Materials Science and Engineering **183** (2017) 012039 doi:10.1088/1757-899X/183/1/012039

Impact of Various Compression Ratio on the Compression Ignition Engine with Diesel and Jatropha Biodiesel

S Sivaganesan[1], M Chandrasekaran[2], M Ruban[3]
[1,3] Assistant Professor, Vels University, Chennai,TN, India
[2] Professor & Director, Dept. of Mechanical Engineering, Vels University, India.

E-mail: [1]sivaganesanme@gmail.com

Abstract. The present experimental investigation evaluates the effects of using blends of diesel fuel with 20% concentration of Methyl Ester of Jatropha biodiesel blended with various compression ratio. Both the diesel and biodiesel fuel blend was injected at 23° BTDC to the combustion chamber. The experiment was carried out with three different compression ratio. Biodiesel was extracted from Jatropha oil, 20% (B20) concentration is found to be best blend ratio from the earlier experimental study. The engine was maintained at various compression ratio i.e., 17.5, 16.5 and 15.5 respectively. The main objective is to obtain minimum specific fuel consumption, better efficiency and lesser Emission with different compression ratio. The results concluded that full load show an increase in efficiency when compared with diesel, highest efficiency is obtained with B20MEOJBA with compression ratio 17.5. It is noted that there is an increase in thermal efficiency as the blend ratio increases. Biodiesel blend has performance closer to diesel, but emission is reduced in all blends of B20MEOJBA compared to diesel. Thus this work focuses on the best compression ratio and suitability of biodiesel blends in diesel engine as an alternate fuel.

1. Introduction

The economy and the development of the country mainly depend on its energy resources. The increasing industrialization and modernization of the world lead to a steep rise in the consumption and demand for the petroleum based fuels every year. At present, the world is confronted with the twin crisis of fossil fuel depletion and environmental degradation which has put a limit in the use of conventional fuels. The search for alternate fuels for diesel engines has been intensified during the last few decades as substitute to fossil fuels [1]. Countries like India which has strong agricultural resources can fulfil their energy requirements by focusing on the production of energy using the energy extract techniques from biofuels.

Among the many alternative fuels, biofuels are considered as accessible and economical, which essentially plays a vital role for the development of industrialization. Many researches represented that there can be large potential work available in bio origin fuels. These biofuels can be used directly without any modifications made in the engine. Some of the bio origin fuels are biodiesel, bio oil, bio gas, vegetable oil and alcohols, which can be used in the industries and in the transport sector as a replacement for existing fossil fuel. The bio fuels which are extracted from oil seed, crop and agricultural waste have less sulfur and more oxygen content. This research contributes an improvement for alternate resources, which reduces the emulating pollution.

The objective of this research is to investigate the usage of biodiesel blended in order to reduce the emissions of all regulated pollutants from diesel engine. The analysis is further carried out with the effect of various compression ratio on engine for bio fuels.

2. Effect of compression ratio in diesel engine

This study reveals the natural aspirated, various compression ratio engine with different biodiesel blends (B10, B20, B30, B40, B60, B80 and B100). The test was conduct in no load to full load in all blends. All emission parameter such as HC, CO, and NOx were measured. The compression ratio was varied from 14 to 18 with equal interval of two. In higher compression ratio 18 the BTE was increased and lower in SFC. The results shown B20 biodiesel was superior in both performance and emission parameter in higher compression ratio of 18 compare to other compression ratio and other biodiesel blends [2].

Investigated about the methyl ester of cotton seed oil production by transesterification process and blends with four different blends B5 to B20 in equally raised by 5%. An experiment is studied at constant speed in 1500 rpm with variable compression ratio (CR). The compression ratio of 15 and 17 were found better in performance characteristics with 5% blends. The oxides of nitrogen emission are higher in compression ratio 17 with B20 blend. Also reduction in carbon monoxide and smoke in all loads in higher compression ratio of 17 with B20 blend. They concluded B20 with higher compression ratio is suitable for diesel engine [3].

The performance of DI compression ignition engine with blends of 5-carbon alcohol, rubber seed biodiesel and diesel at various compression ratio (CR) viz.16: 1, 17.5: 1, 19: 1. It proves that CR 17.5 is the best compression ratio for all the blends. The result shows that the emission is reduced except forNOx when compared with the diesel fuel. Finally concluded that emission was reduced in R10P10 blend at CR 17.5 and it is an eco friendly alternative fuel [4].

This experiment was tested in uneven firmness ratio engine (VCR) with methyl ester of jatropha oil (MEOJ) and diesel. The compression ratio was varied for 14, 16, 18 and 20. The BTE and BSFC were measured for performance parameter and CO, CO2, HC, NOx and smoke for emission parameter. It was noticed that higher CR, performance of engine increased apparently with less BSFC for biodiesel blends fuel. The emission was higher for CO and CO2, HC, NOx and smoke density emission is lower when CR is higher. The higher compression ratio of the combustion parameter is improved in biodiesel blends [5].

3. Experimental Setup

In this experimental investigation was analysis in three different compression ratio in various load at rated speed at 1500 rpm. It is a single cylinder, four stroke, vertical, water cooled DI diesel engine. The engine details were mention in table 1. The both diesel and biodiesel fuel was injected at 23ºBTDC. Biodiesel was extracted from Jatropha oil used in 20% (B20) concentration with diesel in all compression ratios [6]. The physical properties in B20 MEOJ are provided in table 2. Smoke reading was identified by smoke meter and hydrocarbon, carbon monoxide and oxides of nitrogen emission were collected from engine tail pipe. The engine cooling was made by water at constant flow rate for entire experiments. Engine load was varied by adjusting knob, it connected to the eddy current dynamometer.

Table 1. Details of Experimental Engine

Manufacturer	Kirlosker TV – I
Category	Vertical cylinder, DI diesel engine, VCR engine
Number of cylinder	1
Bore X Stroke	87.5 mm X 110 mm

Compression ratio	17.5
Speed	1500 rpm
Rated brake power	5.2 kW
Cooling system	Water cooling
Injection timing	23°BDTC

The fuel utilization was measured by burette with periodic time. During this interval of time, the utilization of fuel was measured, with the help of the stopwatch. In each load the performance parameter and emission parameter were measured [7]. The experimental system line sketch is plotted in figure 1. The inlet and exhaust gas temperature were measured by Chromel-Alumel K-type thermocouples. The combustion parameter was measured by combustion analyzer.

Table 2. Physical Properties in B20 MEOJ

Test Property	B20 MEOJ
Density at 15° C kg/m3	868.2
Kinematic Viscosity at 40°C	4.36
Flash Point (PMCC) °C, (min)	138
Pour point °C	5
Gross Colorific value k.cal/kg	10280

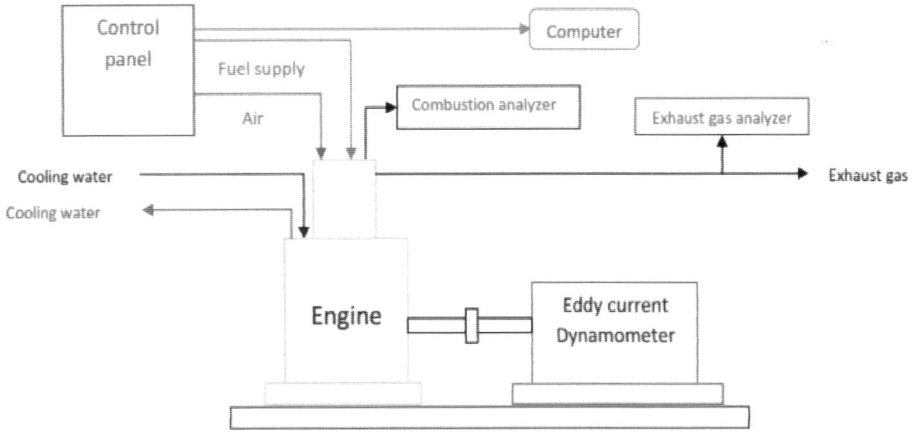

Figure 1. Line Sketch of Experimental System

4. Results and Discussion

4.1 Performance characteristics

The different compression ratio on SFC for the blend of B20 MEOJ and diesel fuel is reported in figure 2. Among the three compression ratio tried, it was observed that specific fuel consumption is least at 17.5 for both diesel and B20 blends. This is due to improvement in physical properties of MEOJ oil. It is concluded that the higher compression ratio shows better results [8].

The difference of BTE with different loads in various compression ratio with blend of B20MEOJ is shown figure 3. From the figure, higher BTE is obtained in higher compression ratio of 17.5 in diesel and B20MEOJ blends the maximum BTE was 28.18 and 28.04% respectively, hence it was predicted that there was a marginal improvement on performance while increasing the compression ratio [9].

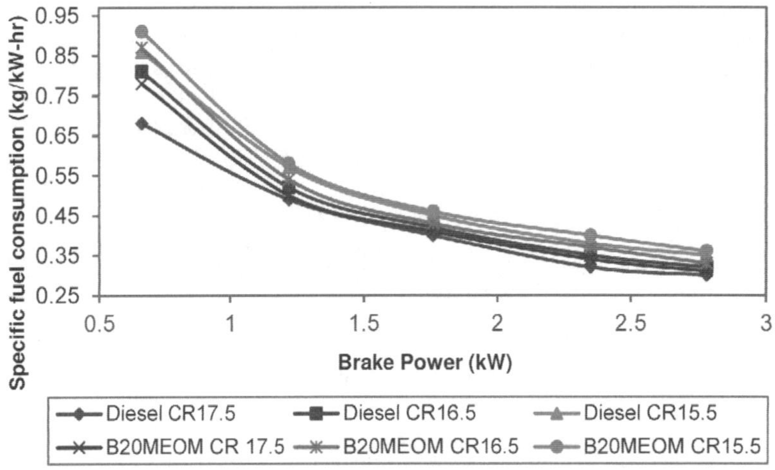

Figure 2. Specific fuel consumption with Brake Power (B20MEOJ-CR)

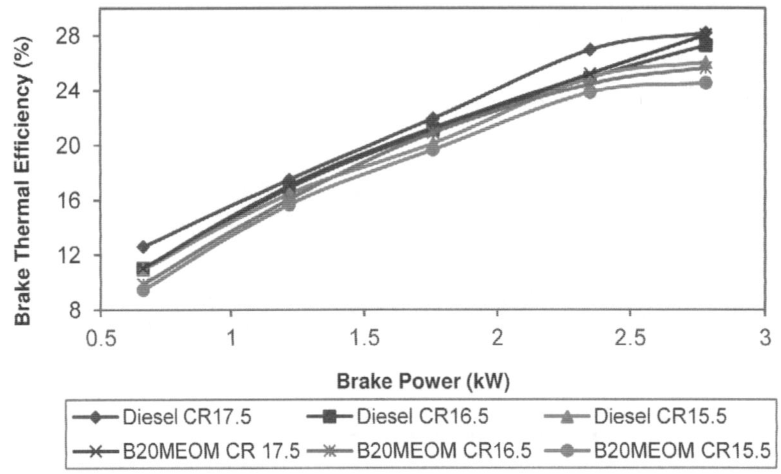

Figure 3. Brake thermal efficiency with Brake Power (B20MEOJ-CR)

4.2 Emission characteristics

The difference in smoke density in diesel and B20 blends of MEOJ oil at compression ratio of 17.5, 16.5 and 15.5 is depicted in figure 4. For the B20 blends of MEOJ the 17.5 compression ration

resulted in the lowest smoke density over the entire output range. All biodiesel blends lower smoke density in all compression ratio is compared with diesel [10].
The differences in NOx emission with different loads at different compression ratio is exhibited in figure 5. For all the blends NOx emission is raising with rising of loads. In compression ratio of 15.5, the NOx emission is optimum over the entire output range with diesel. The corresponding NOx values at maximum output are 846 ppm. Diesel at 15.5 compression ratio is the overall optimum combination for lower NOx emission.
The Hydro Carbon emission for B20 blends of MEOJ and diesel at various compression ratio (CR) effects plots in figure 6. The HC emission increase or decrease is based on the CR value. Normally if the CR value is increased, the HC emission reduced. At full load with the CR of 17.5, the HC for B20 is 33ppm while compared to other compression ratio. During the end of the compression stoke the air temperature gets higher, this enhance the complete burning and reduction HC emission [11].
The difference of Carbon monoxide emission for B20 blends of MEOJ and diesel fuel with various CR is indicated in figure 7. It is noted that the increase of CR value with decrease of Carbon monoxide emission. At full load with the CR value of 17.5, the Carbon monoxide emission for B20MEOJ is 0.07 % and 53% decreasing of Carbon monoxide emission is achieved. The reason is entire burning of the fuel takes place during combustion (Eliana Weber de Menezeset al. (2006)).

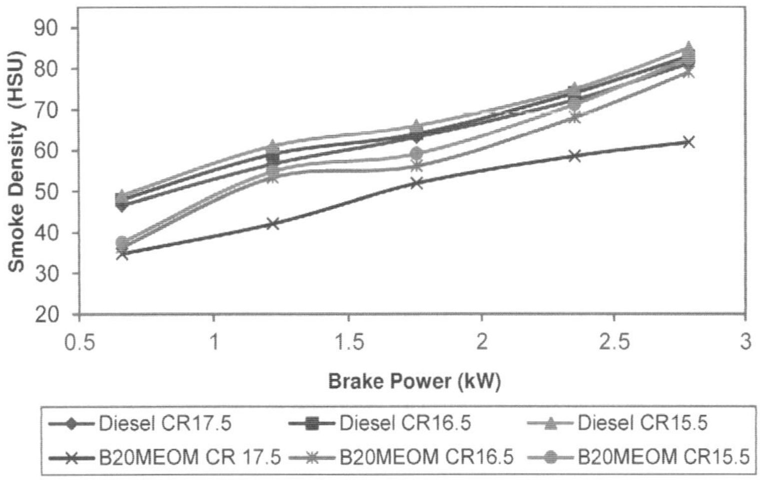

Figure 4. Smoke density with Brake Power (B20MEOJ-CR)

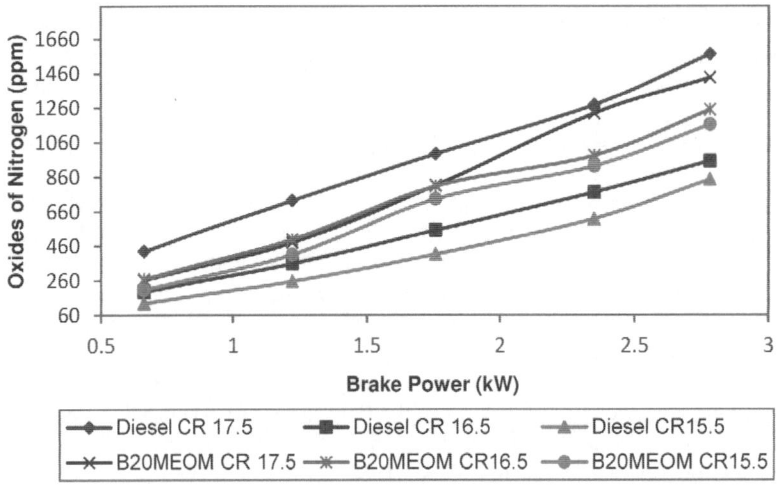

Figure 5. Oxides of Nitrogen with Brake Power (B20MEOJ-CR)

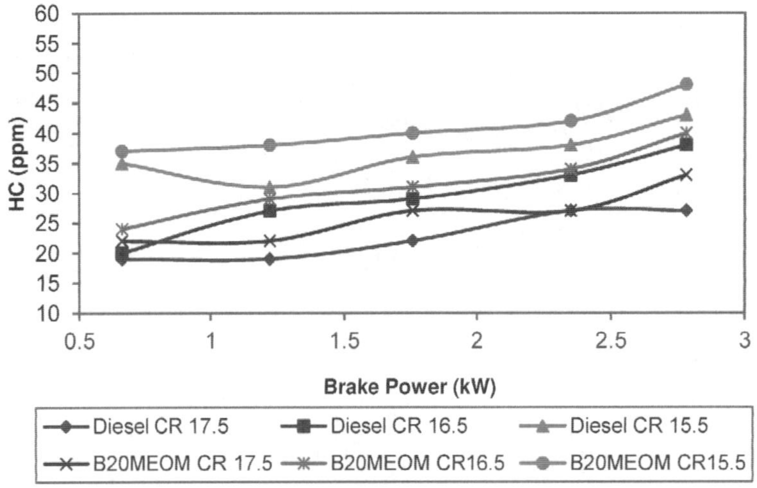

Figure 6. Hydro carbon with Brake Power (B20MEOJ-CR)

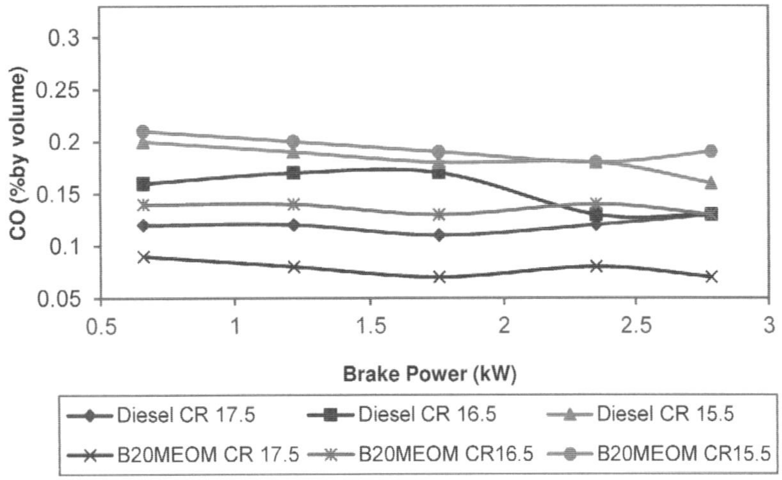

Figure 7. Carbon monoxide with Brake Power (B20MEOJ-CR)

4.3 Combustion characteristics

Cylinder pressure with crank angle of B20 blend of MEOJ and diesel fuel with different compression ratio at maximum load is depicted in figure 8. The maximum cylinder pressure is obtained in diesel compared to B20 blend of MEOJ for entire range of load operation. During the increase in the compression ratio the combustion that occurs before the top dead center as a result rapid pressure rise will take place [12].

Comparison of heat release at different compression ratio with B20 blend of MEOJ, and diesel fuel at maximum load is plotted in figure 9. At compression ratio of 17.5, the heat release rate of B20 blend MEOJ and diesel fuel are 57.75 J/degree and 49.87 J/degree respectively. However the heat release rate is increased in diesel fuel while compared to B20MEOJ. The reason for this is due to the increase in the combustion that takes place during the diffusion phase [13].

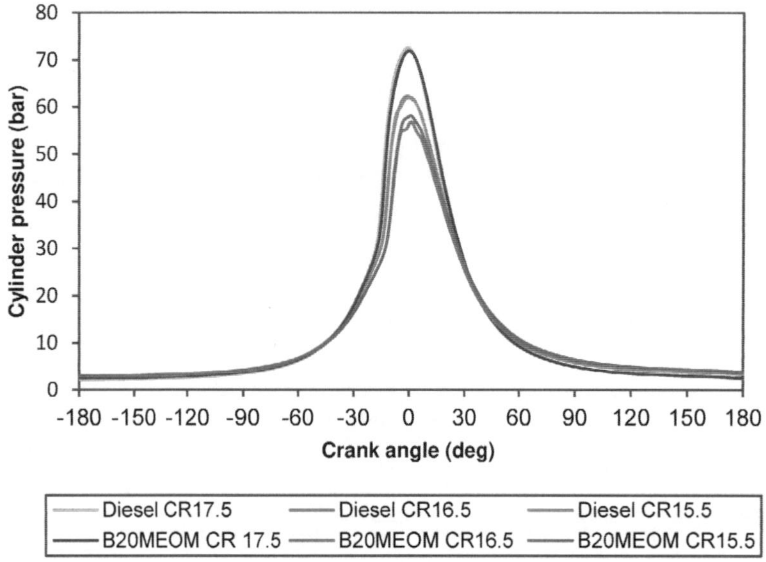

Figure 8. Cylinder pressure with Crank angle(B20MEOJ-CR)

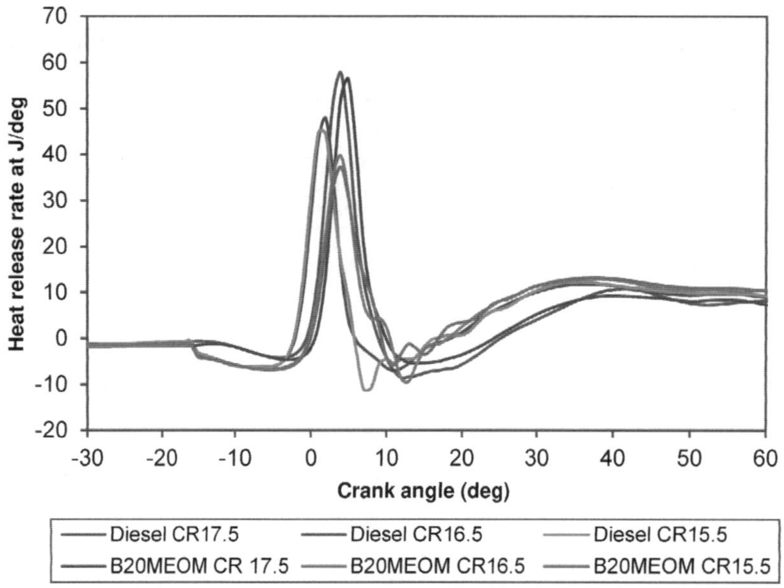

Figure 9. Heat Release rate with Crank angle (B20MEOJ-CR)

5. Conclusions

It was observed that among the three compression ratio specific fuel consumption is least at 17.5 for both diesel and B20 blends. The higher BTE is obtained in higher compression ratio of 17.5 in diesel and B20MEOJ blends. The compression ratio 17.5 resulted in the lowest smoke density over the entire output range. All biodiesel blends emit lower smoke density in all compression ratio when compared with diesel. The NOx emission is reduced in compression ratio of 15.5. When the CR value is increased, the HC and Carbon monoxide emission reduced. The maximum cylinder pressure is obtained in diesel compared to B20 blend of MEOJ and the heat release rate is increased in diesel fuel while compared to B20MEOJ. The results concluded that the compression ratio 17.5 with B20MEOJ biodiesel blend is the best compression ratio in terms of performance, emission and combustion when it is compared with baseline diesel.

References

[1] Osmano Souza Valente, Marcio Jose da Silva, Vanya Marcia Duarte Pasa , Carlos Rodrigues Pereira Belchior, Jose Ricardo Sodre, 2010. Fuel consumption and emissions from a diesel power generator fuelled with castor oil and soybean biodiesel, *Fuel* **89**:3637–3642.

[2] Sejal Narendra Patel and Ravindra Kirar, 2012. An Experimental Analysis of Diesel Engine Using Biofuel at Varying Compression Ratio, *International Journal of Emerging Technology and Advanced Engineering*, **2**:385-391.

[3] Anand R, Kannan G R, Rajasekhar Reddy K and Velmath S, 2009. The Performance and Emissions of a Variable Compression Ratio Diesel Engine Fuelled With Bio-Diesel From Cotton Seed Oil, *ARPN Journal of Engineering and Applied Sciences*, **4(9)**:72-87.

[4] Pillai N. Subramonia, Kannan P. Seeni, and Suresh S, 2016 Experimental Investigation on Emission Characteristics of Variable Compression Ratio CI Engine Fuelled by Combined Biodiesel, *Asian Journal of Research in Social Sciences and Humanities*, **6(9)**:1419-1429.

[5] Venkateswara Rao P, 2015. Compression ratio effect on Diesel Engine working with Biodiesel (JOME) Diesel blend as Fuel", *Research Journal of Chemical Sciences*, **5(7)**:48-51.

[6] Sivaganesan. S and Chandrasekaran. M, 2016. Performance and Emission Analysis of Compression Ignition Engine with Methyl Ester of Jatropha and Diesel, *Indian Journal of Science and Technology* **9(26)**: 1 – 6.

[7] Karthick. D, Dwarakesh. R, and Premnath, 2014. Combustion and Emission Characteristics of Jatropha Blend as a Biodiesel for Compression Ignition Engine with Variation of Compression Ratio, *International Review of Applied Engineering* Research **4**:39-46.

[8] Xiangang Wang, Cheung Yage Di and Zuohua Huang, 2012. Diesel engine gaseous and particle emissions fueled with diesel–oxygenate blends, Energy **94**:317–323.

[9] Yi Ren, Zuohua Huang, Haiyan Miao, Yage Di, Deming Jiang, Ke Zeng, Bing Liu, and Xibin Wang, 2008. Combustion and emissions of a DI diesel engine fuelled with diesel-oxygenate blends, Fuel **87**:2691–2697.

[10] Zannis, Hountalas D, and Kouremenos D, 2004. Experimental Investigation to Specify the Effect of Oxygenated Additive Content and Type on DI Diesel Engine Performance and Emissions, *SAE Technical Paper* 01-0097.

[11] Metin Guru, Atilla Koca, Ozer Can, Can Cinar, and Fatih Sahin, 2010. Biodiesel production from waste chicken fat based sources and evaluation with Mg based additive in a diesel engine, *Renewable Energy* **35(3):**637–643.

[12] Balakrishnan. N and Mailsamy. K, 2014. Effect of compression ratio on compression ignition engine performance with biodiesel and producer gas in mixed fuel mode, *Renewable and sustainable energy* **6**:23-103.

[13] Sunilkumar. R, Kumbhar and Dange. H. M, 2014. Performance Analysis Of single Cylinder Diesel Engine, Using Diesel Blended with Thumba Oil, *International Journal of Soft Computing and Engineering (IJSCE)* **4**:24-30.